Lecture Notes in Energy

Volume 94

Lecture Notes in Energy (LNE) is a series that reports on new developments in the study of energy: from science and engineering to the analysis of energy policy. The series' scope includes but is not limited to, renewable and green energy, nuclear, fossil fuels and carbon capture, energy systems, energy storage and harvesting, batteries and fuel cells, power systems, energy efficiency, energy in buildings, energy policy, as well as energy-related topics in economics, management and transportation. Books published in LNE are original and timely and bridge between advanced textbooks and the forefront of research. Readers of LNE include postgraduate students and non-specialist researchers wishing to gain an accessible introduction to a field of research as well as professionals and researchers with a need for an up-to-date reference book on a well-defined topic. The series publishes single- and multi-authored volumes as well as advanced textbooks.

Jonghyun Kim · Seungjun Lee · Poong Hyun Seong

Autonomous Nuclear Power Plants with Artificial Intelligence

Springer

Jonghyun Kim
Department of Nuclear Engineering
Chosun University
Gwangju, Korea (Republic of)

Seungjun Lee
Department of Nuclear Engineering
UNIST
Ulsan, Republic of Korea

Poong Hyun Seong
Department of Nuclear and Quantum Engineering
KAIST
Daejeon, Republic of Korea

ISSN 2195-1284 ISSN 2195-1292 (electronic)
Lecture Notes in Energy
ISBN 978-3-031-22385-3 ISBN 978-3-031-22386-0 (eBook)
https://doi.org/10.1007/978-3-031-22386-0

This Springer imprint is published by the registered company Springer Nature Switzerland AG
The registered company address is: Gewerbestrasse 11, 6330 Cham, Switzerland

Preface

The top priority of a nuclear power plant (NPP) is safety. Indeed, safety is a ubiquitous concern over the whole life cycle of an NPP, from its design to decommissioning. But in a complex large-scale system with a huge number of components such as an NPP, it is not easy to identify all vulnerabilities and achieve a perfect level of safety. As the TMI-2, Chernobyl, and Fukushima severe accidents demonstrated, an NPP emergency does not occur by a single cause but rather by a complicated combination of hardware, software, human operators, decision-makers, and so on. Research has revealed that human error takes up more than half of core damage frequency, a common risk metric, and it is also known that human factors were the direct or indirect cause of most nuclear accidents in history.

One of the approaches to enhance the safety of NPPs is to develop operator support systems to reduce latent human errors by optimizing the operators' required workload or automating some portion of their tasks. In fact, research into the application of artificial intelligence (AI) to NPPs has been performed for decades, but few developments have actually been reflected in real NPPs. Despite this though, AI technology is now being particularly highlighted again due to increases in data processing and advancements in hardware design, graphics processing units, and related methods. This has led to an explosive growth in recent years of research related to AI techniques in the nuclear industry. Such global efforts toward applying AI techniques to NPPs may result in better performance than conventional methods based on the characteristics of NPPs, such as complexity in operation, dynamic behaviors, and high burden in decision-making.

This book is divided into nine chapters. In the first part, Chaps. 1 and 2, the framework of the autonomous NPP and the fundamentals of AI techniques are introduced. Chapter 1 suggests a high-level framework for an autonomous NPP including the functional architecture of the autonomous operation systems. This chapter defines multiple levels and sub-functions necessary for automating operator tasks. The framework consists of monitoring, autonomous control, the human–autonomous system interface, and the intelligent management of functions. Chapter 2 provides fundamental explanations of the various AI techniques appearing in the book as an overview.

In Chaps. 3–6, the essential methods necessary for developing operator support systems and autonomous operating systems are introduced. Chapters 3, 4, and 5 apply various AI-based techniques for the monitoring of NPPs including signal validation, diagnosis of abnormal situations, and prediction of plant behavior using both supervised and unsupervised learning methods. Chapter 6 presents AI applications for autonomous control using reinforcement learning for both normal and emergency situations along with domain analysis methods.

Lastly, in Chaps. 7 and 8, applications are introduced. Chapter 7 provides an example of an integrated autonomous operating system for a pressurized water reactor that implements the suggested framework. Lastly, Chap. 8 deals with the interaction between operators and the autonomous systems. Even though the autonomous operation systems seek to minimize operator interventions, supervision and manual control by human operators are inevitable in NPPs for safety reasons. Thus, the last chapter addresses the design of the human–autonomous system interface and operator support systems to reduce the task burden of operators as well as increase their situational awareness in a supervisory role.

This book is expected to provide useful information for researchers and students who are interested in applying AI techniques in the nuclear field as well as other industries. Various potential areas of AI applications and available methods were discussed with examples. Traditional approaches to recent applications of AI were examined. In addition, the specific techniques and modeling examples provided would be informative for beginners in AI studies. While the focus of this book was the autonomous operation of NPPs with AI, the methods addressed here would also be applicable to other industries that are both complex and safety-critical.

Gwangju, Korea (Republic of) — Prof. Jonghyun Kim
Ulsan, Republic of Korea — Prof. Seungjun Lee
Daejeon, Republic of Korea — Prof. Poong Hyun Seong
October 2022

Acknowledgements

This book includes various researches that were supported by National Research Foundation of Korea (NRF) grants funded by the Korean government (MSIT) from 2016 to 2022 (NRF-2016R1A5A1013919). The authors appreciate the assistance and support of the Korean national fund that initiated the research toward the application of artificial intelligence to nuclear power plant operation. The authors would also like to acknowledge the sincere contribution and effort by students and graduates from three universities: Dr. Daeil Lee, Hyojin Kim, Younhee Choi, and Subong Lee from Chosun University, Jae Min Kim, Jeeyea Ahn, Jeonghun Choi, Junyong Bae, and Ji Hyeon Shin from the Ulsan National Institute of Science and Technology (UNIST), and Dr. Young Ho Chae and Dr. Seung Geun Kim from the Korea Advanced Institute of Science and Technology (KAIST).

Contents

Acronyms

1D	One-dimensional
2D	Two-dimensional
A3C	Asynchronous advantage actor-critic
ADAGRAD	Adaptive gradient
ADAM	Adaptive moment estimation
ADAS	Advanced driver-assistance system
ADS	Abstraction-decomposition space
AE	Autoencoder
AI	Artificial intelligence
ANN	Artificial neural network
AOO	Anticipated operational occurrence
AOP	Abnormal operating procedures
APR	Advanced power reactor
AUC	Area under receiver operating characteristic curve
BOL	Beginning of life
CCW	Component coolant water
CD	Complex decision
CDN	Condenser
CDS	Condenser vacuum abnormality
CHRG	Charging water system abnormality
CIA	Concealed intelligent assistant
CNN	Convolutional neural network
CNS	Compact nuclear simulator
COSIE	Comparison of safety impact evaluation
CPN	Colored Petri-Net
CSF	Critical safety function
CST	Condensate storage tank
CTMT	Containment
CWS	Circulating water system abnormality
DBA	Design basis accident
DeepLIFT	Deep learning important features

DL	Deep learning
DNB	Departure from nucleate boiling
DNBR	Departure from nucleate boiling ratio
DRL	Deep reinforcement learning
EA	Expected action
EID	Ecological-interface design
EOL	End of life
EOP	Emergency operating procedure
ERG	Emergency response guideline
ESDE	Excess steam demand event
FC	Fully connected
FRP	Functional recovery procedure
GAN	Generative adversarial network
GCN	Graph convolution neural network
GNN	Graph neural network
Grad-CAM	Gradient-weighted class activation mapping
GRU	Gated recurrent unit
GRUD	Gated recurrent unit-decay
HASI	Human–autonomous system interface
HP	High pressure
HSI	Human–system interface
HX	Heat exchanger
I&C	Instrumentation and control
IAEA	International Atomic Energy Agency
KAERI	Korea Atomic Energy Research Institute
KNN	K-nearest neighbor
LC	Loss of component cooling system seal damage
LCO	Limiting conditions for operation
LH	Level high
LN	Water line leakage
LOAF	Loss of all feedwater
LOCA	Loss of coolant accident
LP	Low pressure
LSTM	Long short-term memory network
LTDN	Letdown water system abnormality
MAE	Mean absolute error
MAPE	Mean absolute percentage error
MCR	Main control room
MFP	Main feedwater pump
MGGAN	Manifold-guided generative adversarial network
MICE	Multivariate imputation with chained equation
MIMO	Multi-input multi-output
ML	Machine learning
MLP	Multilayer perceptron
MSE	Mean squared error

MSIV	Main steam isolation valve abnormality
MSLB	Main steam line break
MSLE	Mean squared logarithmic error
MSS	Main steam system abnormality
NCP	Normal charging pump
NPP	Nuclear power plant
NRC	Nuclear Regulatory Commission
OLM	Online monitoring technique
ORP	Optimal response procedure
PC	Principal component
PCA	Principal component analysis
PCC	Procedure compliance check
PID	Proportional-integral-derivative
PM	Pump trip
PORV	Power operated relief valve
POSRV	Pilot operated safety relief valve
PRT	Pressurizer relief tank
PSV	Pressurizer safety valve
PWR	Pressurized water reactor
PZR	Pressurizer
RA	Real action
RBF	Radial basis function
RCP	Reactor coolant pump
RCS	Reactor coolant system
RE	Reconstruction error
ReLU	Rectified linear unit
RGAN	Recurrent generative adversarial network
RL	Reinforcement learning
RMSE	Root mean squared error
RMW	Reactor makeup water tank valve abnormality
RNN	Recurrent neural network
ROC	Receiver operating characteristic
RPM	Turbine revolutions per minute
RPS	Reactor protection system
RV	Reactor vessel
SBO	Station blackout
SD	Reactor coolant pump seal damage
SDM	Shutdown margin
SFP	Spent fuel pool
SG	Steam generator
SGTL	Leakage of tubes inside steam generators
SGTR	Steam generator tube rupture
SHAP	Shapley additive explanations
SI	Safety injection
SR	Surveillance requirement

SVM	Support vector machines
TBN	Turbine
TCS	Turbine control system abnormality
Tech Spec	Technical specification
TMI	Three Mile Island
TSM	Two-stage model
TSMS	Tech Spec Monitoring System
VAE	Variational autoencoder
VV	Valve abnormality

Chapter 1
Introduction

1.1 Background

Autonomous is a term referring to the power of self-government. By this definition, control systems with high degrees of autonomy should have the capacity for self-governance, which allows them to perform their necessary control functions without external intervention over extended time periods (Antsaklis and Passino 1993). Recently, interest in autonomous systems has increased in a variety of fields, from manufacturing to the development of unmanned space, atmospheric, and ground vehicles, based on the belief that the technology can improve safety, reliability, and efficiency.

The area in which autonomous systems have been the most actively studied and applied is the mobility industry. At present, advanced driver-assistance systems (ADASs) are being installed even in commercial vehicles to assist drives in a variety of ways. ADASs provide important information about road traffic and can suggest more efficient routes based on the current traffic conditions. They can also perform functions like basic control, parking, etc., without driver intervention. Furthermore, the technology is currently advancing to detect driver fatigue or distraction, alert him or her, and even take control of the vehicle from the human driver automatically (Joseph and Mondal 2021).

The extent of automation (or autonomy) is often described with levels of automation. Sheridan and Verplanck defined 10 such levels for supervisory control systems (Sheridan and Verplanck 1978). Their levels of automation represent a continuum from low automation (Level 1), in which a human performs a given task manually, and full automation (Level 10), in which the system is fully autonomous, requiring no human intervention. Similarly, Endsley and Kaber formulated a 10-level taxonomy that suggests roles for the human and system in the human information processing stages, i.e., monitoring, generating, selecting, and implementing (Endsley and Kaber 1999). Billings also proposed a comprehensive taxonomy by assigning functions to automatic systems and humans, as shown in Table 1.1 (Billings 1997).

J. Kim et al., *Autonomous Nuclear Power Plants with Artificial Intelligence*, Lecture Notes in Energy 94, https://doi.org/10.1007/978-3-031-22386-0_1

Table 1.1 Level of automation proposed by Billings

Management mode	Automation functions	Human functions
Autonomous operation	Fully autonomous operation. Human not usually informed. System may or may not be capable of being disabled	Human generally has no role in operation and monitoring is limited
Operation by exception	Essentially autonomous operation unless specific situations or circumstances are encountered	Human must approve of critical decisions and may intervene
Operation by consent	Full automatic control under close monitoring and supervision	Human monitors closely, approves actions, and may intervene
Operation by delegation	Automatic control when directed by a human to do so	Human provides supervisory commands that automation follows
Shared control	Automatic control of some functions/tasks	Manual control of some functions/tasks
Assisted manual control	Primarily manual control with some automation support	Human manually controls with assistance from partial automation
Direct manual control	No automation	Human manually controls all functions and tasks

It should be noted here that an "autonomous system" and an "automatic system" do not exist in different worlds. The Billings taxonomy in Table 1.1 uses the term *autonomous* in the two highest levels of automation, i.e., Autonomous Operation and Operation by Exception, while *automatic* is used in the lower levels. Therefore, we can say that an autonomous system is placed on the continuum of automation, but at the higher levels.

Nuclear power plants (NPPs) have been automated with the following purposes: (1) achieving and maintaining a high level of safety, (2) reducing operators' physical and mental burden, and (3) improving production performance and reliability. To achieve the high safety standards required of the industry, in commercial NPPs, reactor trips and the actuation of safety systems are required to be automated by law or regulation. In the power operation mode, the use of automatic, closed loop control for process parameters such as temperature, pressure, flow, etc. is common (IAEA 1992). In many plants, automation has been extended to include the automatic control of plant start up, mode change, and shutdown. The support of operators' decision-making activities is also an area of automation (NRC 2002); examples of such applications include fault diagnosis, safety function monitoring, plant performance monitoring, and core monitoring.

At present, it can be said that the level of automation in most commercial NPPs is around the Shared Control management mode in the Billings taxonomy. That is, some functions are performed by automatic systems and others are performed by human operators. For instance, the actuation of a reactor trip and the engineered safety features is generally performed by an automatic system, while the termination and resetting of these functions is assigned to operators. In addition, the authority of control can be shared along with the plant mode. For instance, the steam generator (SG) water level is generally controlled by operators in the low power and shutdown modes, but it is automatically controlled in the power operation mode.

Recently, the number of studies attempting to achieve higher levels of automation in the control of NPPs is increasing with the help of artificial intelligence (AI) technology. Researchers have investigated the feasibility of applying fuzzy logic, neural networks, genetic algorithms, and expert systems to conventional control methods to raise the level of automation in different aspects of operations such as reactor start-up, shutdown in emergency situations, fault detection and diagnosis, reactor alarm processing and diagnosis, and reactor load-following operations. For instance, Upadhyaya et al. designed an autonomous control system for a space reactor system using a proportional-integral-derivative (PID) controller (Upadhyaya et al. 2007). Oak Ridge National Laboratory suggested an autonomous decision-making framework for small modular reactors (Cetiner et al. 2014), and Boroushaki et al. proposed an intelligent reactor core controller for load-following operations using AI techniques, i.e., a recurrent neural network (RNN) and fuzzy logic systems (Boroushaki et al. 2003).

The purpose of this book is to introduce one of these efforts to build NPPs to become *autonomous* systems through the use of AI techniques. Although current nuclear plants are already highly automated to reduce human errors and guarantee the reliability of system operations, the term *autonomous* is, at the time of this writing, still not popular in the industry. However, the use of AI techniques and autonomous operation itself seem unavoidable considering the great advantages they can provide, in particular for advanced reactors and small modular reactors. Novel approaches with practical examples for autonomous NPPs that minimize operator intervention are covered in this work.

This book comprises nine chapters. The next section in this chapter introduces a functional architecture of an autonomous operation system for NPPs. A framework is defined that consists of multiple levels and sub-functions necessary for automating operator tasks. The framework includes monitoring, autonomous control, human–autonomous system interfaces (HASIs), and the intelligent management of plant functions. Chapter 2 introduces various AI methods, focusing on the particular techniques adopted in this book. Chapter 3 considers particular approaches to signal validation, which is a prerequisite for a successful autonomous operation system. Chapter 4 covers various AI-based techniques for the diagnosis of abnormal situations, including supervised and unsupervised learning approaches and AI methods. In Chap. 5, a couple of methods to predict plant behavior are presented, and in Chap. 6, AI applications for autonomous control using reinforcement learning (RL) are presented for both normal and emergency situations along with domain analysis methods. Chapter 7 introduces some techniques for monitoring whether the plant is in a stable state and whether the autonomous operation is managing the situations correctly. Chapter 8 focuses on the interaction between operators and autonomous systems; even though autonomous operation minimizes operator interventions, supervision and manual control by operators remain inevitable in NPPs for safety reasons. In addition, this chapter discusses how an autonomous operation system selects operational strategies relevant to the situation. Lastly, Chap. 9 concludes this book.

1.2 A Framework of Autonomous NPPs

The functional architecture for an autonomous controller for space vehicles suggested by Antsaklis et al. (Antsaklis and Passino 1993; Antsaklis et al. 1991) can be considered relevant for NPPs. Figure 1.1 shows a functional framework consisting of three levels for an autonomous NPP, modified from that of Antsaklis. At the lowest level, referred to as the execution level, controllers and sensors are used to directly control and monitor the plant. This level provides automatic control, surveillance, and diagnostic functions for the NPP with the main function to generate control actions as dictated by the higher levels of the system. The execution level also senses the response of the NPP, processes it to identify the relevant parameters, estimates the plant state, and detects system failures, information which is then passed to the higher levels. Accordingly, it is necessary to validate the integrity of the signals from the NPP and detect any signal failures at this level.

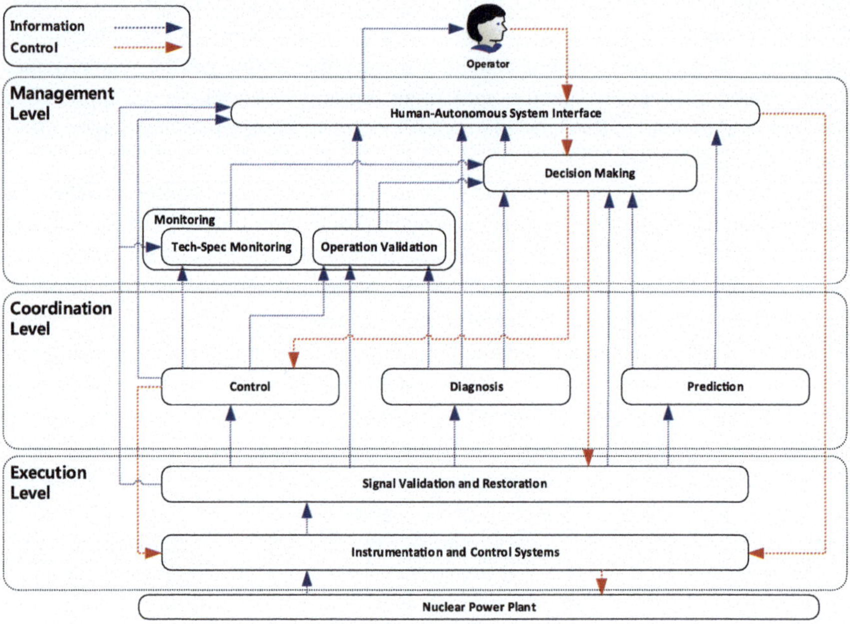

Fig. 1.1 Functional framework for an autonomous NPP

The second level, called the coordination level, receives commands from the top level and generates a sequence of controls as well as an identification algorithm for the execution level. This level has accident management capability to deal with certain component or system failures, for example by detecting and identifying a failure and switching to an alternative control method, along with methods to maintain performance or a certain degree of safety during abnormal or emergency operations. The capability to predict plant behaviors and provide the information to the top level is also included in the coordination level.

The top level or management level oversees and directs all of the activities at both the coordination and execution levels. It evaluates the current situation (i.e., performs monitoring) and predicts what can reasonably be expected within a certain time, as well as generates attainable goals to be accomplished by the autonomous system. The management level also provides for the interaction between the operators and the autonomous system. Through the HASI, operators may intervene in the plant operations by monitoring the status of the NPP and the autonomous system itself, taking over authority, and performing control actions as necessary.

Brief explanations about the functions performed in the levels are as follows, starting with the execution level.

Signal Validation

Signal validation refers to monitoring the signals coming from the NPP through the instrumentation and control (I&C) system and detecting any signal failures. In principle, signals transmitted from the sensors to the operators must be valid and correct in order for the NPP to be operated safely and efficiently. Faulty signals and sensors have the potential to degrade the performances of control systems and operators, which may lead to undesirable situations that compromise NPP safety. When considering an autonomous system, the reliability of the signals becomes even more crucial because the signals are inputs to the system. If the signals are wrong, even a well-developed autonomous system cannot function correctly.

The signal validation function receives signals from the I&C systems as inputs, which take various forms such as parametric values from sensors (e.g., temperature, pressure, flow rate), alarms (e.g., high-pressure alarm), the statuses of systems or components (e.g., pump operating state and valve position), and processed information from computers (e.g., the trip signal and actuation signals from the plant protection system). With such inputs, the signal validation function continuously monitors the signals and validates whether they are normal under the current circumstances. If one is determined to be faulty or missing, this function can generate the expected (or normal) signal. Specific techniques regarding signal validation are covered in Chap. 3.

Diagnosis

In the event of an abnormal or emergency situation, the diagnosis function identifies the initiating event. This function receives the plant parameters, alarms, and system status information selected for diagnosis from the execution level and generates a diagnostic result along with the uncertainty of the result in real time. The result of the diagnosis about the initiating event is then used by the strategy selection function. Several approaches to diagnosis are introduced in Chap. 4.

Prediction

The prediction function forecasts the future states of the plant, systems, and components. This function includes the prediction of the following:

- plant parameters, such as pressure, temperature, and flow rate;
- system and component statuses, such as system actuation and component auto-start/stop;
- system and component health, such as remaining useful life.

In addition to predicting these primary factors, the prediction function has other utilities in autonomous operation. First, it can predict when an abnormal or emergency situation may occur, which can be used to provide a pre-trip alarm so that operators can intervene before the situation actually occurs. Second, the prediction function can be employed by the control and strategy selection functions to choose among operational options. For instance, if several control options are available, the prediction function can evaluate the effectiveness of applying each option, thereby supporting the control function to choose the best one. Chapter 5 discusses methods for the prediction function.

Control

The control function facilitates the autonomous control of systems or components under the given operating condition, which includes normal, abnormal, and emergency situations. This function receives validated NPP signals as inputs, and then based on the inputs, it generates control signals to achieve the desired state as provided by the strategy selection function. These control signals are transmitted to the corresponding systems or components at the execution level. Approaches to the control function are presented in Chap. 6.

Monitoring

The monitoring function covers two aspects of autonomous NPP operation. First, it checks whether the plant is being successfully managed by the autonomous operation system. There are many operational constraints to be monitored in an NPP; for instance, under emergency operation, the critical safety function (CSF)s must be monitored to confirm that they are working correctly. In all operation modes, the limiting conditions for operation (LCOs) should be maintained to ensure the safety of the NPP. Second, the monitoring function keeps track of the health or functionality of the autonomous system itself. In other words, it checks whether the functions of the autonomous system are working correctly, and if any abnormality is detected, an alternative is chosen or operator intervention is requested. Chapter 7 presents some examples of related studies.

Human–Autonomous System Interface

The human–autonomous system interface or HASI provides the means of interaction between the operators and autonomous system. Similar to the main control room

(MCR) of current commercial NPPs, the HASI includes alarms, information displays, and controllers for manual operation. In addition, the HASI conveys the monitoring information regarding the status of the NPP and the autonomous operations, e.g., the autonomous control activities. Alerts to request operator intervention and the means to transfer control authority between the operators and the autonomous system are also presented through this interface. The HASI is discussed further in Chap. 8.

Strategy Selection
The strategy selection function makes the decisions for the operation as the brain of the autonomous system. The main role of this function is to determine whether the operational strategy that is currently applied is adequate and to monitor whether all the functions of the autonomous systems are working normally. To perform decision-making, the strategy selection function utilizes information from the other functions, namely signal validation, diagnosis, prediction, monitoring, and control. If the current strategy is evaluated as ineffective, this function generates a new strategy as an output, which can be an alternative autonomous operation mode (or algorithm) or a request for operator intervention if there is no applicable autonomous operation mode. In addition to discussing the HASI, Chap. 8 also provides an example of the strategy selection function.

References

Antsaklis PJ, Passino KM (1993) An introduction to intelligent and autonomous control. Kluwer Academic Publishers

Antsaklis PJ, Passino KM, Wang SJ (1991) An introduction to autonomous control systems. IEEE Control Syst Mag 11(4):5–13

Billings C (1997) Aviation automation: the search for a human-centered approach

Boroushaki M, Ghofrani MB, Lucas C, Yazdanpanah MJ (2003) An intelligent nuclear reactor core controller for load following operations, using recurrent neural networks and fuzzy systems. Ann Nucl Energy 30(1):63–80

Cetiner SM, Muhlheim MD, Flanagan GF, Fugate DL, Kisner RA (2014) Development of an automated decision-making tool for supervisory control system. ORNL/TM-2014/363 (SMR/ICHMI/ORNL/TR-2014/05), Oak Ridge National Laboratory, Oak Ridge, TN

Endsley MR, Kaber DB (1999) Level of automation effects on performance, situation awareness and workload in a dynamic control task. Ergonomics 42(3):462–492

IAEA (1992) The role of automation and humans in nuclear power plants. Int At Energy Agency

Joseph L, Mondal AK (2021) Autonomous driving and Advanced Driver-Assistance Systems (ADAS): applications, development, legal issues, and testing. CRC Press

NRC (2002) Human-system interface design review guideline (NUREG-0700 Rev. 2). US Nuclear Regulatory Commission

Sheridan T, Verplanck W (1978) Human and computer control of undersea. Teleoperators, 1, pp 1.2

Upadhyaya BR, Zhao K, Perillo SR, Xu X, Na M (2007) Autonomous control of space reactor systems. Univ. of Tennessee, Knoxville, TN (United States)

Chapter 2
Artificial Intelligence and Methods

2.1 Definitions of AI, Machine Learning, and Deep Learning

2.1.1 AI

AI is a broad classification that includes all artificially implemented intelligence. While artificially implemented intelligence may or may not engage in learning, all types of AI that mimic human behavior are generally classified as AI. Research into AI has been actively conducted since the mid-1900s.

Algorithms included in the AI classification range from spam filters with a simple structure and explicit condition provided that checks a specific string and classifies it as spam to current complicated AI models. Algorithms that are AI but not included in machine learning (ML) include algorithms with explicit conditions.

2.1.2 ML

ML is an AI methodology in which the artificially implemented intelligence can learn from previous experience. Algorithms belonging to this category can set and update their criteria through learning even if there are no explicit conditions. Representative methodologies that are ML but not deep learning (DL) include regression analysis, Bayes classifiers, K-means clustering, support vector machine (SVM)s, and single-layer artificial neural networks.

J. Kim et al., *Autonomous Nuclear Power Plants with Artificial Intelligence*, Lecture Notes in Energy 94, https://doi.org/10.1007/978-3-031-22386-0_2

2.1.3 DL

DL is a specific AI category that includes artificial neural networks with multiple layers. Multiple single-layer perceptrons can configure deep neural network (DNN) structures. While research on single-layer perceptrons has been widely conducted since the mid-1900s, expansion to DL research was difficult due to the problems such as data quantity, quality, and operation speed. Figure 2.1 illustrates the relation among AI, ML, and DL.

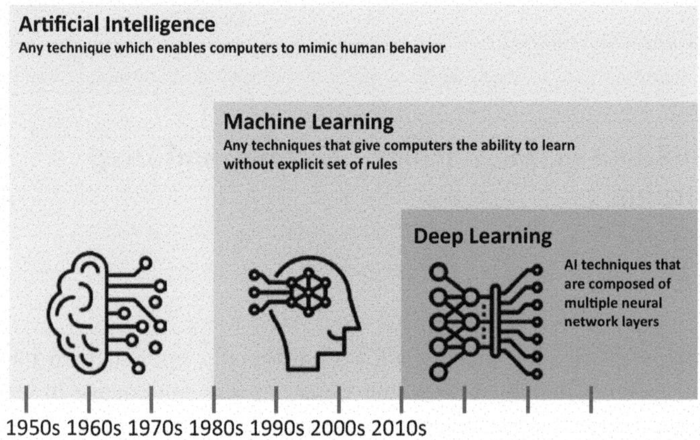

Fig. 2.1 AI, ML, and DL

2.2 Classification of ML Methods Based on Learning Type

ML methodologies are largely divided into supervised learning, unsupervised learning, and reinforcement learning according to the dataset used for training and the learning method of the algorithm; see Fig. 2.2. In this chapter, we discuss applications of each learning method, and specifically, what problems they can help solve in the nuclear field. In this chapter, the input data are expressed as vector X and the output data are expressed as vector Y, as in Eq. (2.1).

$$Input\ data\ X = \begin{bmatrix} x_{11} & \cdots & x_{1n} \\ \vdots & \ddots & \vdots \\ x_{n1} & \cdots & x_{nn} \end{bmatrix}$$

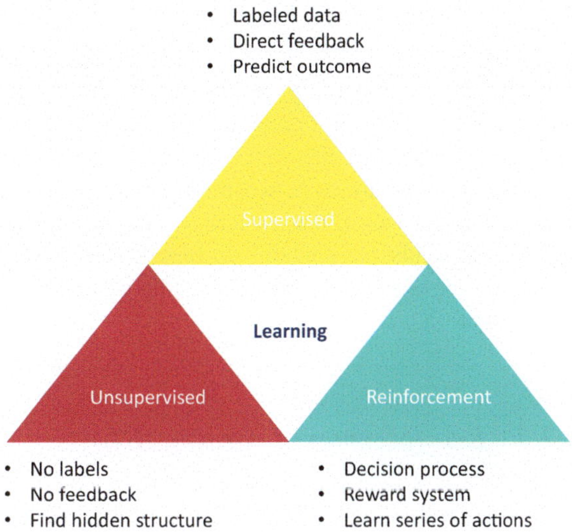

Fig. 2.2 ML methods based on learning type

$$Output\ data\ Y = \begin{bmatrix} y_1 \\ \vdots \\ y_2 \end{bmatrix} \quad (2.1)$$

2.2.1 *Supervised Learning*

Supervised learning is a learning method that can be applied to datasets with pre-assigned input–output pairs. In general, supervised learning applications assume the following structure: a pair of vectors X corresponding to the input, and vector Y corresponding to the output, where the algorithm learns to estimate Y by receiving X. In this case, the algorithm is guided by actual data for Y; it is this characteristic that led to the name "supervised learning". The learning criterion is based on the difference between the actual Y and the estimated Y vectors using the algorithm, Eq. (2.2).

$$Error\,function = f\left(Y_{real} - Y_{estimate}\right) \quad (2.2)$$

Suitable fields for applying supervised learning include classification problems, where Y is classified based on the characteristics of X, and regression problems, where Y is inferred based on the history from X.

2.2.2 Unsupervised Learning

Unlike supervised learning, unsupervised learning utilizes a dataset that does not include the output in the dataset configuration. In this case, the algorithm is unguided because there no explicit correct answer is provided. Instead, the algorithm mainly identifies the similarities between the X vectors received as inputs or the relatively important properties of the X vectors with various characteristics.

Suitable fields for applying unsupervised learning include clustering, feature extraction for the identification and compression of data properties, and dimensionality reduction.

2.2.3 Reinforcement Learning (RL)

Unlike supervised and unsupervised learning schemes, RL involves an agent and an environment. The agent is an entity that operates based on a specific algorithm, and the environment refers to the situation the agent is in. In a fixed space, the agent's behavior affects the environment. Thus, the scores of actions by the agent are differentiated according to how the environment changes by the behavior. For RL, the agent, actions of the agent, reward, environment, and state of the environment should all be defined.

A suitable field for applying RL is strategy finding.

2.3 Overview of Artificial Neural Networks (ANNs)

2.3.1 History of ANNs

Figure 2.3 shows of a general timeline of ANN research. In 1943, a mathematical model of an ANN was proposed by McCulloch and Pitts (McCulloch and Pitts 1943). Based on this, Rosenblatt proposed a single layer perceptron structure in 1957, which received high academic praise for its characteristic of mimicking the operation of the human nervous system (Rosenblatt 1957). However, the single layer perceptron framework, limited to constructing a linear boundary, could not be applied to the XOR problem requiring nonlinear boundaries. Minsky and Papert later revealed that multilayer perceptrons are difficult to apply owing to the inefficiency of learning (Minsky and Papert 1969), which initiated a period of stagnation in ANN research. But in 1975, Werbos proposed a backpropagation algorithm that can efficiently train multilayer neural networks (Werbos 1974). Based on his research, a seminal study applying the backpropagation algorithm to ANN learning was conducted by Rumelhart (Rumelhart et al. 1986a).

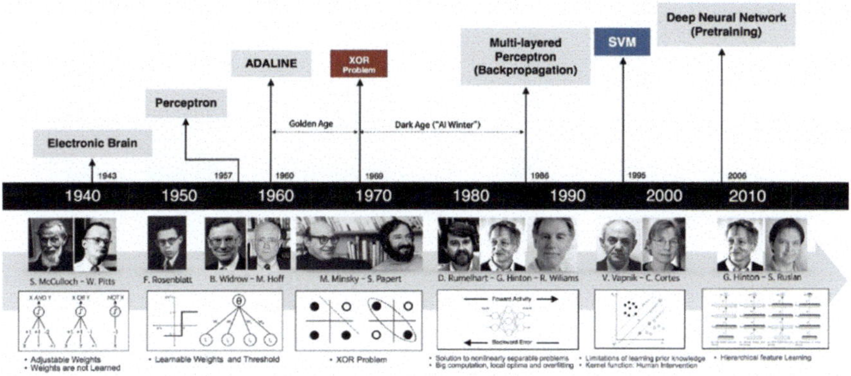

Fig. 2.3 History of ANN development

Since then, through steady research, various techniques and algorithms to improve the learning capabilities of neural networks have been developed, such as dropout, initialization, optimization, and others. Moreover, computing power has significantly improved, as evidenced by a computer program winning a Go match over a world champion in 2016, a feat long considered highly difficult for machines.

2.3.2 Overview of ANNs

2.3.2.1 Basic Calculations of ANNs

The general structure of an artificial neuron is illustrated in Fig. 2.4. After receiving an input, the neuron framework multiplies the received input by a preset weight and passes it to the transfer function. The transfer function performs a predefined operation and delivers the result to the activation function. Finally, the activation function outputs the result corresponding to the input. For input data X and output data Y, the calculation result of an ANN is as follows:

The mathematical expression of the calculation process in an artificial neuron is written as in Eqs. (2.3) and (2.5).

$$v(network\ input) = x_1 w_1 + x_2 w_2 + \cdots + x_i w_i + x_0 w_0 \tag{2.3}$$

$$\begin{gathered} y_{estimate}(output\ of\ neuron) = f(v) \\ Where, f\ is\ an\ activation\ function \end{gathered} \tag{2.4}$$

Finally, the generated error (E) can be defined as below.

Fig. 2.4 ANN architecture

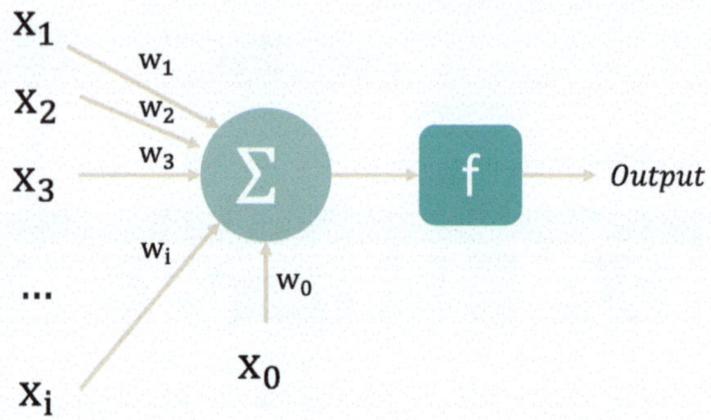

$$E = g(y - y_{estimate})$$
$$where, g \ is \ a \ loss \ function \quad (2.5)$$

Learning of ANNs is accomplished by adjusting the weights of the neurons based on errors. The learning process of an artificial neuron based on the calculation results is defined as in Eq. (2.6).

$$w_{updated,i} = w_i - \alpha \nabla_{w_i} E$$
$$Where, \alpha \ is \ a \ learning \ rate \quad (2.6)$$

Weights are updated based on the contribution of a corresponding weight to the overall error. In the next section, the design of an actual ANN following this calculation process is described.

2.3.2.2 How to Construct an ANN?

ANNs are designed according to the learning method of the specific network. In other words, Eq. (2.6) defines the basic ANN design process. The first step in constructing Eq. (2.6) is to define an error function, which requires a value for $y_{estimate}$ according to Eq. (2.5). To define $y_{estimate}$, it is necessary to define the activation function. As shown in Table 2.1, various functions can be used as the activation function. More specifically, any function that can simulate a neuron's threshold can be used as the activation function.

To reiterate, after calculating $y_{estimate}$ via the defined activation function, the error function can be defined. This function can take various forms as well; that is, any error function suitable for the problem at hand can be chosen. For a regression problem, suitable error functions include mean squared error (MSE), mean squared logarithmic error (MSLE), and mean absolute error (MAE). For performing binary classification

Table 2.1 Activation functions

Activation function	Equation	Graph
Unit step	$f(z) = \begin{cases} 0, z < 0 \\ 1, z \geq 0 \end{cases}$	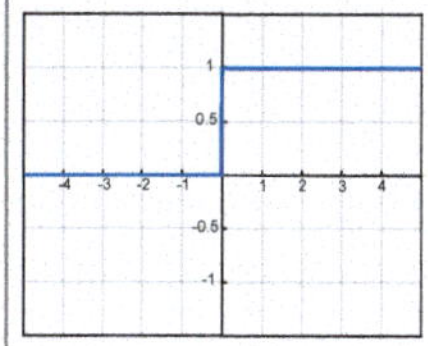
Signum	$f(z) = \begin{cases} -1, z < 0 \\ 0, z = 0 \\ 1, z > 0 \end{cases}$	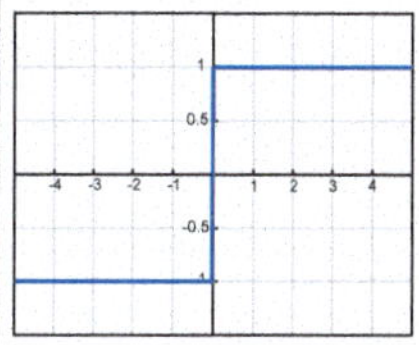
Linear	$f(z) = z$	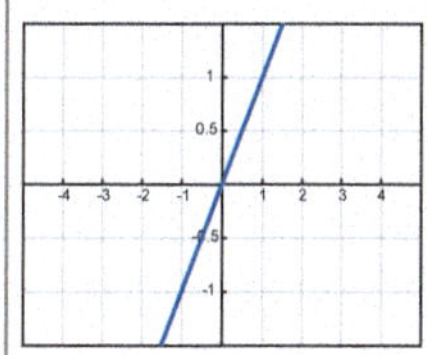
Sigmoid	$f(z) = \frac{1}{1+e^{-z}}$	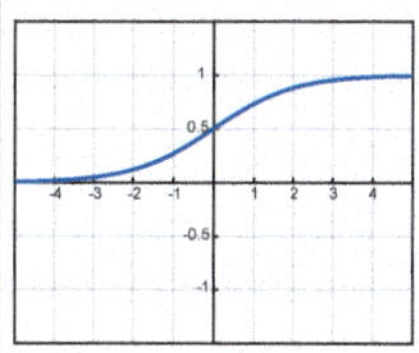
Hyperbolic tangent	$f(z) = \frac{e^{z}-e^{-z}}{e^{z}+e^{-z}}$	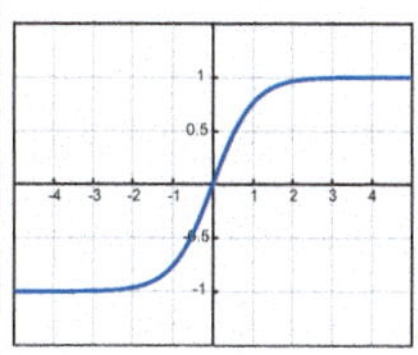
ReLU (Rectified Linear Unit)	$f(z) = max(0, z)$	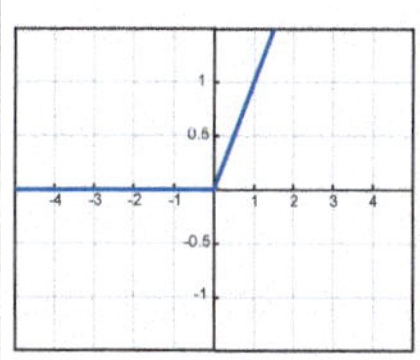

Table 2.2 Error functions

Regression		
	MSE	$\frac{1}{n}\sum_{i=1}^{n}(y_i - y_{i,estimated})^2$
	MSLE	$\frac{1}{n}\sum_{i=1}^{n}\begin{pmatrix}\log(y_i+1)- \\ \log(y_{i,estimated}+1)\end{pmatrix}^2$
	MAE	$\frac{1}{n}\sum_{i=1}^{n}\lvert y_i - y_{i,estimated}\rvert$
Classification		
	Binary cross entropy loss	$-\frac{1}{n}\sum_{i=1}^{n}\begin{pmatrix}y_i * \log(y_{i,estimated})+ \\ (1-y_i) * \log(1-y_{i,estimated})\end{pmatrix}$
	Categorical cross entropy loss	$-\sum_{i=1}^{n} y_i * \log(y_{i,estimated})$
	Kullback–Leibler divergence loss	$-\sum_{i=1}^{n} y_i * \log\left(\frac{y_{i,estimated}}{y_i}\right)$

problems, an appropriate error function is binary cross-entropy loss. Error functions suitable for multi-class classification include categorical cross-entropy loss and Kullback–Leibler divergence loss (Table 2.2).

After determining the appropriate activation and error functions, it is necessary to set up a gradient descent optimization algorithm that determines the update method for the weights according to the error. Equation (2.6) expresses a stochastic gradient descent algorithm, one type of gradient descent optimization algorithm. The update rule of the stochastic gradient descent algorithm is intuitive. However, it converges more slowly than other gradient descent optimization algorithms (which will be described later) and may converge to a local minimum. Therefore, various gradient descent optimization algorithms have been developed to secure fast convergence to global minima. Three such algorithms are described as follows.

Momentum (Qian 1999)

The momentum algorithm is a method of providing inertia when weights are updated through gradient descent. The final update value is calculated by reflecting not only the current gradient value but also the direction it moved in the past. The momentum algorithm is expressed as follows.

$$\begin{aligned} v_{updated} &= \gamma v + \alpha \nabla_{w_i} E \\ w_{i,updated} &= w_i - v_{updated} \\ &Where,\ \alpha\ is\ a\ learning\ rate \\ &\gamma\ is\ a\ momentum\ rate \end{aligned} \tag{2.7}$$

In Eq. (2.7), v is the exponential average of the previous gradients. Therefore, by reflecting the movement direction of the previous gradient in the current update, the momentum algorithm is relatively free from oscillations compared with the stochastic gradient descent method.

Adaptive Gradient (Adagrad) (Duchi et al. 2011)
The Adagrad algorithm is a method of updating rarely changed variables with a large step size and frequently changed variables with a small step size, for the following reasons. As frequently updated variables are more likely to be located near the optimum value, the weights are updated with a small step size to determine the optimum value. Variables that are rarely updated are likely to be far from the optimum, and thus a large step size is applied. The Adagrad algorithm is defined as below.

$$
\begin{aligned}
G_{updated} &= G + \left(\nabla_{w_i} E\right)^2 \\
w_{i,updated} &= w_i - \frac{\eta}{\sqrt{G_{updated} + \epsilon}} * \nabla_{w_i} E \\
&Where,\ \eta\ is\ step\ size \\
&\epsilon\ is\ a\ constant\ t
\end{aligned}
\tag{2.8}
$$

Adaptive Moment Estimation (Adam) (Kingma and Ba, 2015)
The Adam algorithm possesses the characteristics of both the Adagrad and momentum algorithms. Here, the exponential average of the slope and the exponential average of the square of the slope are used simultaneously. The Adam algorithm is defined as follows.

$$
\begin{aligned}
m_{updated} &= \beta_1 m + (1 - \beta_1) \nabla_{w_i} E \\
v_{updated} &= \beta_2 v + (1 - \beta_2) \left(\nabla_{w_i} E\right)^2 \\
w_{i,updated} &= w_i - \frac{\eta}{\sqrt{v_{updated} + \epsilon}} * m_{updated}
\end{aligned}
\tag{2.9}
$$

In sum, a basic artificial neuron can be created by defining a threshold function, defining a loss function according to the problem, and selecting an appropriate optimization algorithm. Through the aforementioned sequences, the ANN updates the weights of the neurons in a series of processes called error backpropagation. The operation of repeatedly correcting the weights of a neural network until an optimum value is obtained is called training. In general, training, test, and validation datasets are separated to evaluate the fitness of the model.

In the next section, various ANN structures are explained.

2.4 ANN Algorithms

An ANN model can be configured in various forms depending on the data preprocessing method, the computational process of the ANN, and the arrangement of the ANN. Among numerous models, this section briefly describes the ANN algorithms that will be discussed later in detail.

2.4.1 Convolutional Neural Networks (CNNs)

The first study on modern convolutional neural networks (CNNs) was conducted by LeCun (LeCun et al. 1998). This work suggested a general structure of a CNN and successfully implemented the algorithm for handwritten character recognition. More recently, Krizhevsky et al. suggested a CNN-based image classification model, called AlexNet (Krizhevsky et al. 2017), that demonstrated state-of-the-art results in image recognition.

A deep CNN has a feature that reflects the Euclidean characteristics of the input data by attaching a convolution layer that performs a preprocessing function before the multilayer neural network. The only difference between CNNs and conventional ANNs is the convolution layer. What the convolution layer does and how it works can be understood as follows.

In general, for image data, the left and right pixels are important. However, the properties of the adjacent pixels are important as well. To reflect the properties of the adjacent data, the convolution layer performs a convolution operation between the input data and a filter. Through this, the data is compressed, including information regarding the adjacent data. The following example demonstrates how the convolution layer functions.

Input data typically take the form of two-dimensional (2D) data (3×3 matrices). The filter is also a 2D matrix (2×2 matrix), as below.

$$X = \begin{bmatrix} 1 & 2 & 1 \\ 0 & 1 & 2 \\ 3 & 2 & 1 \end{bmatrix}, \quad \text{filter} = \begin{bmatrix} 1 & 2 \\ 2 & 1 \end{bmatrix} \tag{2.10}$$

The filter performs a convolution operation on the input data.

$$\boldsymbol{X} = \begin{bmatrix} \boxed{\begin{matrix} 1 & 2 \\ 0 & 1 \end{matrix}} & \begin{matrix} 1 \\ 2 \end{matrix} \\ \begin{matrix} 3 & 2 \end{matrix} & 1 \end{bmatrix}, \text{filter} = \begin{bmatrix} 1 & 2 \\ 2 & 1 \end{bmatrix} \qquad \text{(2.10 Description 1)}$$

From the convolution operation between the data $\begin{bmatrix} 1 & 2 \\ 0 & 1 \end{bmatrix}$ and filter $\begin{bmatrix} 1 & 2 \\ 2 & 1 \end{bmatrix}$, a result of 6 is calculated. Repeating this process, the calculation result from the convolution

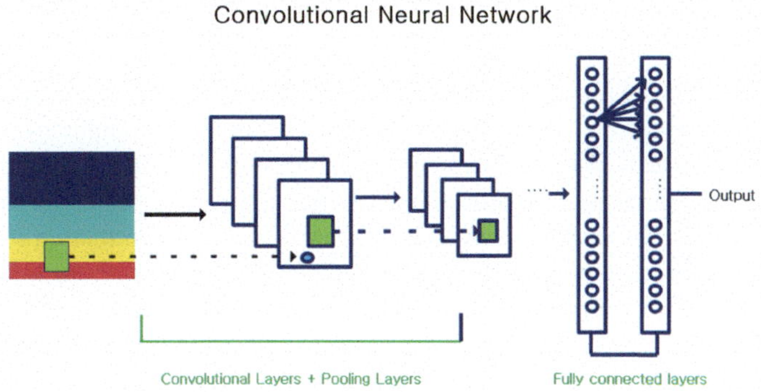

Fig. 2.5 Schematic diagram of a CNN

layer can be stated as follows:

$$
\begin{aligned}
&ouput\,from\,convolution\,layer = \begin{bmatrix} 6 & 8 \\ 10 & 10 \end{bmatrix} \\
&output\,from\,flattening\,layer = \begin{bmatrix} 6 \\ 8 \\ 10 \\ 10 \end{bmatrix}
\end{aligned}
\tag{2.11}
$$

The existing 3 × 3 data are compressed into 2 × 2 data using adjacent information. Subsequently, the result of the convolution operation is flattened through a flattening layer, the result of which is input to the ANN; see Fig. 2.5.

2.4.2 Recurrent Neural Networks (RNNs), Long Short-Term Memory Networks (LSTMs), and Gated Recurrent Units (GRUs)

While a CNN performs an operation that reflects the spatial characteristics of the data, RNNs, long short-term memory networks (LSTMs), and gated recurrent units (GRUs) all perform operations that reflect the temporal characteristics of the data. The basic idea behind the RNN was suggested by Rumelhart (Rumelhart et al. 1986b). The basic structure of both LSTMs and GRUs is the same as that of an RNN; however, they are optimized models with improved calculation methods in the neurons to solve problems specific to RNNs. In this section, the basic structure of an RNN and the problems that occur in the structure are discussed. Then a method to solve the problems that occur in LSTMs and GRUs is provided.

2.4.2.1 RNNs

An RNN, which is the most basic model that reflects data's temporal characteristics, sends not only the calculation results obtained in a neuron to the output layer but also to the input of the next neuron. This structure enables the $i + 1$-th neuron to consider not only the data from x_{i+1} but also the data from x_i indirectly by using the output of the *i*-th neuron. The output from the *i*-th neuron contains information from x_0 to x_i directly or indirectly. In other words, through the RNN structure, the influence of the data from the previous time instance can be considered in the next layer.

However, RNNs have a critical problem, called the vanishing gradient problem. In Fig. 2.6, we assume that error E_2 is calculated. The backpropagation result for each hidden layer can be written as follows.

$$\frac{\partial E_2}{\partial H_2} = \frac{\partial E_2}{\partial H_2}$$
$$\frac{\partial E_2}{\partial H_1} = \frac{\partial E_2}{\partial H_2}\frac{\partial H_2}{\partial H_1}$$
$$\frac{\partial E_2}{\partial H_0} = \frac{\partial E_2}{\partial H_2}\frac{\partial H_2}{\partial H_1}\frac{\partial H_1}{\partial H_0} \tag{2.12}$$

Backpropagation from the last to the first neuron inevitably results in multiple multiplication operations. In general, the gradient value of the hidden layer is less than 1, and therefore, the error is diluted as the information is transmitted during repeated multiplication operations. Consequently, while adjacent neurons are trained properly, neurons far from the result are trained improperly, resulting in the vanishing gradient problem. In other words, the algorithm can tackle short-term memory problems with the information of the nearby neurons but is vulnerable to problems requiring long-term memory.

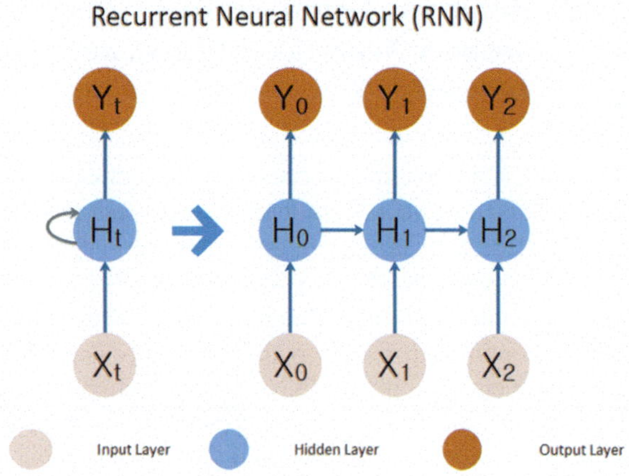

Fig. 2.6 Schematic diagram of an RNN

2.4.2.2 LSTMs

First suggested by Hochreiter and Schmidhuber (1997), the basic structure of an LSTM is the same as that of an RNN. The LSTM likewise sends the calculation results obtained in the neuron to the output layer as well as to the input layer of the next neuron. However, LSTMs differ from RNNs in that they utilize an LSTM cell instead of a simple neuron. The LSTM cell consists of a forget gate, input gate, and output gate. As shown in Fig. 2.7, the cell state passed by the LSTM cell to the next cell can be expressed as follows.

$$c_t = f_t * c_{t-1} + i_t * \widehat{c_t} \tag{2.13}$$

The partial derivation of c_t over c_{t-1} can be written as below.

$$\frac{\partial c_t}{\partial c_{t-1}} = \frac{\partial f_t}{\partial c_{t-1}} * c_{t-1} + f_t + \frac{\partial i_t}{\partial c_{t-1}} * \widehat{c_t} + \frac{\partial \widehat{c_t}}{\partial c_{t-1}} * i_t \tag{2.14}$$

It should be observed that the gradient between the cell states is not composed of multiplication operations only. Accordingly, it is possible to prevent the vanishing gradient problem thanks to the addition operation in the error backpropagation process. As a result, both short-term and long-term memory tasks can be tackled relatively well compared to naïve RNNs.

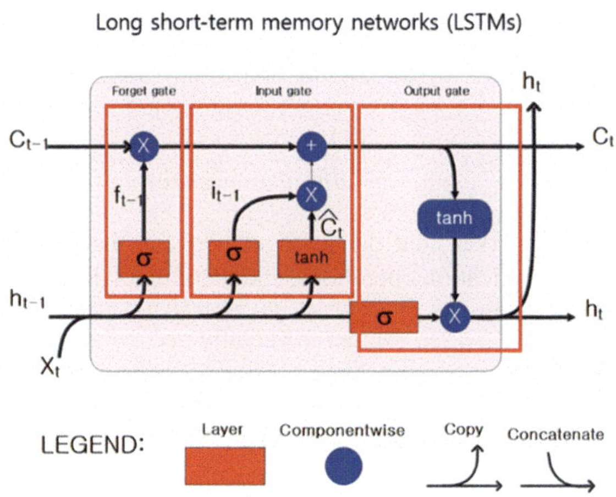

Fig. 2.7 Schematic diagram of an LSTM cell

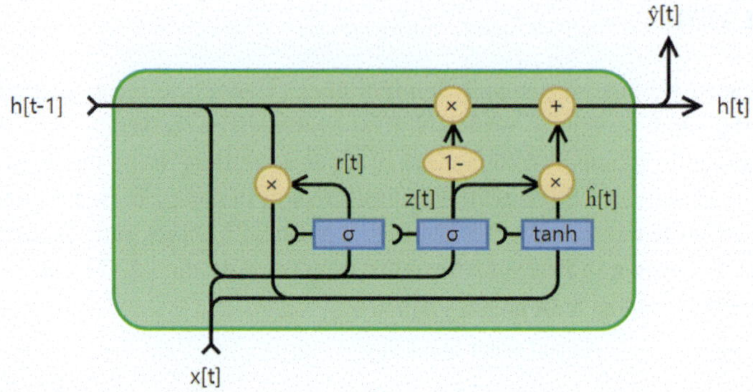

Fig. 2.8 Schematic diagram of a GRU

2.4.2.3 GRUs

The GRU structure was first suggested by Cho et al. (2014). Figure 2.8 represents a GRU structure. GRUs operate in a similar manner as LSTMs, but unlike LSTMs, GRUs only have two gates. In an LSTM, the forget and input gates operate independently, while in a GRU model, the total amount of information is fixed and the input is forgotten as soon as it is processed. This leads to improved operation speed compared with that of an LSTM because GRUs can operate with fewer parameters.

The output of a GRU and the partial derivation of h_t over h_{t-1} can be expressed as follows.

$$h_t = z_t * h_{t-1} + (1 - z_t) * \widehat{h_t} \tag{2.15}$$

$$\frac{\partial h_t}{\partial h_{t-1}} = \frac{\partial z_t}{\partial h_{t-1}} * h_{t-1} + z_t + \frac{\partial (1 - z_t)}{\partial h_{t-1}} * \widehat{h_t} + \frac{\partial \widehat{h_t}}{\partial c_{t-1}} * (1 - z_t) \tag{2.16}$$

Again, it can be observed that the gradient between the hidden layers is not composed of only multiplication operations, and therefore, GRUs can also avoid the vanishing gradient problem.

2.4.3 *Variational Autoencoders (VAEs)*

The VAE structure was first suggested by (Kingma and Welling (2013), which is based on a autoencoder (AE) structure. The VAE is a generative model that consists of two paired ANNs to learn the data-generation process. For example, suppose we need to create a distribution that is as similar as possible to a given distribution, using the distribution of math scores in a class as an input. If the math scores follow a normal distribution, their distribution can be replicated using the mean and standard

Fig. 2.9 Schematic diagram of a VAE

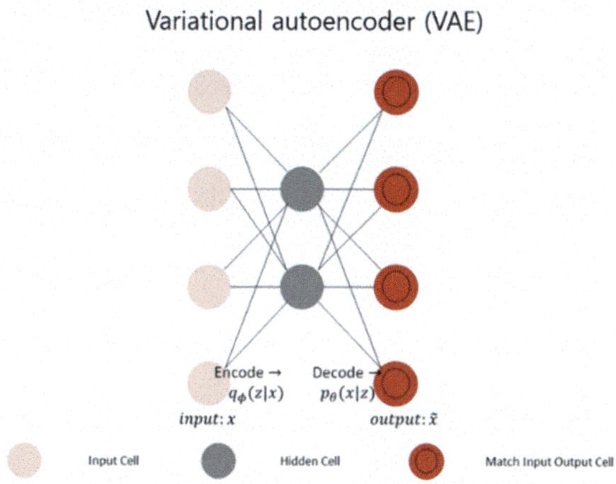

deviation of the math score distribution. In other words, the distribution of the math scores can be compressed into two variables: mean and standard deviation. Data that follow a normal distribution can be encoded statistically following this process; but in general, it is difficult to understand the distribution of the data. A model that both encodes and decodes based on ANNs is called a VAE.

In Fig. 2.9, the ANN corresponding to the front part acts as an encoder, while the ANN corresponding to the rear part acts as a decoder. The encoder neural network receives input data x and creates a latent vector z. The encoder extracts the potential features from the data and compresses them. Conversely, the decoder neural network receiving the latent vector z imitates the input data of the encoder, x, as similarly as possible.

The performance of a VAE depends on the performance of the decoder neural network. Therefore, in designing the decoder neural network loss function, the output of the network should be similar to the existing data x [maximize $p_\theta(x)$]. This can be expressed as follows.

$$p_\theta(x) = \int p_\theta(z)p_\theta(x|z)\mathrm{d}z \tag{2.17}$$

Maximizing Eq. (2.17) should be learned by the network, but it is basically impossible to calculate for the latent vector z as there can be countless cases. To solve this problem, a variational inference technique is used. By defining an additional encoder network $q_\phi(z|x)$ that tracks $p_\theta(x|z)$, the equation can be rewritten as follows with the evidence lower bound.

$$p_\theta(x) \geq L(x^i, \theta, \phi) = \mathrm{E}_z\left[\log\left(p_\theta\left(x^i|z\right)\right)\right] - \mathrm{D_{KL}}(q_\phi\left(z|x^i\right)||p_\theta(z)) \tag{2.18}$$

Changing Eqs. (2.17)–(2.18) has two advantages. The Kullback–Leibler divergence between the distributions can be calculated easily, and thus the meaning of loss becomes clear. The first term on the right-hand side of Eq. (2.18) corresponds to the reconstruction loss. As mentioned above, the encoder takes data x and creates a latent vector z, while the decoder receives the latent vector z created by the encoder and restores the original data x; the first term on the right side of the above expression indicates the cross-entropy between the two. The second term on the right-hand side of Eq. (2.18) corresponds to the Kullback–Leibler divergence regularizer. In other words, z sampled from the posterior is as diverse as possible (to prevent mode collapse), while simultaneously, the amount of information in the posterior and the prior should be similar.

2.4.4 Graph Neural Networks (GNNs)

The basic structure of the modern GNN model was first suggested by Gori (Gori et al. 2005) and then later refined by Scarselli (Scarselli et al. 2008). Figure 2.10 shows a schematic diagram of a GNN.

While a CNN is a network that has been preprocessed to reflect the spatial characteristics of the input data (Euclidean space), a GNN is a network model that has been preprocessed to reflect the characteristics of non-Euclidean space data. For photo-type data, adjacent data are very important, but for graph-type data such as social networks, the existence of relationships among the data is more important than physical distance.

In a GNN, graph information to be provided to the input layer can be extracted using a graph Laplacian. The graph Laplacian operation is defined as follows.

$$\nabla^2 f = \sum_{v_j \sim v_i} [f(v_i) - f(v_j)]$$

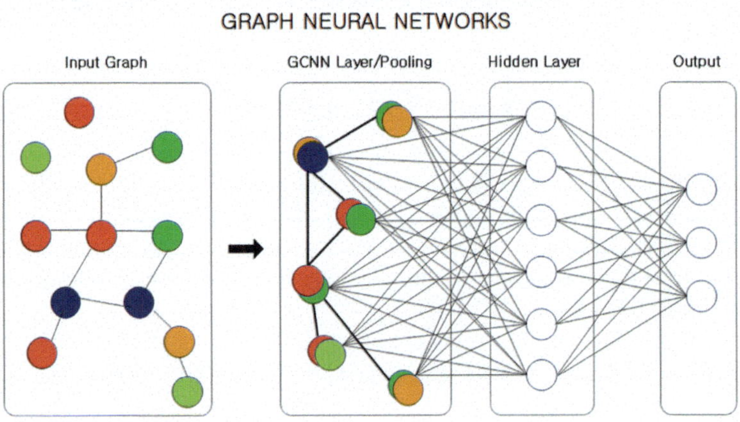

Fig. 2.10 Schematic diagram of a GNN

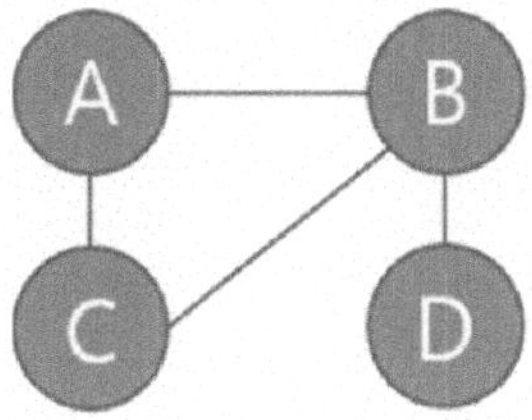

$$\text{A (Adjacent matrix)} = \begin{matrix} 0 & 1 & 1 & 0 \\ 1 & 0 & 1 & 1 \\ 1 & 1 & 0 & 0 \\ 0 & 1 & 0 & 0 \end{matrix}$$

$$\text{L (Laplacian matrix)} = \begin{matrix} 2 & -1 & -1 & 0 \\ -1 & 3 & -1 & -1 \\ -1 & -1 & 2 & 0 \\ 0 & -1 & 0 & 1 \end{matrix}$$

Fig. 2.11 Example of a graph Laplacian

$$Where,\, i\, and\, j\, are\, nodes \tag{2.19}$$

Figure 2.11 shows the result of taking a graph Laplacian on an example graph structure

In the adjacency matrix, only the connection relationship between each node is expressed. However, as a result of transforming into a Laplacian matrix, the number of edges exiting from each node is expressed by checking the diagonal element as well as the connection relationship between nodes. In addition, when the graph Laplacian is performed, the matrix becomes diagonalizable, as in Eq. (2.20).

$$L = D - A = D^{-\frac{1}{2}} L D^{-\frac{1}{2}} = U \Lambda U^T \tag{2.20}$$

The diagonalized matrix U contains the results of the graph information. If matrix U is reflected in the neuron operation in the form of Eq. (2.21), then the neuron operation reflecting the graph information can be performed.

$$\begin{aligned} &H^k (Output\, from\, k - th\, hidden\, layer) = \sigma\left(\sum_{i=1}^{f_{k-1}} U W_{i,j}^k U^T H_{:,i}^{k-1}\right) \\ &Where,\, W\, is\, weight \\ &\sigma\, is\, the\, activation\, function \end{aligned} \tag{2.21}$$

2.4.5 Generative Adversarial Networks (GANs)

The basic structure of the GAN model was first suggested by Goodfellow (Goodfellow et al. 2020). Figure 2.12 represents a schematic diagram of a GAN. A GAN, similar to a VAE, is a structure that combines two ANNs. In this case, the two parts are a generator network and a discriminator network. The generator network is a model that generates data similar to the received data, following a similar mechanism as that of the VAE model. The discriminator network receives the generated data as an input and tries to distinguish whether the received data are simulated by the generator network or actual real-world data. If the discriminator well distinguishes between the data simulated by the generator and actual data, the generator network receives a large loss and the discriminator receives a small loss; in this way, the two networks simultaneously learn in an adversarial manner. When the generator well simulates the real data, the discriminator network reaches a state of random guessing where it can no longer provide the correct answer. Learning is stopped at this point and the generator network is removed, as a generator that effectively simulates real data has been achieved. Owing to this operation, the loss function of a GAN can be described in two ways, as shown in Eq. (2.22).

$$\begin{aligned}
&\textit{Generator loss} = \min_G L(D, G) \\
&\quad = \mathrm{E}_{\mathrm{x}\sim \mathrm{p}_{\mathrm{data}}(\mathrm{x})}\left[\log(D(x))\right] + \mathrm{E}_{\mathrm{z}\sim \mathrm{p}_{\mathrm{z}}(z)}(\log(1 - D(G(\mathrm{Z})))) \\
&\textit{Discriminator loss} = \max_D L(D, G) \\
&\quad = \mathrm{E}_{\mathrm{x}\sim \mathrm{p}_{\mathrm{data}}(\mathrm{x})}\left[\log(D(x))\right] + \mathrm{E}_{\mathrm{z}\sim \mathrm{p}_{\mathrm{z}}(z)}(\log(1 - D(G(\mathrm{Z})))) \\
&\text{where, D is a discriminator network} \\
&\text{G is a generator network} \\
&\mathrm{x} \sim \mathrm{p}_{\mathrm{data}}(x) \textit{ is a distribution of the real data} \\
&\mathrm{z} \sim \mathrm{p}_{\mathrm{z}}(z) \text{is a distribution from the latent vector}
\end{aligned} \tag{2.22}$$

If the generator network perfectly imitates real data, the discriminator judges it to be real data with a high probability, and D(G(Z)) converges to 1. In this case, the generator network has a very small loss by the last term in Eq. (2.22).

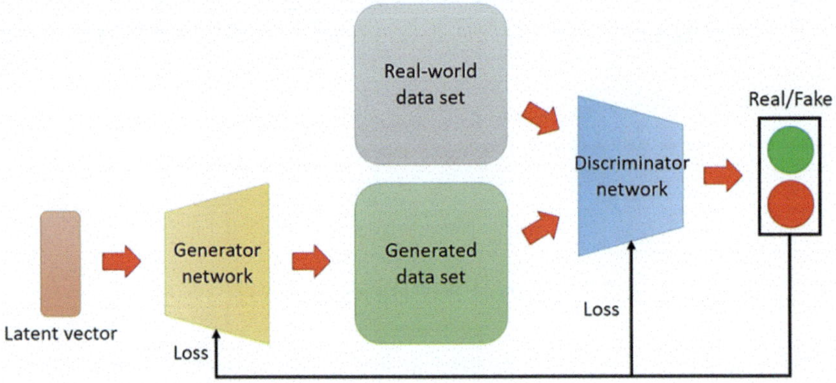

Fig. 2.12 Schematic diagram of a GAN

2.5 Model-Based and Data-Based Approaches

Detection and diagnosis problems can be solved using two different approaches, namely model-based and data-based methods. Both can be classified as quantitative or qualitative (Alzghoul et al. 2014), and thus these approaches can be arranged into analytical methods (quantitative model-based methods), knowledge-based methods (qualitative model-based and data-based methods), and data-driven methods (quantitative data-based methods). In this book, for convenience of understanding, quantitative model-based methods are called *model-based methods*, and quantitative data-based methods are referred to as *data-driven methods*.

A model-based approach uses explicit rules (e.g., a mathematical model of the system), and therefore a deep understanding of the system is required. As a result of using explicit rules, fast and accurate inference and extrapolation are possible. The drawback, though, is that designing explicit rules requires a huge amount of time and expense. Moreover, it is difficult to model multi-dimensional systems.

A data-driven approach uses data to derive decisions. The data-driven methods do not require rules, and thus a deep understanding of the system is not required. As a result of using data, time and cost can be saved in understanding the target system. Also, multi-dimensional complex systems can be analyzed through the use of data. Consequently, the success of a data-driven method is highly dependent on the quantity and quality of the data.

For example, suppose that an agent that distinguishes between a cat and a human is separately designed using a model-based method and a data-driven method. To design the agent following a model-based method, first, explicit rules should be defined. We can simply define such a rule as follows: a thing that walks on four legs is a cat and a thing that walks on two legs is a human. To design the agent following a data-driven method, plenty of human and cat pictures should be provided. With such agents, potential issues may arise as follows. For the agent following the model-based methodology, a problem in diagnosing a cat may be encountered when a person is drunk and crawls on all fours. And although the data-driven methodology may distinguish between humans and cats well, it may not be clear why the particular distinctions are made. Therefore, hybrid approaches are often developed to compensate for the shortcomings of each approach.

References

Alzghoul A, Backe B, Löfstrand M, Byström A, Liljedahl B (2014) Comparing a knowledge-based and a data-driven method in querying data streams for system fault detection: a hydraulic drive system application. Comput Ind 65(8):1126–1135

Cho K, Van Merriënboer B, Bahdanau D, Bengio Y (2014) On the properties of neural machine translation: Encoder-decoder approaches. arXiv preprint arXiv:1409.1259

Duchi J, Hazan E, Singer Y (2011) Adaptive subgradient methods for online learning and stochastic optimization. J Mach Learn Res 12(7)

Goodfellow I, Pouget-Abadie J, Mirza M, Xu B, Warde-Farley D, Ozair S, Courville A, Bengio Y (2020) Generative adversarial networks. Commun ACM 63(11):139–144

Gori M, Monfardini G, Scarselli F (2005) A new model for learning in graph domains. In: Proceedings. 2005 IEEE international joint conference on neural networks, vol 2, no 2005, pp 729–734

Hochreiter S, Schmidhuber J (1997) Long short-term memory. Neural Comput 9(8):1735–1780

Kingma DP, Welling M (2013) Auto-encoding variational bayes. arXiv preprint arXiv:1312.6114

Kingma DP, Ba JL (2015) Adam: a method for stochastic gradient descent. In: ICLR: international conference on learning representations, pp 1–15

Krizhevsky A, Sutskever I, Hinton GE (2017) Imagenet classification with deep convolutional neural networks. Commun ACM 60(6):84–90

LeCun Y, Bottou L, Bengio Y, Haffner P (1998) Gradient-based learning applied to document recognition. Proc IEEE 86(11):2278–2324

McCulloch WS, Pitts W (1943) A logical calculus of the ideas immanent in nervous activity. Bull Math Biophys 5(4):115–133

Minsky M, Papert S (1969) An introduction to computational geometry. Cambridge tiass, HIT, 479, pp 480

Qian N (1999) On the momentum term in gradient descent learning algorithms. Neural Netw 12(1):145–151

Rosenblatt F (1957) The perceptron, a perceiving and recognizing automaton Project Para, Cornell Aeronautical Laboratory

Rumelhart DE, Hinton GE, Williams RJ (1986a) Learning representations by back-propagating errors. Nature 323(6088):533–536

Rumelhart DE, Smolensky P, McClelland JL, Hinton G (1986b) Sequential thought processes in PDP models. Parallel Distrib Process Explor Microstruct Cogn 2:3–57

Scarselli F, Gori M, Tsoi AC, Hagenbuchner M, Monfardini G (2008) The graph neural network model. IEEE Trans Neural Networks 20(1):61–80

Werbos P (1974) New tools for prediction and analysis in the behavioral sciences. Ph. D. dissertation, Harvard University

Chapter 3
Signal Validation

In an NPP, there are more than 10,000 sensors installed to measure all the various plant parameters. Clearly, valid and precise signals from sensors are a prerequisite for both the safety and the efficient operation of the plant. In the case of faulty sensors and signals, the performance of the operators and control systems may be degraded. This may lead to an undesirable situation that compromises NPP safety as learned from the Three Mile Island (TMI) and Fukushima Daiichi NPP accidents, operator error resulting from incorrect signals represents one of the main contributors to historical severe accidents. During the TMI accident, due to wrong indications, the operating staff did not build up a correct situational awareness about the plant status.

The validity of plant signals is even more important for the successful operation of autonomous systems. As alluded to above, the performance of various plant functions largely relies on the correctness and reliability of the related signals. In an autonomous NPP, signals from sensors are used as inputs to the functions that are implemented with AI techniques. Wrong inputs therefore increase the probability that the autonomous system generates wrong outputs.

Typical failure modes of sensors in NPPs can be categorized into bias, drift, and stuck failures, as illustrated in Fig. 3.1. Among them, bias is a signal failure in which constant values are added to or subtracted from the normal and intact signals, while a drift failure is a time-correlated permanent offset failure. A stuck failure is, as its name implies, an incorrect constant value. Typical stuck failures in NPPs include "stuck at the highest value" (stuck-high), "stuck at the lowest value" (stuck-low), or "stuck at the current value at the time of failure" (stuck-as-is).

Based on the essential nature of plant signals, there has been wide research into methods for signal validation and signal reconstruction. Approaches to signal validation can be classified into model-based and data-driven approaches (see Sect. 2.5). The model-based approaches (Gertler 1997), which were typically applied in earlier studies, start from an understanding of the physical mechanisms within the given system and their accurate modeling. Accordingly, some applications may require a large number of models to represent all possible failures, which could result in high computational costs. Moreover, while a higher-order parity space yields improved

J. Kim et al., *Autonomous Nuclear Power Plants with Artificial Intelligence*, Lecture Notes in Energy 94, https://doi.org/10.1007/978-3-031-22386-0_3

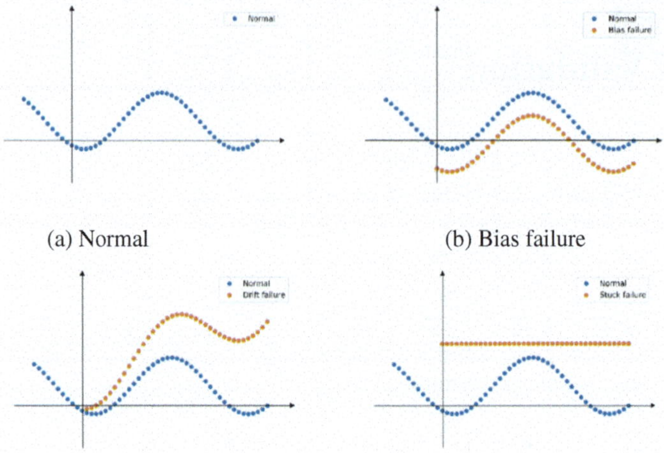

(a) Normal (b) Bias failure

(c) Drift failure (d) Stuck failure

Fig. 3.1 Three typical types of sensor failures in NPPs. Reproduced with permission from Choi et al. (2021)

performance, it also leads to higher computational costs (Hwang et al. 2009). In contrast, data-driven approaches (Fantoni and Mazzola 1996; Choi and Lee 2020; Xu et al. 1999; Hines et al. 1998; Kim et al. 2019b; Di Maio et al. 2013; Yoo et al. 2006; Li et al. 2018; Kaistha and Upadhyaya 2001; Baraldi et al. 2010; Albazzaz and Wang 2004; Zavaljevski and Gross 2000; Zio and Di Maio 2010) use empirical operational data without references to accurate model representations (Li et al. 2018). In this way, data-driven approaches appear to be better suited to complex and nonlinear systems such as NPPs considering the difficulties associated with developing accurate physical models of all the myriad mechanisms.

The recently increasing availability of large signal-measurement datasets provides a key advantage in the application of data-driven approaches to reconstruct signals (Zúñiga et al. 2020). Typical examples of data-driven approaches include ANNs (Fantoni and Mazzola 1996; Xu et al. 1999; Hines et al. 1998; Kim et al. 2019a), auto-associative kernel regression (Di Maio et al. 2013), principal component analysis (PCA) (Yoo et al. 2006; Li et al. 2018; Kaistha and Upadhyaya 2001; Baraldi et al. 2010), independent component analysis (Albazzaz and Wang 2004), the multivariate state estimation technique (Zavaljevski and Gross 2000), SVMs (Zavaljevski and Gross 2000), and fuzzy similarity (Zio and Di Maio 2010).

In contrast to the above works, relatively few studies on signal reconstruction have been conducted to date in the nuclear field. Lin and Wu (2019) applied the multivariate autoregressive method, which is a model-based approach, to the reconstruction of signals in the case of bias and drift failures. Otherwise, a few studies employing data-driven approaches have been reported: the denoised auto-associative sensor model

(Shaheryar et al. 2016), iterative PCA (Li et al. 2018), and CNNs (Lin et al. 2021; Yang et al. 2022).

This chapter introduces three approaches, namely two for signal validation and one for signal reconstruction. Section 3.1 presents a method to detect sensor faults using an LSTM, which is a supervised learning method. Section 3.2 describes a signal validation approach using an unsupervised learning method, and Sect. 3.3 introduces a GAN-based signal reconstruction method.

3.1 Sensor Fault Detection Through Supervised Learning

State awareness in complex systems like aircraft systems, oil and gas facilities, and NPPs derives from a multitude of sensors installed across numerous locations. In terms of maintaining the safety and stability of NPP operation, I&C systems are installed that collect the plant parameters from the sensors and process the data. In a nuclear accident in particular, all parameters are expected to undergo complex changes with potentially dramatically different features from normal operation. Operators in the MCR depend on the displayed sensor values to provide reasoning for their mitigation actions. Thus, the integrity of the sensor data is a critical factor to the safety of NPP systems, in which case information about the conditions of the sensors is a valuable feature for proper accident responses.

Since a 1995 US Nuclear Regulatory Commission (NRC) report concluded that online monitoring (OLM) techniques are able to monitor field sensors and signals (Hines 2009), OLM techniques have been designed and developed to monitor the performance of instrument channels so as to extend the calibration intervals required by technical specifications. Instrument conditions are tracked by OLM via the application of both hardware redundancy and analytical redundancy approaches. Applications of developed OLM techniques are mainly seen in several data-driven approaches (Simani et al. 1999; Fantoni et al. 2004; Zavaljevski and Gross 2000). However, previous data-driven methods only considered the steady-state condition with small uncertainty in the calibration of the target sensors. The general process involved reconstructing sensor signals and performing a residual analysis that compared the estimated signals to measured parameters (Baraldi et al. 2015).

The previous data-driven methods for sensor state monitoring mostly adopted unsupervised learning-based methods to reconstruct the input signal data, after which the sensor state was determined based on residual analysis as mentioned above. Unsupervised learning is advantageous in terms of its versatility; however, in this case the parameter optimization process is a key factor because the trained model should express the relations between the sensor values across diverse situations. Therefore, to monitor the sensor state in an emergency situation, a supervised learning-based sensor fault monitoring model is introduced in this chapter. The application of typical unsupervised learning-based methods showed insufficient monitoring performance when handling the nonlinear changes of NPP accident data. To overcome this limitation, supervised learning was applied with suggested data labels. With the higher performance of supervised learning structures, the sensor fault monitoring model can generate labels that indicate the sensor state.

3.1.1 Sensor Fault Detection System Framework with Supervised Learning

Conventional methods that reconstruct sensor signals and calculate the residuals from the measurements provide advantages in both error detection and sensor value estimation under normal operating conditions. But as mentioned above, such methods are not suitable for NPP emergency situations because a reactor trip and subsequent actuation of safety systems result in highly unstable plant parameters, which presents an issue as the previously applied data-driven models including several kernel-based approaches cannot handle nonlinear parameter changes. In this context, the research aim became to construct a framework for detecting faulty sensors during NPP emergency situations. Applying a neural network, the model monitors for sensor error during the rapid changes of all plant parameters following a reactor trip by considering multivariate inputs, or in other words, a large number of sensor conditions. Model training is conducted with numerous time-series data that include the uncertain and complex variables involved in the early phases following a reactor trip. This way, the sensor error detection model is able to monitor the soundness of the target sensors, which it expresses with a *consistency index* of the sensors (see Sect. 3.1.1.1). While the system does not estimate the plant parameters, it directly displays the current state of the sensors during an emergency situation. In other words, the model output mainly focuses on detecting any deviation of the measured sensor values from the real values.

An overview of the data processing flow for the neural network model training involved in the developed system is shown in Fig. 3.2. Raw data matrices from a compact nuclear simulator (CNS) are first processed by injecting errors, and then the consistency index is labeled on every time-series sensor parameter. These labeled data are then used to train the ML model, which in this case is an LSTM. Following training, the model generates consistency values for every sensor in the test data. Considering all the dynamic features of NPPs, developing a sensor fault monitoring technique for accident sequences represents a challenge when compared to other steady states. Relations between various sensor parameters are difficult to clarify as all input parameters are differently influenced by the reactor trip. To monitor sensor states in such an unstable situation, the developed model requires sufficiently high pattern recognition performance to produce case-specific signal validations.

As mentioned in Sects. 2.3 and 2.4, RNNs feature special structural connections between nodes, where each node has memory to process inputs from the present state as well as from connected nodes. This structure enables the system to consider the temporal context of data (Boden 2002). Among the various ANNs, for tasks

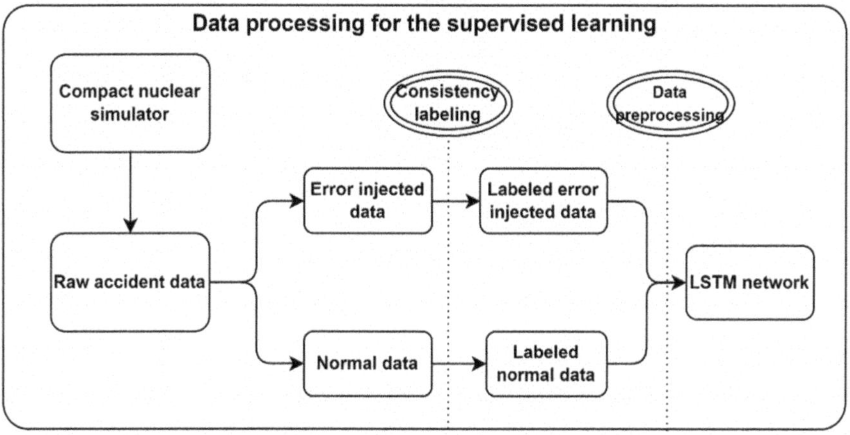

Fig. 3.2 Overview of the data processing for neural network model training

involving time-series data, LSTMs have been preferred for their ability to solve the vanishing gradient problem from the long-term dependencies of the data.

Prior to the selection of this model, a pilot study was conducted in which a deep multilayer perceptron model and a CNN model were tested. Both network models demonstrated poor performance in the present analysis as compared with the LSTM. Similarly, in a previous comparison between multilayer perceptron and LSTMs for plant parameter prediction in an NPP emergency situation, the LSTM showed much more accurate results (Bae et al. 2020).

3.1.1.1 Consistency Index

Introduced for the sensor fault monitoring model, the consistency index expresses the soundness of the measurements. Initially, it may take values of 0 or 1 to respectively represent faulty and sound sensors; however, such binary classification of sensor states would result in a lowering of the threshold for the judgment of sensor failure. This is because ML models may easily designate a normal sensor with no error as a faulty sensor based on small deviations such as from sensor oscillation (Hashemian 2010) or untrained sensor values. This leads to the need for a separate evaluation of the sensor measurements in order to avoid false detections of sensor error for those particular measurements close to the real value. In this case, consistency index values are derived from the relative measurement accuracy; around an error of approximately 10%, the consistency index is calculated from the square of the relative measurement accuracy. Previously, it has been reported that relative measurement error or accuracy may be adopted to quantify instrumentation performance or data quality (Rabinovich and Rabinovich 2010). Equations (3.1) and (3.2) provide the relative measurement error (ε) and the consistency index (C) as follows.

$$\varepsilon = \frac{\tilde{A} - A}{A} \tag{3.1}$$

$$C_{i,t} = \begin{cases} (1 - \varepsilon_{i,t})^2 = 1 - \left|\frac{\tilde{A}_{i,t} - A_{i,t}}{A_{i,t}}\right|^2, & 0 \le \varepsilon_{i,t} \le 0.1 \\ 0, & \varepsilon_{i,t} > 0.1 \end{cases} \tag{3.2}$$

In these expressions, Ã and A are the measured and real values, respectively, i is the parameter number, and t is time. As Eq. (3.2) shows, the consistency index C is taken from the relative measurement error ε. A graphical example of consistency labeling is depicted in Fig. 3.3.

A consistency index is applied to every time-series parameter. The trained model processes all normal data and sensor error-injected data during the emergency situation as inputs and generates consistency outputs from the sensor parameter inputs. With no accident information, the model is able to estimate the states of the sensors by considering the analytical relation between other sensors and the temporal relations between the early sequences of a reactor trip. Normal sensor trends are trained through normal data, while parameter deviations are trained through error-injected data. In the case of an anticipated sensor error or accident, the system directly lowers the consistency index. High computing power enables the use of a large number of parameter

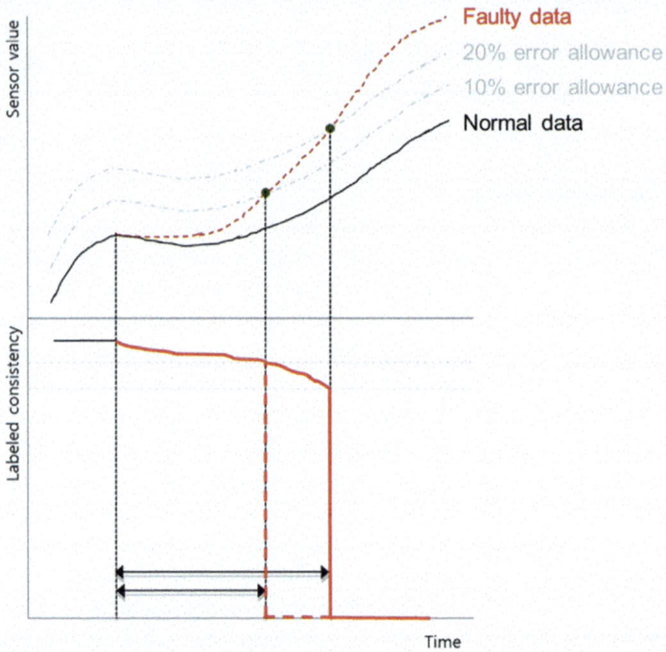

Fig. 3.3 Example of the consistency index labeling rule. The sensor parameter trend with drift error is plotted along with the corresponding consistency index trend

inputs and training for complex changes. As a result, the model output, i.e., consistency index values, provides a simple solution to replace parameter reconstruction and residual analysis.

3.1.1.2 Sensor Error Modes

Sensor failures or sensor uncertainties can be the result of either external or internal environmental causes. In previous applications of OLM techniques, typical uncertainty sources were considered to determine the allowance of channel uncertainties. Such sources include temperature and pressure (external), and systematic structure and sensor type (internal) (Chen et al. 2010).

Among the possible error modes introduced in Fig. 3.1, two modes are selected considering their influence on human error during diagnosis tasks: drift error, which is a time-correlated permanent offset error, and stuck at constant error. The reasons for choosing these parameters are as follows. In an NPP accident situation, diagnosis is conducted mainly by checking the trends of various plant parameters, specifically whether the parameters exceed their thresholds. Drift and stuck at constant errors present a similar behavior as during error-free operation, and so they are good candidates for analysis as the main target errors for detection. The other error modes, for example the stochastic offset and stuck at zero sensor errors, can be easily recognized by the operators due to their discrepancies from real trends.

To implement the two selected error modes, a uniform error injection method is applied to all sensors regardless of sensor type or environment. The stuck at constant error is implemented by fixing the sensor value to that at the point of error injection, while the drift error is classified as slow or rapid drift according to the slope of the change, respectively implemented with rates of change of two and five times the real trends. In the case that there is no change in the real values, the drift types are implemented with 0.4% and 4% of the fixed values. Other than slow or rapid, each drift is also divided into upward and downward by the direction of the drift. All together, the five injected errors are as follows: stuck at constant, slow upward drift, slow downward drift, rapid upward drift, and rapid downward drift. Figure 3.4 shows examples of injected error data plotted against normal parameter trends.

3.1.1.3 Data Preprocessing

Two data preprocessing steps are performed for an efficient training of the LSTM, the first of which is a Gaussian smoothing of the time-series parameters. Parameter oscillation is frequently observed in Loss of Coolant Accident (LOCA) data, where it is estimated to result from coolant vaporization. As oscillations can lead to the false detection of sensor error, a one-dimensional (1D) Gaussian smoothing filter is applied, where the Gaussian distribution function used for smoothing is given as follows.

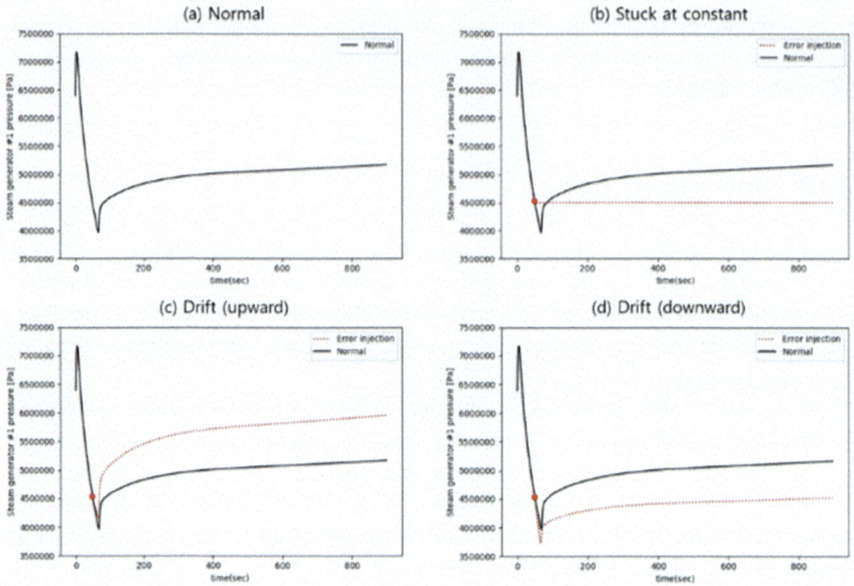

Fig. 3.4 Examples of error injection. **a** Normal, **b** stuck at constant, **c** slow drift (upward), and **d** slow drift (downward) sensor error during a loss of coolant accident. Reproduced with permission from Choi and Lee (2020)

$$G(x) = \frac{1}{\sigma\sqrt{2\pi}} \exp^{-\frac{x^2}{2\sigma^2}} \tag{3.3}$$

From the value of σ (standard deviation), the Gaussian distribution width, or in other words the effect of the smoothing, is determined (Ito and Xiong 2000). Here, Gaussian smoothing with σ = 100 is applied to the oscillating parameters; as the example in Fig. 3.5 shows, the oscillation is removed while the parameter trend is maintained.

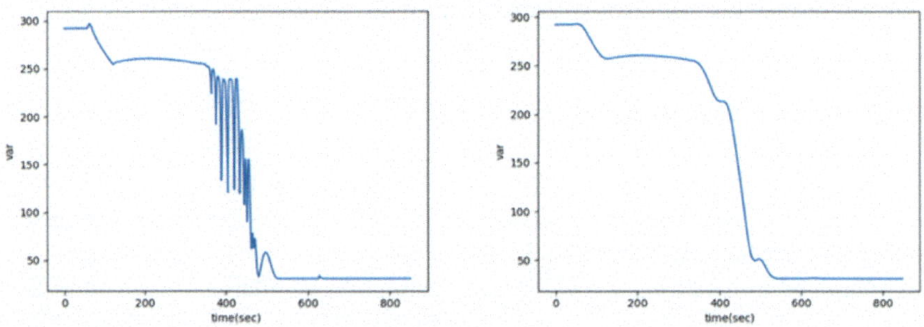

Fig. 3.5 Example parameter oscillation (left) and smoothing result (right). Reproduced with permission from Choi and Lee (2020)

The second preprocessing step involves a min–max normalization of all the parameters. As seen in Table 3.1, various parameter scales are used for training the LSTM network, requiring normalization to reflect all plant parameters. The minimum and maximum parameter values along the entire accident sequence are gathered for the normalization, following Eq. (3.4).

$$x_{norm} = \frac{x - x_{\min}}{x_{\max} - x_{\min}} \quad (3.4)$$

Table 3.1 Selected plant parameters from the CNS

	Plant parameter (units)
1	Pressurizer (PZR) level (m)
2	Reactor vessel (RV) water level (m)
3	Containment (CTMT) radiation (mRem/hr)
4	COLD-LEG #1 temperature (°C)
5	HOT-LEG #1 temperature (°C)
6	Core outlet temperature (°C)
7	SG #1 level, wide range (m)
8	SG #2 level, wide range (m)
9	SG #3 level, wide range (m)
10	Secondary system radiation (mRem/hr)
11	SG #1 pressure (Pa)
12	SG #2 pressure (Pa)
13	SG #3 pressure (Pa)
14	PZR pressure (Pa)
15	Feed water line 1 flow (kg/sec)
16	Feed water line 2 flow (kg/sec)
17	Feed water line 3 flow (kg/sec)
18	CTMT sump water level (m)
19	Steam line 1 flow (kg/sec)
20	Steam line 2 flow (kg/sec)
21	Steam line 3 flow (kg/sec)

3.1.2 *Case Study*

3.1.2.1 Data Description

Reactor trips, a process in which control rods are inserted into the reactor, are initiated to ensure NPP safety in response to the detection of plant parameters deviating from predetermined set points. Control rod insertion causes a rapid decrease in reactivity, which in turn leads to a decrease in reactor thermal power and rapid changes in various plant components such as turbine trips, valve openings and closings, the actuation of safety systems, and so on. The situations that necessitate reactor trips are called emergency situations in the nuclear field. Once a reactor trip has been alerted, operators in the NPP MCR follow emergency operating procedures (EOPs) to mitigate the factors that caused the plant parameters to exceed the set points of the reactor protection system (RPS) or the engineered safety features, or other established limits. By conducting the relevant EOP, operators are able to cope with the symptoms in the early trip phase and diagnose the accident.

It is clear that accident diagnosis depends on the process parameter values as well as their trends. Regarding the time required for accident diagnosis in an NPP emergency situation, the International Atomic Energy Agency (IAEA) published a safety report recommending that operators complete the diagnosis within 15 min from the first indication of the accident (IAEA 2002). Based on the diagnosis results, operators move on to the appropriate optimal response procedure (ORP) providing the tasks required to mitigate the particular accident considering the current symptoms. Example ORPs include those covering a reactor trip, LOCA, SG tube rupture (SGTR), excess steam demand event (ESDE), loss of all feedwater (LOAF), loss of forced circulation, loss of off-site power, and station blackout (SBO) (IAEA 2006).

Each ORP includes different contextual tasks, which means that an accident misdiagnosis could, with critical consequences, result in an omission of the appropriate mitigation action or an error of commission. One of the lessons learned from the TMI accident is that wrong displays of plant parameters during an accident sequence can significantly increase the risk of human error, which could potentially result in even graver consequences following the diagnosis steps. Even a single sensor parameter can be a critical factor in accident diagnosis; for example, an increase in the secondary system radiation is a strong indicator of an SGTR. As such, an error in such a critical sensor during an emergency situation may lead to a misdiagnosis. It has previously been shown that indicator failure is among the factors influencing misdiagnosis error (Lee et al. 2009).

From the emergency situations covered by the ORPs, the following four accidents are selected for the sensor fault monitoring model: LOCA, SGTR, ESDE, and LOAF. Reactor trips with no specific symptoms or accidents from a loss of power loss are not considered due to their lack of distinguishable symptoms in the CNS. The relevant malfunctions of the selected accidents are injected to normal CNS operation, after which the reactor trips by an automatic signal. From the point of reactor trip, time-series data of certain plant parameters are collected for 15 min, corresponding to the recommended accident diagnosis time limit. Among the plant parameters, 21 are selected based on the diagnosis procedures and the importance of the parameters to the estimation of the accident. The time interval of data collection is 1 s, leading to 900 collected time points per dataset. For data variation, nine break sizes and three break locations are included in the accident data for LOCA, SGTR, and ESDE.

In the validation of the system, the following target sensors are considered for their diagnostic importance: sensors monitoring the PZR pressure; CTMT radiation; secondary system radiation; SG #1, #2, #3 pressure and water level; RV water level; cold-leg #1 temperature; hot-leg #1 temperature; and core outlet temperature. As the PZR pressure is affected by all accident types, operators typically use it for a rough diagnosis. Data from the other 13 target sensors are important to diagnose or estimate LOCA, SGTR, ESDE, and LOAF accidents. Errors are injected at six time points during the early phase of the reactor trip at intervals of 60 s for the training and validation datasets, while the test set has error injection at three different time points with the same time interval. To train the single failure of each of the target sensors, 7488 training sets, 2223 validation sets, and 3159 test sets are used.

3.1.2.2 Case Study Results

Figure 3.6 plots examples of the output, or consistency trend. The yellow lines in the panels depict the real consistency trend of the target sensor obtained by labeling both normal and error-injected raw data with consistency values, and the blue lines show the estimated consistency values of the target sensor from the LSTM network test. With a normal sensor in Fig. 3.6a, the estimated result clearly reflects the sensor soundness. In Fig. 3.6b with error injection, the estimated consistency effectively follows the drop in the real consistency by detecting the injected error. But in Fig. 3.6c with a normal sensor, the estimated consistency shows dips in the early phase of the reactor trip owing to the drastic parameter changes as well as the error injection located in the front of the accident sequence. Then Fig. 3.6d with error-injected data shows a slow decrease in the estimated consistency. Each parameter has various rates of change and trends, and likewise the consistency trends also vary. Yet even in this case, the estimation drops in the later stages, mimicking the real consistency value.

3.1.2.3 Error Criteria

The test error datasets for which the real consistency labels did not drop to zero are not considered because they are not perceived to be faulty sensors in the current system. Similarly, for all test data, both success and fault criteria for the sensors should be identified. Note that while setting higher criteria would result in the rapid detection of sensor error, this may be accompanied by increases in false detections of sensor error. Table 3.2 gives the number of datasets in terms of the minimum consistency index location from the test outputs of the trained LSTM. From the results, the consistency between normal and error cases can be clearly discriminated. With no sensor errors

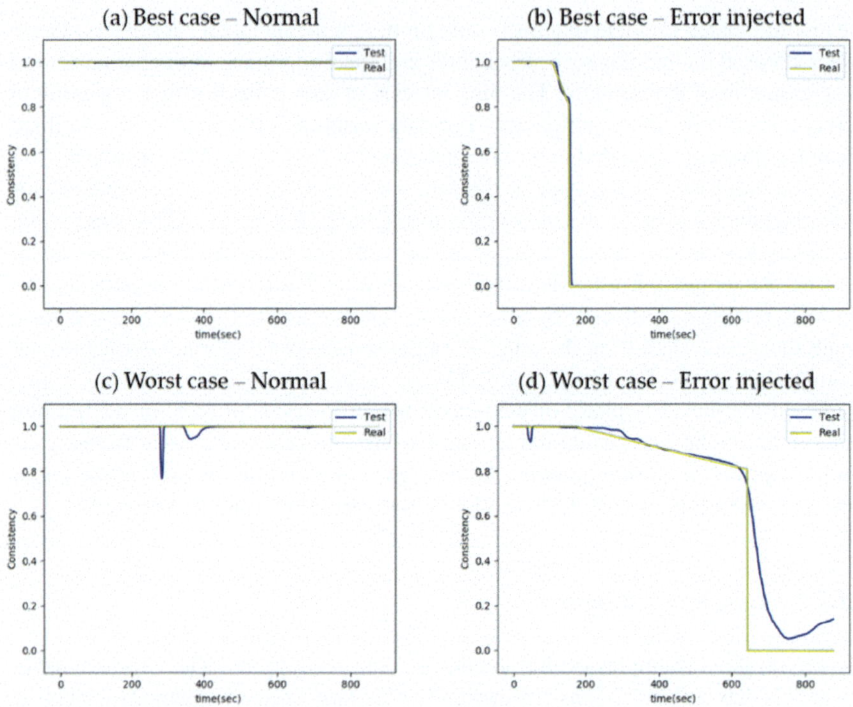

Fig. 3.6 a–d Consistency trends of the best and worst cases from normal and error-injected tests. Reproduced with permission from Choi and Lee (2020)

Table 3.2 Location of minimum consistency index

C Index	< 0.1	0.1–0.2	0.2–0.3	0.3–0.7	0.7–0.8	0.8–0.9	> 0.9	Total
Normal	0	0	0	0	17 (0.79%)	37 (1.72%)	2098 (97.49%)	2152
Error	6751 (72.77%)	2514 (27.10%)	12 (0.13%)	0	0	0	0	9277

(normal case), 97.49% of the data maintains an index value of over 0.9, while during the accident sequence, the other 0.79 and 1.72% of the data have index values over 0.7 and 0.8. As for the error case, 72.77% of the error data are distributed under an index of 0.1, 27.10% are between 0.1 and 0.2, and 0.13% are between 0.2 and 0.3. Determining the error criteria should be based on the ability to distinguish between normal and error states of sensors. According to the results here, the 0.3–0.7 range is where the error criteria should be set. Based on the steady output of the normal case, the higher error criterion, 0.7, is selected for sensor error detection.

Table 3.3 Average time to reach the C index in the error test data

C Index	0.9	0.8	0.7	0.6	0.5	0.4	0.3	0.2	0.1	0
Detection time (s)	60.20	98.52	116.11	130.13	132.14	134.02	135.58	137.55	140.01	670.31

3.1.2.4 Error Detection Time Analysis

Table 3.3 shows the average time to attain the consistency indexes from the error-injected test data. From the results, the average analysis time for the error data was 116.11 s. After reaching a consistency value of 0.8, the decrease of the index accelerates.

The average times required for sensor error detection are given in Tables 3.4 and 3.5 for the 13 target sensors. The system produces outputs of the true values in all the normal cases, and thus the normal data results can be classified as “success” or “failure” based on whether the index value is maintained above the error criteria over all time steps. Accordingly, all normal cases are classified as “success”. Looking at the stuck-at-constant error cases, no meaningful results can be seen for some test data (marked with dashes) because the real values remained constant over the course of the accident sequences. The tables show that the detection time varied by accident type, error mode, and target sensor; the parameter trends or features completely differ in each accident. While it can be seen that the rapid drift errors are detected earlier than the others, the stuck and slow drift error present unstable results. For the CTMT radiation and secondary radiation sensors, the errors are detected in a relatively short time when compared with the other sensor errors, likely based on their relatively simple parameter trends and the fact that these parameters can be clearly distinguished from other parameters.

The results given in Table 3.2 exhibit the performance of the model in distinguishing between normal and error data for all test sets. The error criteria value, 0.7 as mentioned above, was determined based on the observed uncertainty of these results. With this criterion, the sensor error detection system achieved successful performance. Despite this result, though, the addition of more training data and accident types will likely increase the degree or number of peak points, potentially leading to false detection of sensor error; this possibility is noted as a limitation of the current error detection model. In the case that a false case is observed, more logics to determine the sensor error can be incorporated to prevent recurrence. For this, the application of a cumulative function or a downward adjustment of the error criteria can be considered.

Table 3.4 Average sensor error detection time (1) (units, sec)

Accident	Error mode	PZR pressure	CTMT radiation	Secondary radiation	SG #1 pressure	SG #1 level	SG #2 pressure	SG #2 level
LOCA	Normal	Success	Success	Success	Success	Success	Success	Success
	Stuck	71.56	16.26	—	218.02	15.17	236.11	138.85
	Slow drift	69.89	41.72	11.15	85.03	21.16	195.33	14.26
	Rapid drift	53.45	23.63	7.93	56.24	13.73	110.09	8.83
SGTR	Normal	Success	Success	Success	Success	Success	Success	Success
	Stuck	46.80	—	206.35	58.67	60.69	15.36	91.83
	Slow drift	45.64	6.70	151.74	112.17	18.45	218.41	19.24
	Rapid drift	34.88	4.63	125.18	86.46	13.13	121.22	11.40
ESDE	Normal	Success	Success	Success	Success	Success	Success	Success
	Stuck	99.71	—	—	173.48	60.17	85.40	118.27
	Slow drift	81.35	5.98	10.34	90.69	81.36	103.10	45.85
	Rapid drift	52.59	4.17	7.98	54.59	39.78	71.03	36.37
LOAF	Normal	Success	Success	Success	Success	Success	Success	Success
	Stuck	24.33	—	—	63.00	13.00	112.00	9.50
	Slow drift	17.83	4.00	11.50	56.00	9.83	119.50	11.92
	Rapid drift	16.17	2.50	8.50	45.17	5.00	80.00	5.58

3.2 Signal Validation Through Unsupervised Learning

Section 3.2 introduces an algorithm for signal validation that can detect the stuck failure of signals in NPP emergency situations that involve rapidly changing signals through the use of unsupervised learning methods. The algorithm combines a VAE, which employs unsupervised learning, and an LSTM. Development of the signal detection algorithm mainly consists of tailoring it to signal validation in the design stage and optimizing it at each step for improved performance. The algorithm is then validated through a demonstration using a CNS that simulates a Westinghouse-type NPP.

Table 3.5 Average sensor error detection time (2) (units, sec)

Accident	Error mode	SG #3 pressure	SG #3 level	RV water level	Cold-Leg #1 temperature	Hot-Leg #1 temperature	Outlet temperature
LOCA	Normal	Success	Success	Success	Success	Success	Success
	Stuck	220.25	15.39	140.09	343.00	—	338.00
	Slow drift	82.35	21.72	129.30	342.67	334.41	355.51
	Rapid drift	15.29	13.79	92.38	128.00	134.72	135.52
SGTR	Normal	Success	Success	Success	Success	Success	Success
	Stuck	17.27	116.30	246.11	—	—	319.50
	Slow drift	125.59	19.54	21.63	311.11	294.01	282.58
	Rapid drift	64.83	14.07	14.06	168.56	175.07	175.45
ESDE	Normal	Success	Success	Success	Success	Success	Success
	Stuck	77.38	149.75	10.00	161.41	320.50	192.28
	Slow drift	111.77	64.55	66.46	221.59	286.41	246.61
	Rapid drift	63.98	40.41	3.37	97.10	178.28	117.52
LOAF	Normal	Success	Success	Success	Success	Success	Success
	Stuck	110.00	12.67	—	279.00	296.00	296.50
	Slow drift	117.50	12.33	66.67	179.33	138.33	139.67
	Rapid drift	79.00	7.00	3.67	77.33	85.00	83.67

3.2.1 *Signal Behaviour in an Emergency Situation*

Signal failures have a number of causes, such as anomalies of the related sensors, transmitters, and cables. These in turn can be caused by internal, external, or environmental problems including pollution, vibrations, extreme temperatures, and aging (Zúñiga et al. 2020). As mentioned in Sect. 3.1, the typical failure modes of sensors in NPPs are classified as bias, drift, and stuck failures (i.e., stuck-high, stuck-low, and stuck-as-is).

In an emergency situation, tasks such as visually checking and detecting anomalous signals from automatic systems become difficult. Under normal circumstances, NPP parameters generally show stable values, meaning that a faulty signal can be readily distinguished from a normal one. In contrast, many parameters change rapidly in an emergency situation, making it highly challenging to determine whether the parameter changes are due to an accident or signal failure.

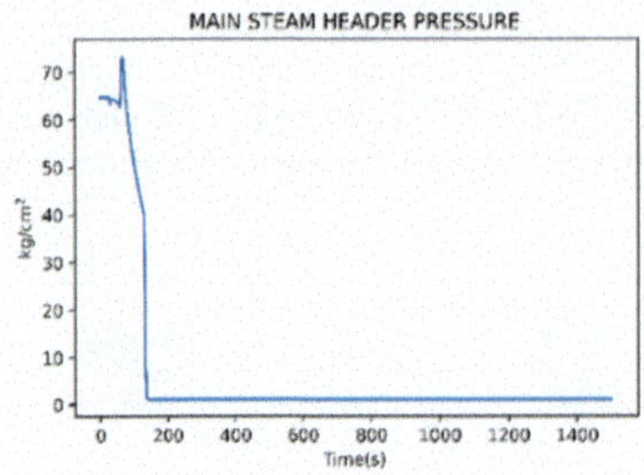

(a) Main steam head pressure, indicating the minimum value

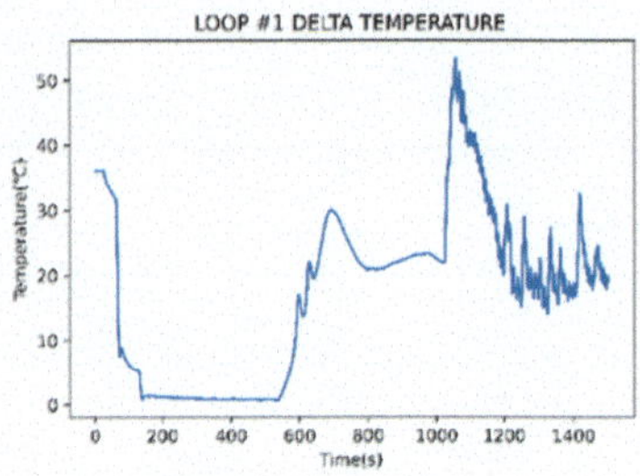

(b) Loop 1 delta temperature

Fig. 3.7 Behavior of different parameters in a LOCA scenario. Reproduced with permission from Choi et al. (2021)

Stuck failures may make it impossible for operators to fully understand the current situation if they wrongly consider a faulty signal to be a normal one. For example, Fig. 3.7 shows the behavior of two different parameters in a LOCA scenario. Like the plot in Fig. 3.7a shows, some parameters may indicate the minimum signal value for a particular measurement, which is a similar behavior as a stuck-low signal failure. Conversely, many other parameters will change quickly over time, as the example in Fig. 3.7b illustrates. Therefore, if an emergency situation includes a stuck failure, the operator may incorrectly recognize the reality of the system, and the emergency response may be adversely affected.

3.2.2 *Signal Validation Algorithm Through Unsupervised Learning for an Emergency Situation*

The developed signal validation algorithm employs a combination of a VAE and an LSTM to detect stuck failures (i.e., stuck-high, stuck-low, stuck-as-is) in an emergency situation in an NPP. As shown in Fig. 3.8, the overall algorithm consists of the signal validation algorithm itself along with optimization, which is meant to improve the performance. This chapter considers a LOCA scenario as a representative example of an emergency situation and demonstrates the algorithm through LOCA simulation using a CNS.

3.2.2.1 Signal Validation Algorithm

The signal validation algorithm comprises four main steps: input preprocessing (Step 1), signal reconstruction (Step 2), reconstruction error (RE) calculation (Step 3), and determining the signal failures (Step 4). The optimization conducted concurrently with these steps is covered in the next section.

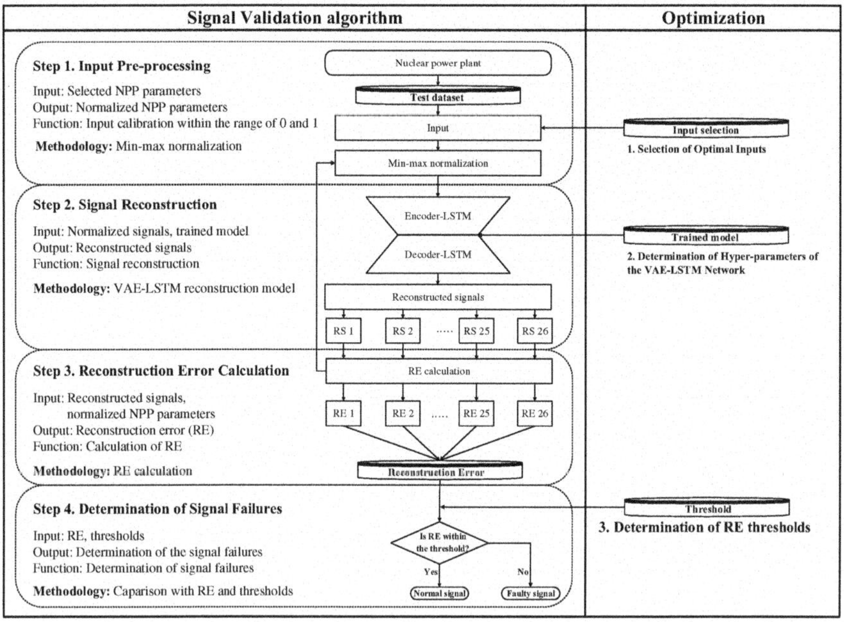

Fig. 3.8 Overview of the signal validation algorithm. Reproduced with permission from Choi et al. (2021)

Step 1 is to pre-process the inputs by using min–max normalization. This step aims to convert each input signal value to a value between 0 and 1. Signals in an NPP have different units and ranges, such as loop 1 cold-leg temperature 290.5 °C, steam line flow 533.4 kg/sec, and valve states of open, closed, or 50%. Generally, variables with higher values will have a larger effect on the network results (Kim et al. 2021; Yang and Kim 2018, 2020). However, including such high values is not proper because it can cause local minima. To address this issue, Step 1 of the algorithm receives the plant parameters as inputs and then outputs normalized plant parameters.

In addition to preventing local minima, the min–max normalization in Step 1 is applied to increase the learning speed of the system. Here, normalization is used to transform the input signal values from the NPP into a value between 0 and 1 following Eq. (3.4). A signal closely related to an interesting signal is selected as an input value by Pearson correlation analysis to obtain the highest performance in the detection of stuck failure.

Step 2 of the signal validation algorithm is to reconstruct the signal using the combined VAE-LSTM network; Fig. 3.9 shows an overview of the network structure, consisting of several layers and nodes. This step tries to generate the same signal value as each pre-processed input from Step 1. To do this, the hyperparameters of the VAE-LSTM network are determined through optimization, as shown in Sect. 3.2.2.2.

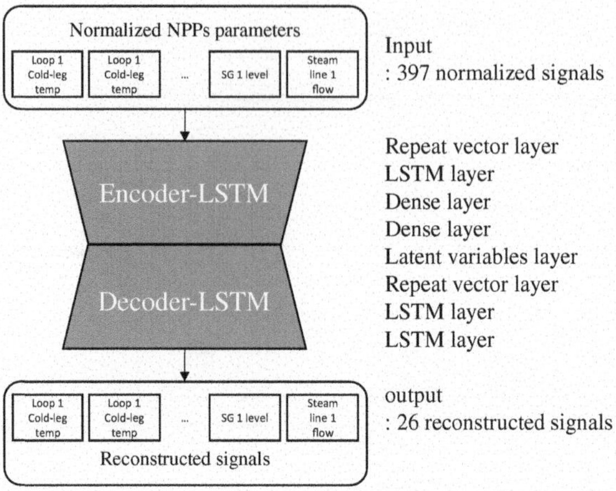

Fig. 3.9 Structure of Step 2 of the signal validation algorithm

The VAE, a Bayesian inference-based variant of the AE (see Sect. 2.4.3), is an unsupervised learning method that forms a network in which the output value resembles the input value (An and Cho 2015). The detection of defect signals by a VAE is based on the probability that the normal signals can be successfully reconstructed. That is, if the VAE successfully reconstructs the input signal, the implication is that the input signal has similar characteristics as the normal trained signal. In contrast, a large difference between the reconstructed signal and the input signal means that the input may not be trained, implying it may be a faulty signal. The LSTM, as discussed in Sect. 2.4.2.2, is a type of RNN that can learn the long short-term dependencies of a dataset (Hwang et al. 2009; Yang and Kim 2018, 2020) and is designed to avoid the problem of long-term dependencies associated with RNNs (Gers et al. 2000). In this step of the algorithm, the LSTM is used to process time-series data.

The VAE-LSTM model receives a normalized signal from Step 1, as shown in Fig. 3.9. At this time, an encoder is trained to extract the characteristics of the input data by using the average and standard deviation of the input data. After that, a decoder generates an output similar to the input based on the extracted characteristics. Both the encoder and decoder of the network adopt LSTM layers to consider the time series data.

Step 3 is to conduct the RE calculation process. In other words, this step calculates the deviation between the reconstructed input signal generated as the output in Step 2 and the normalized input signal. Following this step (i.e., RE calculation), the calculated RE is utilized to determine signal failure in Step 4. The RE is calculated using Eq. (3.5).

$$\mathrm{RE} = (\mathrm{x_t} - \hat{\mathrm{x}})^2 \tag{3.5}$$

where,

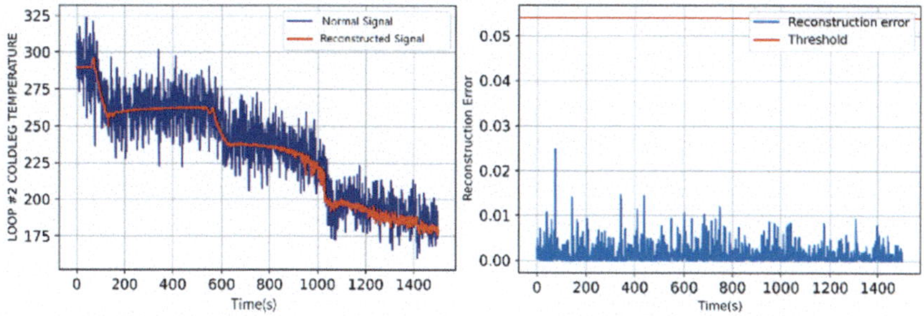

(a) Reconstructed and normal signals

(b) RE calculation results

Fig. 3.10 Reconstruction results for a normal signal of the loop 2 cold-leg temperature. Reproduced with permission from Choi et al. (2021)

x_t normalized value of the original (normal) signal
$\hat{x}$ reconstructed value

Figure 3.10 shows an example of the reconstruction process for a normal signal (without faults) and the corresponding calculated RE. Specifically, Fig. 3.10a plots the original temperature signal of the loop 2 cold-leg (blue line) in the LOCA scenario along with the reconstructed signal (red line, Step 2 results) obtained from the VAE-LSTM network. When the VAE-LSTM correctly trains the normal signal, the RE indicates a small value, as shown in Fig. 3.10b. Figure 3.11 shows an example of the reconstruction process and RE calculation for a faulty signal; Fig. 3.11a plots the faulty temperature signal in the form of a stuck-low failure at 300 s. As shown in Fig. 3.11b, since the network is not trained for the faulty signal, the difference between the reconstructed signal and the faulty input signal is large. By selecting a proper RE criterion (or threshold) that can distinguish between normal and faulty signals, successful signal failure detection can be achieved. The process for the determination of this RE threshold is covered in the optimization step (see Sect. 3.2.2.2).

Finally, Step 4 is to determine whether the input signal is faulty or normal by using the comparison results of the RE calculated in Step 3 with the predefined threshold (i.e., the optimization of Step 3, as discussed in the next section). If the RE of the input signal is lower than the predefined threshold, the signal is labeled as a normal signal. Conversely, if the RE exceeds the predefined threshold, the signal is determined as a faulty signal. This process is shown in Fig. 3.12.

3.2.2.2 Optimization

To improve the performance of the signal validation algorithm, the algorithm undergoes the following optimization processes: (1) selecting the optimal inputs, (2) determining the hyperparameters of the VAE-LSTM network, and (3) determining the RE thresholds. For these optimizations, a CNS is used to simulate various emergency situations. The CNS, developed by the Korea Atomic Energy Research Institute

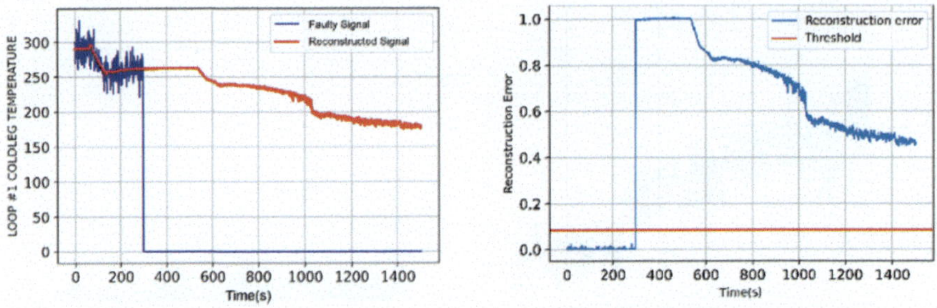

(a) Reconstructed signals (b) RE calculation results

Fig. 3.11 Reconstruction results for a faulty signal of the loop 2 cold-leg temperature. Reproduced with permission from Choi et al. (2021)

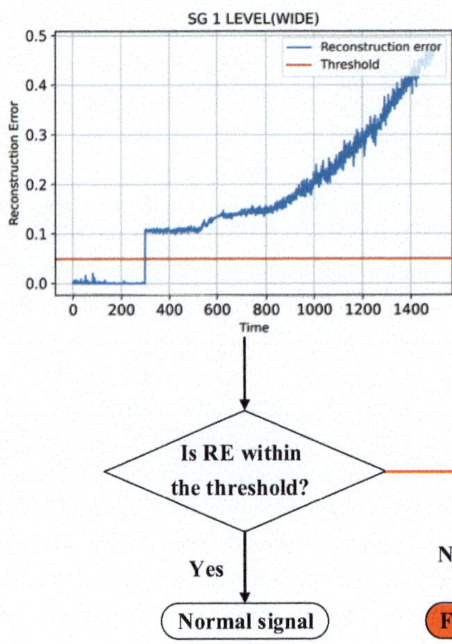

Fig. 3.12 Result of Step 4 of the signal validation algorithm. Reproduced with permission from Choi et al. (2021)

(KAERI), has as a reference plant the Westinghouse 3-loop 990 MW pressurized water reactor (PWR). Figure 3.13 shows the display of the CNS as an overview.

A total of 26 signals are selected for the optimization of the signal validation algorithm, as listed in Table 3.6. In other words, optimization is conducted to detect the stuck failures of these 26 signals.

Data collection is required to conduct the validation and optimization of the signal validation algorithm; Fig. 3.14 provides a breakdown of the data collection. The data is divided into four groups. The first, Data #1, includes normal signals from 49

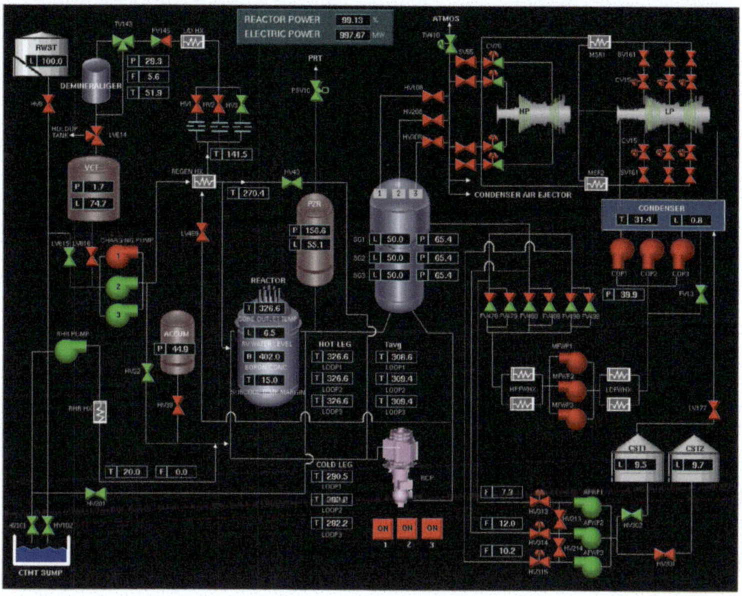

Fig. 3.13 Overview of the CNS components

LOCA scenarios and is used for VAE-LSTM network training. The second group, Data #2 includes normal signal data from five scenarios and is used for Optimization 1 (i.e., selection of the optimal inputs) and Optimization 2 (i.e., determination of the hyperparameters). The data for faulty signals are separated for the purposes of optimization and validation. Data #3 includes the stuck failures of the 26 selected variables for Optimization 3 (i.e., determination of the RE thresholds), as discussed in Sect. 3.2.2.2. It should be noted that the stuck-low dataset includes only the failures of 12 variables because in the scenarios, the other 14 signals indicate the lowest values without any faults and are therefore indistinguishable from stuck-low failures. The final data group, Data #4, is used for validation. A detailed list of all scenarios is given in Table 3.7.

In the first optimization, or Optimization 1, the optimal inputs for the VAE-LSTM network are selected using Pearson correlation analysis. This optimization aims to find the best set of inputs to the VAE-LSTM network for it to reconstruct the normal signals of the 26 selected plant variables. Before this optimization is performed, the VAE-LSTM network is trained to generate normal signals using Data #1. Then Data #2 is used for the optimization.

Different sets of inputs to the VAE-LSTM network would result in different performances during Step 2 of the algorithm (i.e., reconstruction). Correlation analysis is therefore conducted to select the optimal input sets from 2200 variables that can be collected by the CNS. Pearson correlation analysis (Xu and Deng 2017) applies the correlation coefficient given in the following equation.

Table 3.6 The 26 target signals for optimization

NPP parameter	Units
Feedwater pump outlet pressure	kg/cm^2
Feedwater line 1 flow	kg/sec
Feedwater line 2 flow	kg/sec
Feedwater line 3 flow	kg/sec
Feedwater temperature	°C
Main steam flow	kg/sec
Steam line 1 flow	kg/sec
Steam line 2 flow	kg/sec
Steam line 3 flow	kg/sec
Main steam header pressure	kg/cm^2
Charging line outlet temperature	°C
Loop 1 cold-leg temperature	°C
Loop 2 cold-leg temperature	°C
Loop 3 cold-leg temperature	°C
Pressurized temperature	°C
Core outlet temperature	°C
Net letdown flow	kg/sec
PZR level	%
PZR pressure	kg/cm^2
Loop 1 flow	kg/sec
Loop 2 flow	kg/sec
Loop 3 flow	kg/sec
SG 1 level	%
SG 2 level	%
SG 1 pressure	kg/cm^2
SG 1 pressure	kg/cm^2

$$r = \frac{\sum\left(\left(\frac{X_i - \overline{X}}{s_X}\right)\left(\frac{Y_i - \overline{Y}}{s_Y}\right)\right)}{N - 1}. \tag{3.6}$$

Here, N is the number of observations, X_i and Y_i are the values for the i-th observation where X indicates the 26 target variables for signal validation through stuck failure detection and Y indicates all the available variables in the CNS (i.e., 2200 plant variables), and s is the standard deviation. Pearson's coefficient r has a value between –1 and 1, where the larger the absolute value of r, the higher the correlation. An r value approaching 1 means that there is positive linearity, while that approaching –1 means that there is negative linearity. A coefficient of 0 indicates that there is no linear correlation between the two variables.

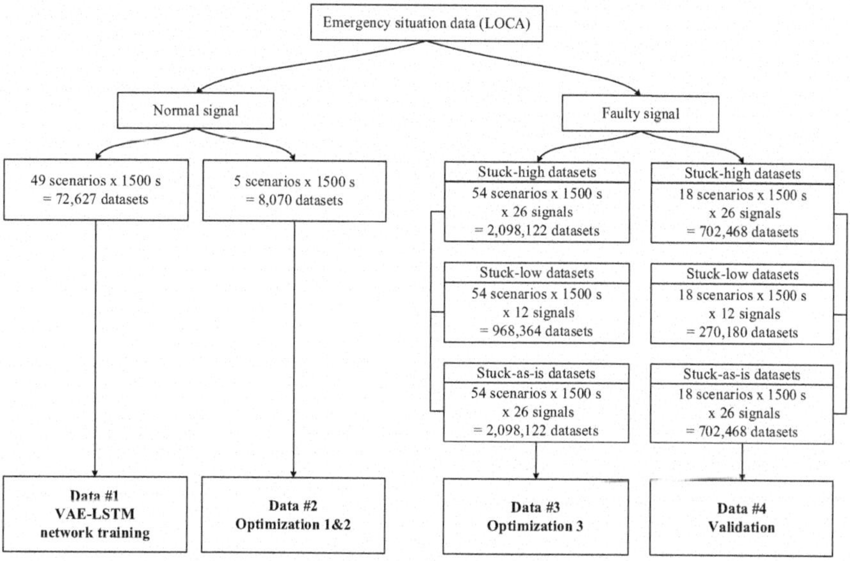

Fig. 3.14 Acquired datasets for the LOCA scenarios. Reproduced with permission from Choi et al. (2021)

As shown in Eq. (3.6), *r* is calculated among the 26 target variables and the CNS-available variables. Figure 3.15 shows a portion of the calculation. Plant variables with correlation coefficients higher than a specific threshold are selected as the optimal input; this threshold is determined here through an experimental approach.

The accuracy achieved in reconstructing the 26 target signals is checked while varying the correlation coefficient. Table 3.8 shows the results of the selected input optimization in Step 1 of the algorithm. The results demonstrate that the reconstruction of the 26 target signals shows the highest accuracy at $r = 0.985$. Accordingly, a total of 397 variables among the CNS parameters having a correlation coefficient higher than $r = 0.985$ are selected as the optimal inputs for the VAE-LSTM network.

In Optimization 2, the hyperparameters—i.e., the number of batches, layers, and nodes of the network—to be used for the signal reconstruction step are determined. In general, hyperparameters influence the performance of a network, in this case reconstruction performance. To optimize the hyperparameters, a trial and error approach is performed until the reconstruction result is as high as possible.

Table 3.9 shows a loss comparison among eight different network configurations. Here, loss is a number that indicates the range of inaccurate network predictions. Perfect network prediction results in a loss of zero, while more incomplete predictions lead to greater loss values. Loss is calculated based on Eq. (3.7) (An and Cho 2015).

$$\mathrm{Loss} = -0.5\sum_{t=1}^{N}\left(1 + \log\left(\sigma_t^2\right) - \mu_t^2 + \sigma_t^2\right) + \frac{1}{N}\sum_{t=1}^{N}\left(x_t - \hat{x}_t\right)^2 \tag{3.7}$$

Table 3.7 Detailed list of the LOCA scenarios

Failure Mode		Scenarios	Purpose
Normal		Hot and cold legs of Loops 1, 2, and 3 with nine different sizes from 10 to 50 cm^2 (every 5 cm^2): 49 scenarios excluding the five used in Optimizations 1 and 2 among 54 total scenarios (2 legs × 3 loops × 9 sizes)	VAE-LSTM network training (Sect. 3.2.2.2)
		Hot-leg of Loop 3 with a 30 cm^2 rupture size Hot-leg of Loop 3 with a 35 cm^2 rupture size Hot-leg of Loop 3 with a 40 cm^2 rupture size Hot-leg of Loop 3 with a 45 cm^2 rupture size Hot-leg of Loop 3 with a 50 cm^2 rupture size	Optimizations 1 & 2 (Sect. 3.2.2.2)
Faulty	Stuck-high	Hot and cold legs of Loops 1, 2, and 3 with nine different sizes from 10 to 50 cm^2 (every 5 cm^2): 54 (2 legs × 3 loops × 9 sizes)	Optimization 3 (Sect. 3.2.2.2)
		Hot and cold legs of Loops 1, 2, and 3 with three different sizes from 2 to 6 cm^2 (every 2 cm^2): 18 (2 legs × 3 loops × 3 sizes)	Validation (Sect. 3.2.3)
	Stuck-low	Hot and cold legs of Loops 1, 2, and 3 with nine different sizes from 10 to 50 cm^2 (every 5 cm^2): 54 (2 legs × 3 loops × 9 sizes)	Optimization 3 (Sect. 3.2.2.2)
		Hot and cold legs of Loops 1, 2, and 3 with three different sizes from 2 to 6 cm^2 (every 2 cm^2): 18 (2 legs × 3 loops × 3 sizes)	Validation (Sect. 3.2.3)
	Stuck-as-is	Hot and cold legs of Loops 1, 2, and 3 with nine different sizes from 10 to 50 cm^2 (every 5 cm^2): 54 (2 legs × 3 loops × 9 sizes)	Optimization 3 (Sect. 3.2.2.2)
		Hot and cold legs of Loops 1, 2, and 3 with three different sizes from 2 to 6 cm^2 (every 2 cm^2): 18 (2 legs × 3 loops × 3 sizes)	Validation (Sect. 3.2.3)

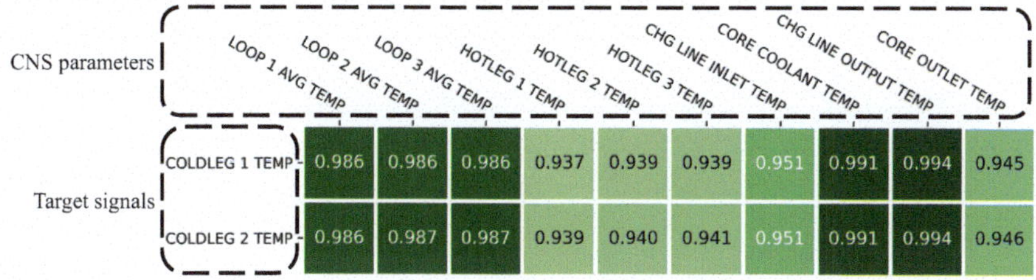

Fig. 3.15 Example of Pearson correlation analysis results. Reproduced with permission from Choi et al. (2021)

Table 3.8 Reconstruction accuracy and inputs as a function of r

r value	Reconstruction accuracy of the target signal (%)	# of inputs
0.995	94.2	157
0.985	99.8	397
0.975	97.5	604

Table 3.9 Performance comparison results for different network hyperparameters

Configuration No.	Batch	LSTM layers	LSTM nodes	Latent nodes	Loss
1	32	2	2	4	1.129E-3
2	32	2	4	8	8.721E-4
3	32	3	2	4	9.017E-4
4	32	3	4	8	5.816E-4
5	64	3	4	8	8.753E-4
6	64	3	8	16	7.139E-4
7	32	4	4	8	1.010E-3
8	64	4	4	8	1.090E-3

In this equation, x_t is the normalized value of the original signal (the output of Step 1), $\widehat{x_t}$ is the reconstructed value (from Step 2), and σ and μ are the mean and deviation values sampled from the latent space z.

Figure 3.16 shows the loss trends of the eight configurations using Data #2. The loss values are calculated for 300 epochs, where an epoch is one loop through the full training dataset consisting of one or more batches of sampling data. In terms of reconstruction performance, Fig. 3.17 compares the result of Configurations 1 and 4 for SG #1 pressure in the cold-leg #1 LOCA scenario as an example. Configuration 4 is shown to reconstruct the original signal more accurately and stably than Configuration 1. Among the configurations, Configuration 4 demonstrates the minimum loss. Thus, through this optimization, it can be determined that the VAE-LSTM network for the signal reconstruction step (i.e., Step 2 of the signal validation algorithm) should include three LSTM layers, four LSTM nods, and eight latent nodes with

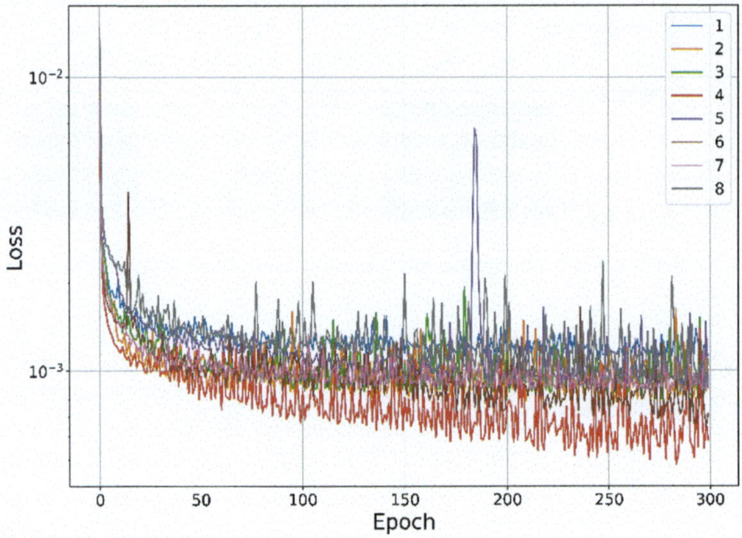

Fig. 3.16 Losses of eight configurations of tested hyperparameters. Reproduced with permission from Choi et al. (2021)

32 batches, as listed for Configuration 4 in Table 3.9. As a result, the loss of the optimized VAE-LSTM network in signal reconstruction is 5.816E-4.

The final optimization, or Optimization 3, determines the particular RE threshold that is the most suitable for detecting the stuck failures of NPP signals. Figure 3.18

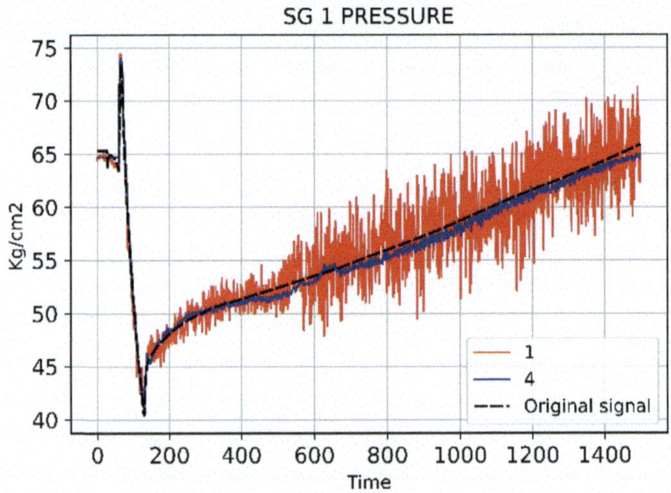

Fig. 3.17 Comparison results of the original signal and reconstructed signals of configurations 1 and 4. Reproduced with permission from Choi et al. (2021)

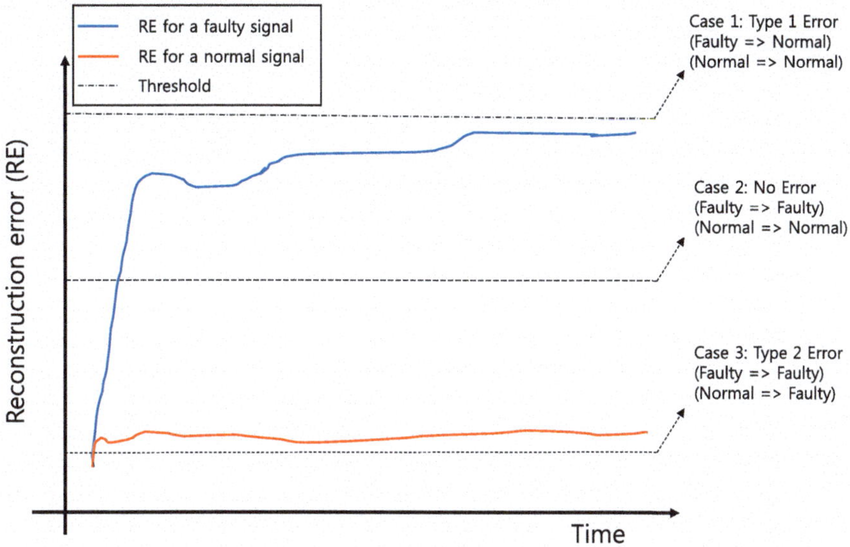

Fig. 3.18 Conceptual determination of RE thresholds for faulty signal detection. Reproduced with permission from Choi et al. (2021)

depicts the process to determine the threshold. In the developed algorithm, RE has a large value for faulty signals because such signals are untrained. Hence, selecting the optimal threshold for discriminating faulty signals is critical for the performance of the signal validation algorithm.

If the threshold is set too high, like in Case 1 in Fig. 3.18, the algorithm determines that both normal and faulty signals are normal. Therefore, a faulty signal is regarded as normal, which is termed here as a Type 1 error. If the threshold is chosen too low, as Case 3 in Fig. 3.18 shows, the algorithm determines that both normal and faulty signals are faulty. In this case, a normal signal is detected as faulty, which is termed a Type 2 error. If the threshold is chosen properly, demonstrated by Case 2 in Fig. 3.18, the algorithm becomes capable of correctly distinguishing between normal and faulty signals.

The RE threshold is determined based on the statistical method proposed by Shewhart (Nazir et al. 2014). Shewhart's control charts are widely used to calculate changes in process features from the in-control state using Eq. (3.8).

$$RE\ threshold = \mu + k\sigma. \tag{3.8}$$

Here, μ and σ represent the mean and standard deviation, respectively, of the RE for each variable in the training data (i.e., Data #1), and k is a constant. This optimization step calculates the results of distinguishing normal and faulty signals for the 26 target variables by entering different k values (i.e., $k = 0.5$, 1, 2, or 3). By testing for Type 1 and Type 2 errors according to the k value, the optimal k can be determined.

Table 3.10 show the results of comparison between Type 1 and Type 2 errors for diverse k values. Based on this comparison, the algorithm adopts $k = 1$, which demonstrates the best performance considering both Type 1 and Type 2 errors. The optimal RE thresholds when $k = 1$ are finally obtained as follows.

$$RE\ threshold = \mu + \sigma \tag{3.9}$$

The μ and σ in this equation are the same as in Eq. (3.8), representing the mean and standard deviation of the RE for each variable in the training data. The right-most column of Table 3.10 shows the thresholds for each signal following Optimization 3.

The conducted optimizations can now be checked for each failure mode. Table 3.11 shows the result of the Optimization 3 using Data #3 from Fig. 3.14. The signal validation algorithm determines 99.81% of the normal signals as "normal" while detecting 97.6% of the signal failures, specifically 100% of the stuck-high, 98.92% of the stuck-low, and 93.88% of the stuck-as-is failures.

3.2.3 *Validation*

Figure 3.19 shows an example of the process by which the signal validation algorithm process detects stuck signal failures. In this illustration, the algorithm receives two signals as inputs from the LOCA scenario. The loop 1 cold-leg temperature signal is faulty, namely a stuck-high failure, while the other signal, PZR pressure, is normal. Step 1 of the algorithm normalizes these signal inputs to a range of 0 to 1. Step 2 attempts to reconstruct the normalized signals similarly to the input signals. Then Step 3 calculates the RE from the difference between the normalized and reconstructed signals. Step 4 compares the calculated RE to the threshold defined in the third optimization.

As shown in Fig. 3.19, the RE of the loop 1 cold-leg temperature is larger than the threshold, and based on the comparison, Step 5 determines that the input signal is faulty.

The signal validation algorithm is now validated with data not used in either the training or optimization, i.e., Data #4. The validation focuses on evaluating whether the proposed algorithm can correctly detect stuck failures of the 26 selected signals. The dataset contains the three failure modes (i.e., stuck-high, stuck-low, and stuck-as-is) and accordingly includes 1152 scenarios with added stuck failures. As shown in Table 3.12, the signal validation algorithm detects 96.70% of all stuck failures and 98.29% of all normal signals, similar to the results achieved through the optimization (Table 3.11).

Table 3.10 Type 1 and Type 2 errors with different *k* values

Parameter	k = 0.5		k = 1		k = 2		k = 3	
	Type 1	Type 2	Type 1	Type 2	Type 1	Type 2	Type 1	Type 2
Feedwater pump outlet press	0	0.07	0	0	0	0.06	0	0
Feedwater line 1 flow	0	0.02	0	0	0	0	0	0
Feedwater line 2 flow	0	0	0	0	0	0	0	0
Feedwater line 3 flow	0	0.001	0	0	0	0	0	0
Feedwater Temp	0	0.11	0	0	0.90	0.09	0.49	0
Main steam flow	0	0.06	0	0	0.93	0.06	0.06	0
Steam line 1 flow	0	0.011	0	0	0	0.10	0	0
Steam line 2 flow	0	0.011	0	0	0	0.10	0	0
Steam line 3 flow	0	0.011	0	0	0	0.11	0	0
Main steam header pressure	0	0.011	0	0	0	0.10	0	0
Charging line outlet temp	0	3.5	0.06	0.06	0.67	0.21	1.28	0
Loop 1 Coldleg Temp	0	3.5	0.06	0.02	0.14	0.38	0.58	0
Loop 2 Coldleg Temp	0	3.5	0.12	0.02	0.03	2.97	0.95	0.005
Loop 3 Coldleg Temp	0	3.5	0.06	0.01	0.23	0.23	0.64	0
PZR temp	0	3.5	0	0.02	0.14	0.14	0.43	0
Core Outlet Tempe	0	3.5	0.35	0.01	0.55	0.12	0.98	0
Net Letdown Flow	0	0.002	0	0	0	0.002	0	0
PZR level	0.75	0.05	0.41	0	1.07	0.04	0.87	0
PZR pressure	0	3.5	0.49	0.02	2.05	0.15	3.04	0
Loop 1 flow	0	3.56	0	0	0	0.13	0	0
Loop 2 flow	0	3.56	0	0	0	0.15	0	0
Loop 3 flow	0	3.56	0	0	0	0.11	0	0
SG 1 level (wide)	0	3.5	0	0.01	0	1.24	0.29	0
SG 2 level (wide)	0	3.5	0	0.01	0	1.90	0.20	0
SG 1 pressure	0	3.5	0.52	0.002	0.20	0.02	1.53	0
SG 2 pressure	0	0.99	0.35	0.001	0.61	0.02	1.56	0
Sum	0.75	47.41	2.40	0.19	7.52	8.44	12.91	0.005

3.3 Signal Generation with a GAN

The majority of the decisions in NPPs, as large-scale, safety–critical systems with high complexity, are made by human operators who monitor numerous instrumentation signals and utilize them while following the various procedures that correspond to the plant states to maintain safe and efficient operation. But in harsh conditions,

Table 3.11 Optimization results of the algorithm for each failure mode

Failure mode		Classification result (%)	
		Faulty	Normal
Failed	Stuck-high	100	0
	Stuck-low	98.92	1.08
	Stuck-as-is	93.88	6.12
	Total	97.6	2.4
Normal		0.19	99.81

multiple instrumentation signals may become unavailable. During the Fukushima accident, for example, most of the instrumentation systems were inoperative. Despite this clear importance, relatively few studies have explored the reconstruction of multiple missing signals in NPP emergency situations.

Most conventional signal reconstruction methods strongly depend on correlations between signals. As signal correlations can vary drastically with the plant condition, traditional methods have limitations in that numerous models need to be developed to cover the various plant conditions, and also prior knowledge of the plant conditions is necessary to select the appropriate model for signal reconstruction. These limitations lead to a type of circular dilemma: normal (reconstructed) signals are necessary to accurately diagnose the plant condition, but a proper reconstruction model for obtaining the signals cannot be identified until the plant condition is diagnosed.

New approaches to signal reconstruction are therefore required to deal with multiple missing signals in a flexible manner under diverse conditions, including emergency situations. Such a method should not depend too strongly on the correlations between signals, nor should it require prior knowledge of the plant condition.

3.3.1 GAN

The basic structure and algorithm of a GAN is described in Sect. 2.4.5. Representative problems that are commonly encountered during the training of a GAN model are as follows.

- Premature convergence: If the generator becomes far superior to the discriminator or vice versa during the training process, then the GAN model may not be properly trained. Typically, it is the discriminator that becomes superior to the generator.
- Mode collapse: If the generator is trained to generate only a single mode or some modes of the data distribution, other modes may be left out.
- Loss oscillation: Instead of converging, the states of the generator and discriminator may oscillate, regardless of the training length.

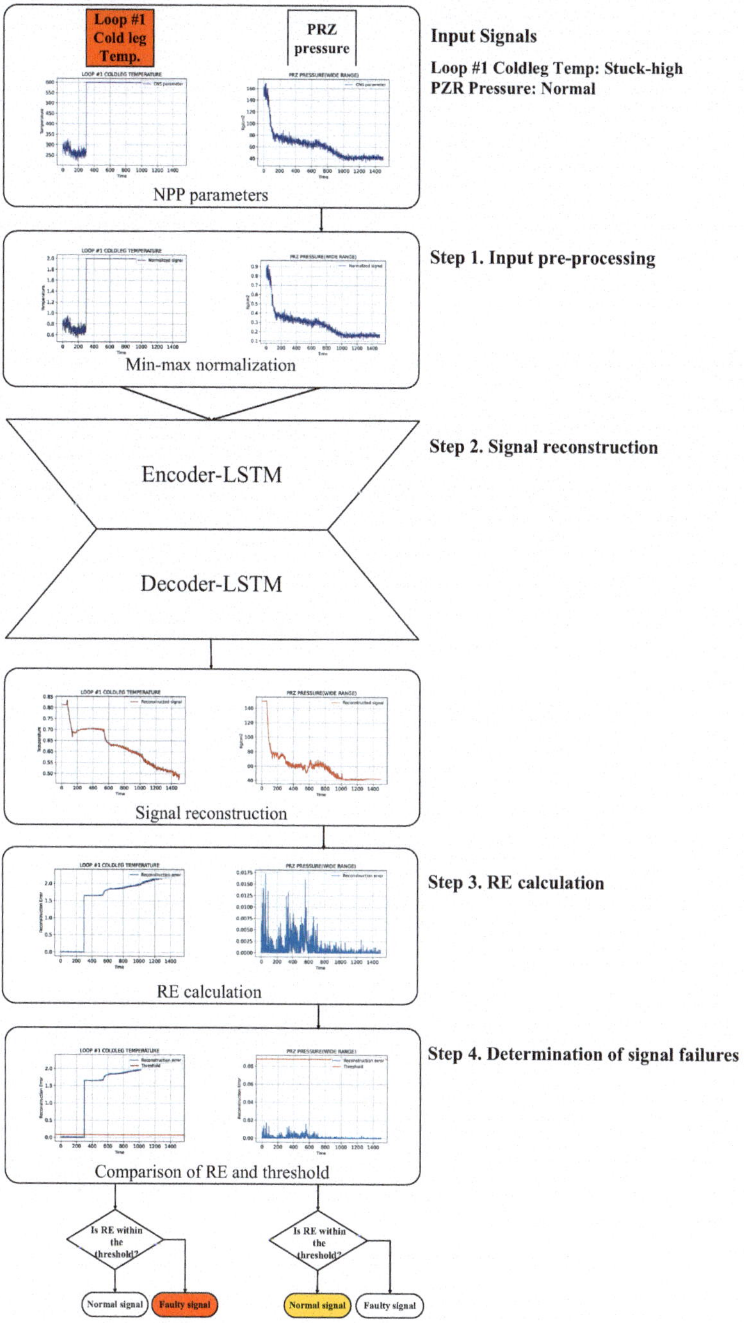

Fig. 3.19 Validation process of the signal validation algorithm

Table 3.12 Validation results of the algorithm for each failure mode

Failure modes		Classification results (%)	
		Failed	Normal
Failed	Stuck-high	100	0
	Stuck-low	97.92	2.08
	Stuck-as-is	92.18	7.82
	Total	96.70	3.30
Normal		1.71	98.29

Different GAN architectures each have unique characteristics and advantages, and thus by merging their concepts, new GAN structures suitable for specific application can be constructed. The first variant selected in the current GAN development to consider the sequential characteristics of instrumentation signals is a recurrent generative adversarial network (RGAN) (Esteban et al. 2017; Press et al. 2017). The RGAN adopts an RNN as its baseline architecture, and as mentioned in Chap. 2, RNNs that include a directed cycle between nodes perform well for sequential or time-series data (e.g., natural language and sound). An LSTM (Hochreiter and Schmidhuber 1997) is also selected for its strength in considering long-term dependencies.

The second GAN variant chosen is the conditional GAN (Mirza and Osindero 2014), which is adopted to more precisely consider sets of instrumentation signals in various NPP conditions. The conditional GAN was originally proposed to address limitations of the vanilla GAN, which is unable to selectively generate samples with the desired attributes only. The overall architecture of the conditional GAN is similar to that of the vanilla GAN, but the conditional GAN's generator and discriminator take as inputs information on specific attributes of the data, or labels, in addition to the latent vector or data. Moreover, while the conditional GAN discriminator outputs the same classification results (i.e., real or fake), it also compares the characteristics or attributes of the input data and with those in the given information or labels. The discriminator of the conditional GAN thus performs two separate functions—classification of realistic and unrealistic data, and estimation of data attributes—and accordingly it can be divided into two networks. The one that performs the attribute or label estimation for the input data is called the classifier network.

The last variant selected is a manifold-Guided generative adversarial network (MGGAN) (Bang and Shim 2021), which is adopted to mitigate the mode collapse problem. The MGGAN was developed by adding additional guidance networks, namely an encoder network and an encoder-discriminator network, to the conventional GAN architecture.

3.3.2 GAN-Based Signal Reconstruction Method

The baseline concepts of the current signal reconstruction method are similar to those of the image in painting method (Yeh et al. 2017). But owing to the many differences between image characteristics (static, high spatial correlations) and instrumentation signals (sequential, low spatial correlations), significant modifications to the underlying GAN architecture, loss functions, and performance metrics are needed before adopting the concepts of the image in painting method in the development of the signal reconstruction method.

The GAN-based signal reconstruction method follows three steps: GAN model training, searching for the optimal latent vector and optimal label, and signal reconstruction. The following sections describe these steps in detail.

3.3.2.1 Training of the GAN Model

The first step of the GAN-based signal reconstruction method involves training the GAN model to prepare it for the subsequent steps. In the training process, the generator is trained to generate a realistic signal set from the given latent vector, while the discriminator is trained to distinguish real-world signal sets and generated signal sets.

To achieve successful signal reconstruction across various NPP conditions, the generator needs to be able to mimic the signal sets that can be observed in the different conditions. It should also be fairly general, meaning that the generator should be capable of mimicking not only the signal sets used for training it but also the signal sets that belong to similar manifold distributions.

As mentioned in Sect. 3.3.1, the current GAN architecture combines the concepts of the RGAN, conditional GAN, and MGGAN. As a result, it has five subnetworks: generator (G), discriminator (D), classifier (C), encoder (E), and encoder-discriminator (D_E). Figure 3.20 shows a schematic of the GAN architecture.

Among the subnetworks, the AE should be trained before the training of the other aspects of the GAN model in order to utilize the encoder, as follows from the adoption of the MGGAN. After training the AE and the rest of the GAN, the model parameters are fixed and used in all subsequent steps.

The loss functions necessary for updating the network parameters of the GAN architecture are defined in Eqs. (3.10) to (3.13), in which expectation terms and normalizing constants are omitted. During the experiments, we attempted to apply the Wasserstein-distance-based loss function and a sigmoid cross-entropy loss function for the generator and discriminator, but empirically found that such loss functions did not perform well, except for the following least-squares loss function.

$$L_{Gen} = (D(G(z, c)) - 1)^2 + (D_E(G(z, c)) - 1)^2 \tag{3.10}$$

$$L_{Dis} = (D(x) - 1)^2 + (D_E(x) - 1)^2 + (D(G(z, c)))^2 + (D_E(G(z, c)))^2 \tag{3.11}$$

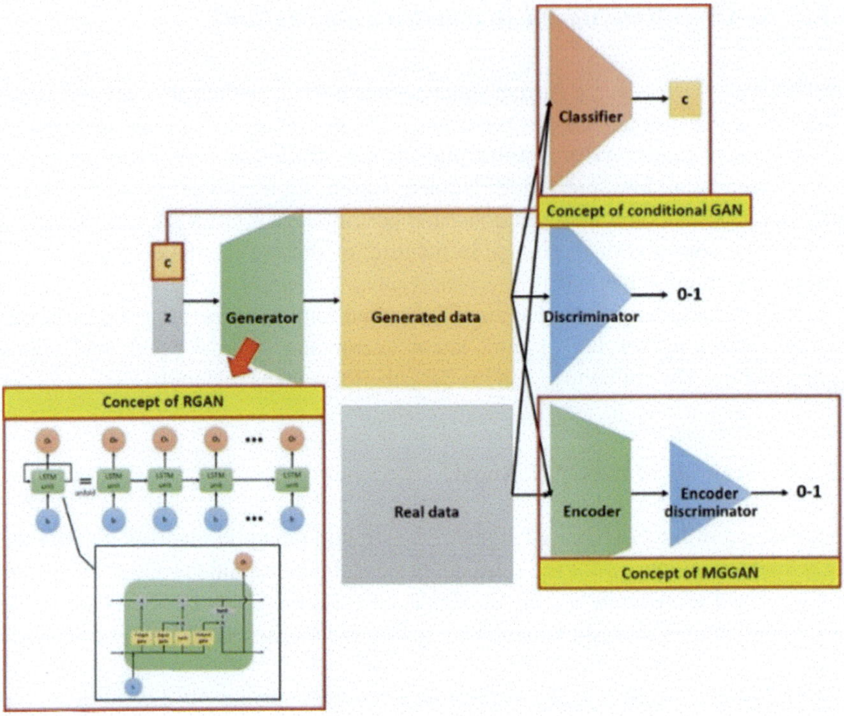

Fig. 3.20 Schematic of the GAN architecture for signal generation. Reproduced with permission from Kim et al. (2020)

$$L_{AE} = \sum_{i}\sum_{j}\left(AE(x)_{(i,j)} - x_{(i,j)}\right)^2 \tag{3.12}$$

$$\begin{aligned} L_{Cla} = &\sum_{n=1}^{N}\left\{\left|C_{cont,n}(x) - c_{cont,x,n}\right| + \left|C_{cont,n}(G(z,c)) - c_{cont,G(z,c),n}\right|\right\} \\ &- \sum_{i=1}^{Q}\sum_{j=1}^{n(Q_i)}\left\{C_{disc,i,j}(x)\log\left(c_{disc,x,i,j}\right) + C_{disc,i,j}(G(z,c))\log\left(c_{disc,G(z,c),i,j}\right)\right\} \end{aligned} \tag{3.13}$$

In these equations, L_{Gen} is the generator loss function, L_{Dis} is the discriminator loss function, L_{AE} is the AE loss function, L_{Cla} is the classifier loss function, c is the label, $c_{cont,x,n}$ is the n-th continuous label of data x, $c_{disc,x,i,j}$ is the value of the discrete label of data x (i-th discrete label, j-th class), D_E is the encoder-discriminator function, AE is the AE function ($AE_{(i,j)}$, i-th row, j-th column element of AE output), $x_{(i,j)}$ is the i-th row, j-th column element of input x, $C_{cont,n}$ is the n-th continuous-label-estimation function of the classifier network, and $C_{disc,i,j}$ is the discrete-label-estimation function of the classifier network (i-th discrete label, j-th class). Otherwise, N, Q, and n(Q_i) are the number of continuous labels, number of discrete labels, and number of classes in the i-th discrete label, respectively.

Compared with other data types, such as image data, it is relatively more challenging to check the quality of the generated signal sets by manually inspecting drawn samples. To address this issue, a simple performance metric called the generative error is introduced to easily confirm the performance of the generator. For a given sample, the generative error is defined as the minimum value of the deviations between the training dataset and the generated dataset. By this definition, the generative error is zero when the generator produces data identical to the training data. The generative error can be mathematically expressed with Eq. (3.14).

$$E_g(z, c) = \min_k \left[\sum_i \sum_j \left(\left| G(z, c)_{(i,j)} - x_{k,(i,j)} \right| \right) \right]. \tag{3.14}$$

Here, E_g denotes the generative error, $G_{(i,j)}$ is the i-th row, j-th column element of the generator output, and $x_{k,(i,j)}$ is the i-th row, j-th column element of the k-th training data. In terms of the performance of the GAN model in generating realistic samples, a large mean generative error over multiple samples indicates that the generator is ill-trained to generate realistic samples as intended, assuming that the training datasets are generally evenly distributed. But in terms of the generality of the model, an overly small mean generative error is not desirable either, as this can indicate an occurrence of the overfitting problem. Despite the fact that the suggested metric is not suitable for fixing precise criteria, it is clear that a certain level of mean generative error is desirable when considering the applicability of the model. We set the mean generative error in this work to be about 1%–3% and trained the GAN model following this criterion.

3.3.2.2 Search for the Optimal Latent Vector and Optimal Label

To reconstruct a damaged signal set, the trained GAN model should generate a signal set that is appropriate for reconstruction. In the second step, which is to find the optimal latent vector and optimal label, a specific latent vector and label are identified and used as inputs for the trained generator. This allows the model to generate suitable appropriate signal sets for reconstruction.

The search for the optimal latent vector and optimal label is performed iteratively, where a specific loss function (differing from those in the GAN model training step) is minimized; Fig. 3.21 shows a schematic of the iterative process used to find the optimal latent vector and label. The loss function for this step is determined based on three loss elements, as described below, that are related to the shared characteristics of the damaged signal set and the generated signal set as well as the quality of the generated signal set.

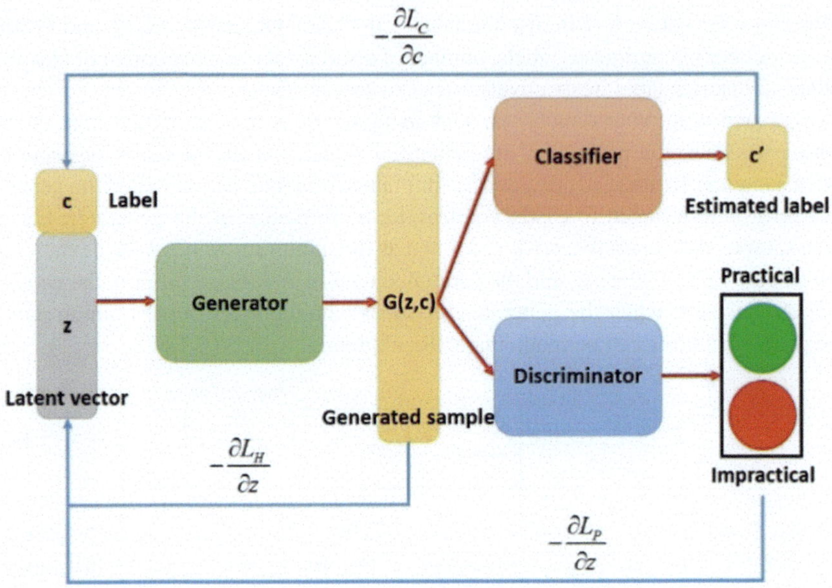

Fig. 3.21 Schematic of the iterative process used for finding the optimal latent vector and optimal label. Reproduced with permission from Kim et al. (2020)

The first loss element is called homogeneity loss, which measures the deviation between the normal aspects of the damaged signal set and the equivalent aspects of the generated signal set. Homogeneity loss is a fundamental loss element, allowing the normal data in the damaged signal set to be utilized as a clue for the signal reconstruction.

The second loss element, classification loss, measures the deviation between the labels estimated from the normal aspects of the damaged signal set and the labels estimated from the equivalent aspects of the generated signal set. While the role of classification loss is similar to that of homogeneity loss, classification loss employs a classifier network instead of a discriminator network.

The third loss element is called practicality loss, which is used to determine whether the generated signal set is realistic or unrealistic. During the iterative process of the search step, LP helps prevent the generation of unrealistic solutions with low values of homogeneity loss and classification loss.

Summing the values of the three loss elements with corresponding weighting factors λ1 (for classification loss) and λ2 (for practicality loss) gives the total loss (Ltotal). The λ1 and λ2 weighting factors are considered hyperparameters, and thus should be tuned to obtain proper values. Mathematically, the above loss elements and total loss can be expressed as in Eqs. (3.15) to (3.18).

In these equations, L_H is the homogeneity loss, L_C is the classification loss, L_P is the practicality loss, L_{total} is the total loss, s is the damaged signal set, m is the mask matrix indicating the missing and normal parts with binary values (0 and

1, respectively), λ_1 is a relative weighting factor for classification loss, and λ_2 is a relative weighting factor for practicality loss. The symbol $\odot$ is an operator for element-wise multiplication.

$$L_H(z, c|s, m) = \sum(m \odot |G(z, c) - s|) \tag{3.15}$$

$$\begin{aligned} L_C(z, c|s, m) = &\sum_{n=1}^{N} |(C_{cont,n}(m \odot G(z, c)) - C_{cont,n}(m \odot s)| \\ &- \sum_{i=1}^{Q} \sum_{j=1}^{n(Q_i)} C_{disc,i,j}(m \odot s)\log(C_{disc,i,j}(m \odot (G(z, c))) \end{aligned} \tag{3.16}$$

$$L_P(z, c) = \log(1 - D(G(z, c))) \tag{3.17}$$

$$L_{total} = L_H(z, c|s, m) + \lambda_1 L_C(z, c|s, m) + \lambda_2 L_P(z, c) \tag{3.18}$$

3.3.2.3 Signal Reconstruction

Once the optimal latent vector and the optimal label needed for signal reconstruction are found through the iterative process in the second step, the trained GAN model generates an 'optimal' signal set. Signal reconstruction then proceeds by replacing the missing parts of the damaged signal set with the corresponding parts of the generated 'optimal' signal set. Expressions for the signal reconstruction are given below.

$$\hat{z}, \hat{c} = \arg\min_{z,c}(L_{total}) \tag{3.19}$$

$$\hat{s} = (m \odot s) + \big((1 - m) \odot G(\hat{z}, \hat{c})\big). \tag{3.20}$$

Here, $\hat{z}$ denotes the optimal latent vector, $\hat{c}$ denotes the optimal label, and $\hat{s}$ denotes the reconstructed signal. It should be noted that the reconstructed signal in Eq. (3.20) may include normalized data, and thus post-processing should be conducted to obtain the actual signal values.

3.3.3 Experiments

Experiments are designed and conducted to validate the GAN-based signal reconstruction method. In the pilot/main experiments, data acquisition and preprocessing are followed by GAN model training, after signal reconstruction is performed. The pilot experiment is conducted to verify the signal reconstruction performance and determine the most suitable parameter sets. The main experiment is conducted using the suitable parameter sets from the pilot experiment and for several conditions. Python 3.6 and TensorFlow (version 1.5) are used for programming and as the ML platform.

3.3.3.1 Data Acquisition and Preprocessing

As the signal reconstruction method is a data-driven method, a sufficient number of realistic signal sets reflecting NPP emergency conditions must be prepared for the method to be effective. For this, a CNS developed by KAERI was used to simulate data for the experiments, based on the lack of real-world data: there have only been a small number of cases of NPP emergencies, and real-world instrumentation signal data in emergency situations are likewise scarce. The CNS was developed using the SMABRE (small break) code and has as its reference plant the Westinghouse three-loop 900 MW PWR (KAERI 1990). Although the employed CNS is not as precise as the simulators typically used in thermal–hydraulic analyses, its performance in simulating the early phases of NPP emergency events has been established (Kurki and Seppälä 2009).

In the experiments, four design basis accidents (DBAs) are simulated with the CNS: cold-leg LOCA (cold-leg LOCA), hot-leg LOCA (hot-leg LOCA), SGTR, and main steam line break (MSLB). For each scenario, 91 break sizes from 10 to 100 cm^2 at 1 cm^2 intervals (multiplied by 10 for the MSLB) are simulated. In all simulations, the initial plant condition is 100% full-power normal operation, and data are collected for over 5 min starting from the actuation of the reactor trip by an anomaly. As for the instrumentation signals, 31 different types of signals (refer to Table 3.15, experiment I) that are important for plant state diagnosis are selected.

Following simulation, the simulated data undergo the following preprocessing steps to effectively train the GAN model and to be used in the experiments.

- All data are processed to have the same time length (300 s of plant operation time)
- Data for all instrumentation signals are linearly interpolated to have the same time interval (1 s of plant operation time)
- Unit input data are generated with a time length of 30 s (plant operation time)
- Measurement values are normalized to be between –1 (minimum) and 1 (maximum)
- Unit input data are labeled according to accident type, break size, and elapsed time (from reactor trip).

In the labeling step, the accident type as a discrete label is one-hot encoded, while the break size and time as continuous labels are labeled with values between –1 (minimum value) and 1 (maximum value).

As a result of the preprocessing, a total of 98,280 unit data (4 scenarios, 91 break sizes, 270 unit data per simulation) are obtained. Each unit data is a 31 × 31 matrix and comprising 31 signal types each over 30 s. From the preprocessed data, 55,080 unit data (4 scenarios, 51 break sizes, 270 unit data per simulation), about 56%, are

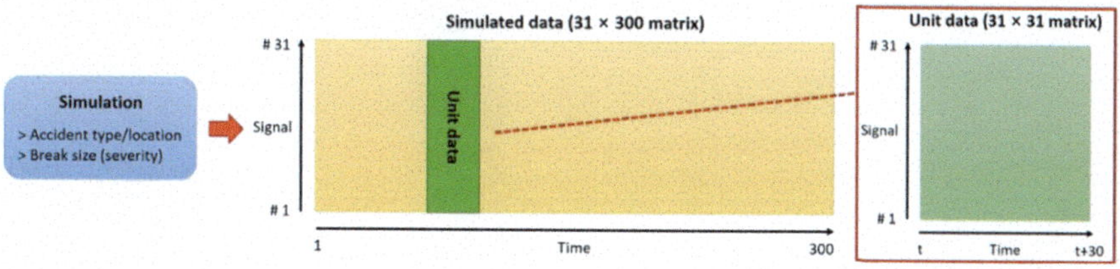

Fig. 3.22 Schematic of the data structure and unit data generation. Reproduced with permission from Kim et al. (2020)

used as training data. Figure 3.22 shows a schematic of the data structure and the unit data generation.

3.3.3.2 GAN Model Training Results with Pilot/Main Experiments

Based on the GAN architecture (Fig. 3.20), the model is repeatedly trained using various sets of hyperparameters. During the AE training process in particular, the encoded size (i.e., the size of the encoder's output), numbers of LSTM layers and LSTM sequences, number of fully connected (FC) layers, number of FC layer nodes, and learning rate are mainly considered. For the training of the other subnetworks of the GAN, the latent vector size, numbers of LSTM layers and LSTM sequences, number of FC layers, number of FC layer nodes, and learning rate are considered. To find the hyperparameter sets that minimize the AE loss and mean generative error in the training process of the AE and other GAN subnetworks, respectively, the grid search method is applied. To prevent excessive training time, hyperparameters excluding the learning rate are first roughly determined, after which the learning rate is determined precisely.

Table 3.13 lists the hyperparameter sets that are selected for training the GAN model. In the subsequent experiments, the corresponding GAN model is trained with these hyperparameters. The Adam optimizer Kingma and Ba (2014) is applied using its default settings ($\beta_1 = 0.9$, $\beta_2 = 0.999$, $\varepsilon = 10^{-8}$) for updating all networks.

Table 3.13 Hyperparameter set selected for GAN model training

	# of LSTM layers	# of LSTM sequences	# of FC layers	# of FC nodes	Learning rate (10 − 7)
Generator	2	31	3	248/124/62	50
Discriminator	2	31	3	248/124/62	10
Encoder (separately trained)	2	31	2	124/62	200
Encoder-discriminator	2	31	3	248/124/62	2
Classifier	2	31	3	248/124/62	10

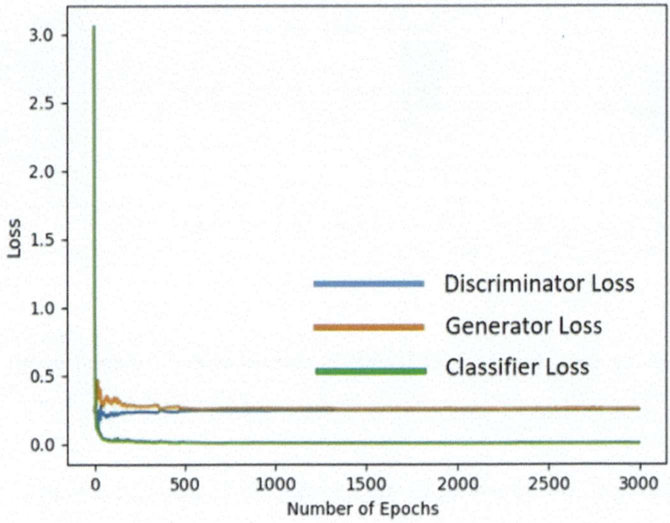

Fig. 3.23 Losses of the generator, discriminator, and classifier networks during model training. Reproduced with permission from Kim et al. (2020)

During training, 200 samples are drawn every 100 epochs to check the performance of the model. After 3000 training epochs with the selected hyperparameter sets, the mean generative error of the drawn samples is 1.10%, the maximum generative error is 3.46%, and the standard deviation of the generative errors is 0.0076. The losses of the generator, discriminator, and classifier networks during training are shown in Fig. 3.23, and the mean generative errors during training are shown in Fig. 3.24. As shown in the figures, all three loss types mostly converge after about 500 epochs, while the mean generative error decreases further after 500 epochs.

Signal Reconstruction

In the last step of the experiments, the trained GAN model is applied for signal reconstruction. Prior to conducting the main experiment, a pilot experiment is first attempted, the results of which are used to make modifications for the main experiment.

Pilot Experiment and Modifications

In the pilot experiment to verify the signal reconstruction performance of the trained GAN model, 1000 unit data are randomly selected from the total 98,280 unit data, and the signal reconstruction sequences are performed after intentionally omitting signals. The types of omitted signals are randomly selected in every trial, and the number of omitted signals varies from 1 to 20. For a quantitative evaluation of the

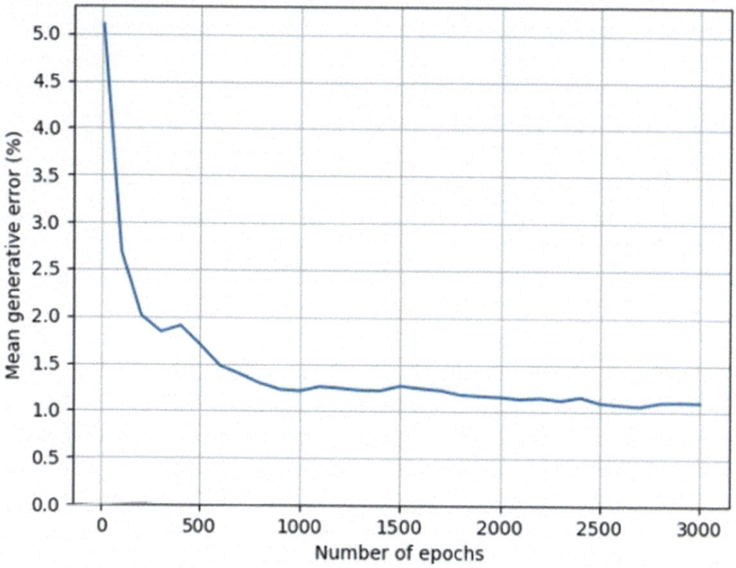

Fig. 3.24 Mean generative error during model training from samples drawn every 100 epochs. Reproduced with permission from Kim et al. (2020)

signal reconstruction performance, the mean RE, maximum RE, and standard deviation of the RE are measured for the reconstruction results of the 1000 intentionally damaged signal sets.

Similar to the GAN model training step, signal reconstruction is repeatedly performed with different sets of hyperparameters. The mainly considered ones are λ_1, λ_2, and the learning rate. The grid search method is again applied to find the proper hyperparameters for signal reconstruction with a minimal mean RE. Section 3.3.1 discusses the detailed process of determining the two weighting factors, λ_1 and λ_2.

Table 3.14 lists the finally selected hyperparameter set, with which the corresponding reconstruction results are used for performance evaluation. As in the GAN model training step, the Adam optimizer Kingma and Ba (2014) is used at its default settings ($\beta_1 = 0.9$, $\beta_2 = 0.999$, $\varepsilon = 10^{-8}$) for the iterative process of searching for the optimal latent vector and optimal label.

The reconstruction performance tends to be similar or better for increasing numbers of iterations, but it is undesirable to spend a significant amount of time on the reconstruction task. A maximum of 750 iterations are therefore performed in

Table 3.14 Selected hyperparameters for the iterative process of signal reconstruction

	Max. no. of iterations	λ_1	λ_2	Learning rate (10^{-2})
Finding the optimal latent vector and optimal label	750	5	10	10

this experiment, requiring about 30 s which is identical to the time length of the unit data. The maximum number of iterations can be varied as long as the reconstruction performance is not excessively degraded or the time required for the reconstruction is not excessively long; any changes to the number of iterations, though, should be made after considering the performance of the underlying hardware.

The reconstruction results are obtained for different numbers of missing signals. Figures 3.25 and 3.26 plot the number of data that exceed 10% RE and the maximum RE according to the number of missing signals, respectively.

As the current method does not produce any RE for the normal signals in the damaged signal set, it can be expected that the RE increases with an increasing number of missing signals. However, multiple performance metrics in the pilot experiment showed contradictory results, for example the reducing trend in the number of data that exceed 10% of the RE and the maximum RE with increasing number of missing signals as seen in Figs. 3.25 and 3.26. Inspecting the signal reconstruction process in the pilot experiment reveals two possible reasons for the contradictory results.

One reason is related to a limitation of the trained GAN model itself, namely that it may poorly express a particular signal in a particular scenario. Such a limitation would result in large REs for specific signal sets, and even for specific signals within a signal set. Accordingly, when reconstruction is attempted in the case that multiple signals are missing, several signals may not be appropriately reconstructed even if most of the missing signals are reconstructed properly. In this case, with an increasing

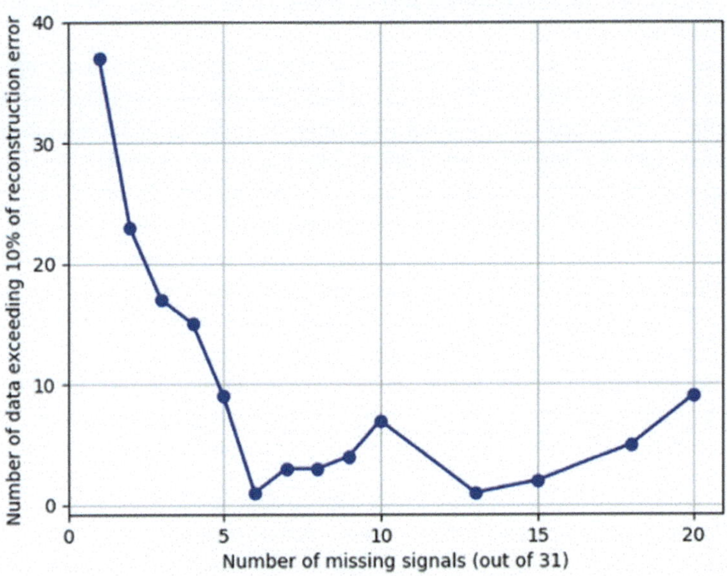

Fig. 3.25 Number of data exceeding 10% RE, pilot experiment. Reproduced with permission from Kim et al. (2020)

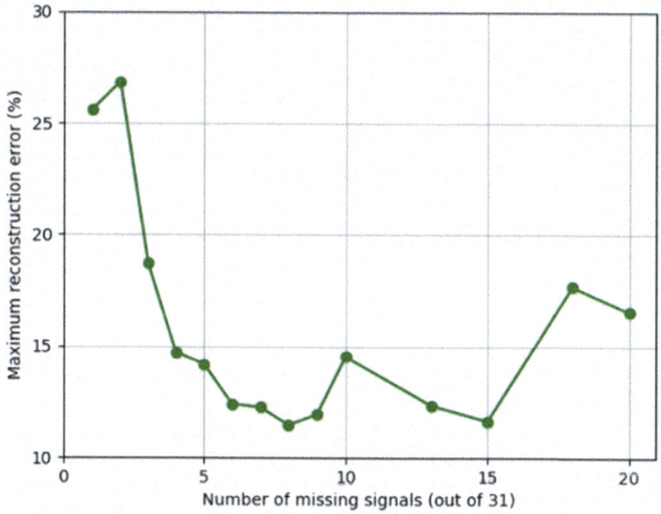

Fig. 3.26 Maximum RE, pilot experiment. Reproduced with permission from Kim et al. (2020)

number of missing signals, the effect that the poorly reconstructed signals have on the RE is reduced because their effect is compensated for by the well-reconstructed signals. This case was indeed found to occur in the pilot experiment. It was also found that specific types of signals, in particular those related to water levels (e.g., RV and PZR water levels), induce higher REs in numerous cases compared to other types of signals.

The other reason for the contradictory results mentioned above, as an extension of the first reason, is that RE alone is insufficient to measure the overall reconstruction performance of the model. Note that RE is obtained by equally considering all missing signals in the signal set. Therefore, it is a valid performance metric for evaluating the model from a macroscopic perspective, but it is inappropriate for evaluating model performance from a microscopic perspective because it does not clearly reveal whether the model successfully reconstructs specific damaged signal sets.

Considering these two reasons, the following modifications are made for the main experiments. The first modification is to conduct additional experiments that exclude several types of signals that induce large REs as candidates for the intentionally omitted signals. The second modification is to adopt a new performance metric, namely mean-max RE, in addition to the conventional RE. As its name implies, mean-max RE is the largest value of the mean REs for all signal types. By defining a reconstruction result as failed when its mean-max RE exceeds a certain threshold, it easier to evaluate whether specific reconstruction cases are successful. Figure 3.27 shows a schematic of the calculation process for both RE and mean-max RE. It can be noted that, in terms of terminology, the term *RE* as used so far would correspond to *mean-mean RE*.

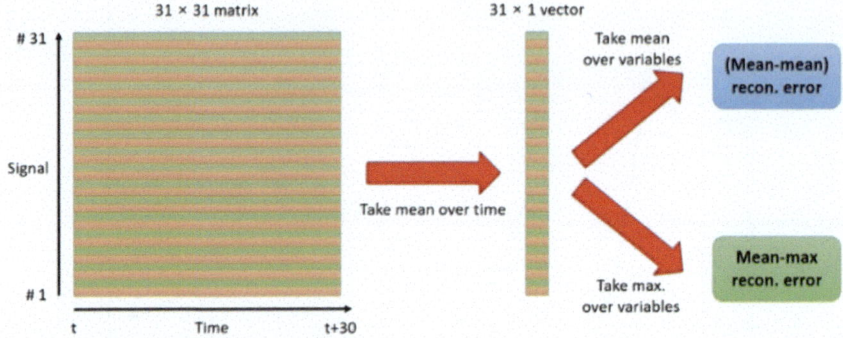

Fig. 3.27 Schematic of the calculation processes for the (mean-mean) RE and mean-max RE. Reproduced with permission from Kim et al. (2020)

Main Experiment

The main experiment is conducted with the modifications determined from the pilot experiment. The general process and configurations of the main experiment mirror those of the pilot experiment, but with several differences as follows.

- Metrics based on the mean-max RE, i.e., the number of data exceeding 5 and 10% of the mean-max RE, are additionally considered.
- The number of missing signals is varied from 1 to 21.
- Additional trials are conducted that exclude 5 or 10 types of signals that induce large REs from the candidates of intentionally omitted signals. In what follows, the specific experiments excluding 0, 5, and 10 types of signals as candidates for intention omission are referred to as Experiments I, II, and III, respectively.

Table 3.15 lists the candidate signals for intentional omission and their units for Experiments I, II, and III. Figures 3.28, 3.29, 3.30, and 3.31 show the number of data exceeding 10% of the mean-max RE, percentages of successful reconstruction, mean REs, and standard deviations of the REs according to the number of missing signals, respectively.

Table 3.15 Candidates of intentionally omitted signals and their units

Signal (Experiment I, 31 types of signals)	Signal (Experiment II, 26 types of signals)	Signal (Experiment III, 21 types of signals)	Unit
CTMT sump level	CTMT sump level	—	m
CTMT radiation	CTMT radiation	—	mrem/h
CTMT relative humidity	CTMT relative humidity	CTMT relative humidity	%
CTMT temperature	CTMT temperature	CTMT temperature	°C
CTMT pressure	CTMT pressure	—	kg/cm^2
Core outlet temperature	Core outlet temperature	Core outlet temperature	°C
Hot leg temperature (loops 1, 2, and 3)	Hot leg temperature (loops 1, 2, and 3)	Hot leg temperature (loops 1, 2, and 3)	°C
Cold leg temperature (loops 1, 2, and 3)	Cold leg temperature (loops 1, 2, and 3)	Cold leg temperature (loops 1, 2, and 3)	°C
Delta temperature (loops 1, 2, and 3)	Delta temperature (loops 1, 2, and 3)	Delta temperature (loops 1, 2, and 3)	°C
PRT temperature	PRT temperature	PRT temperature	°C
PRT pressure	PRT pressure	PRT pressure	kg/cm^2
H2 concentration	H_2 concentration	H_2 concentration	%
RV water level	—	—	%
PZR temperature	PZR temperature	PZR temperature	°C
PZR level	—	—	%
PZR pressure (wide range)	PZR pressure (wide range)	PZR pressure (wide range)	kg/cm^2
SG pressure (loops 1, 2, and 3)	SG pressure (loops 2 and 3)	SG pressure (loops 2, and 3)	kg/cm^2
SG narrow range level (loops 1, 2, and 3)	SG narrow range level (loops 2 and 3)	—	%
Feedwater flow rate (loops 1, 2, and 3)	Feedwater flow rate (loops 2 and 3)	Feedwater flow rate (loops 2 and 3)	t/h

CTMT: containment, PRT: PZR relief tank, RV: reactor vessel, PRZ: pressurizer, SG: steam generator

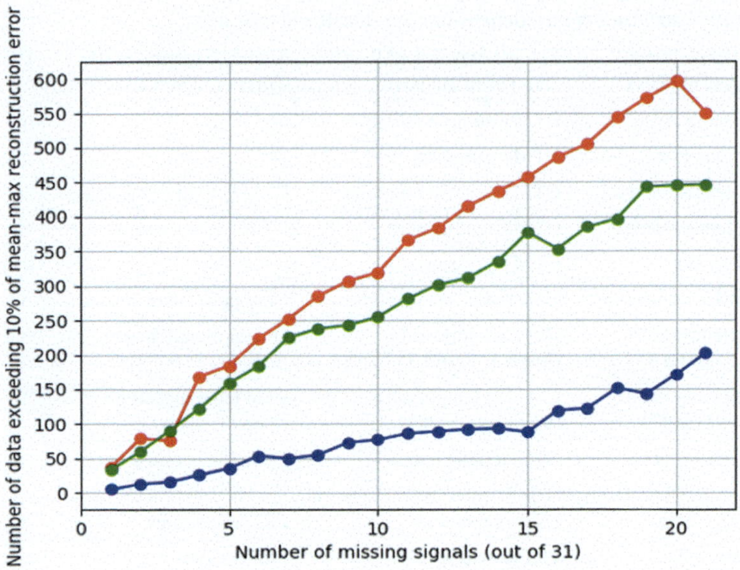

Fig. 3.28 Number of data exceeding 10% of the mean-max RE. Red, green, and blue lines denote the results of Experiments I, II and III, respectively. Reproduced with permission from Kim et al. (2020)

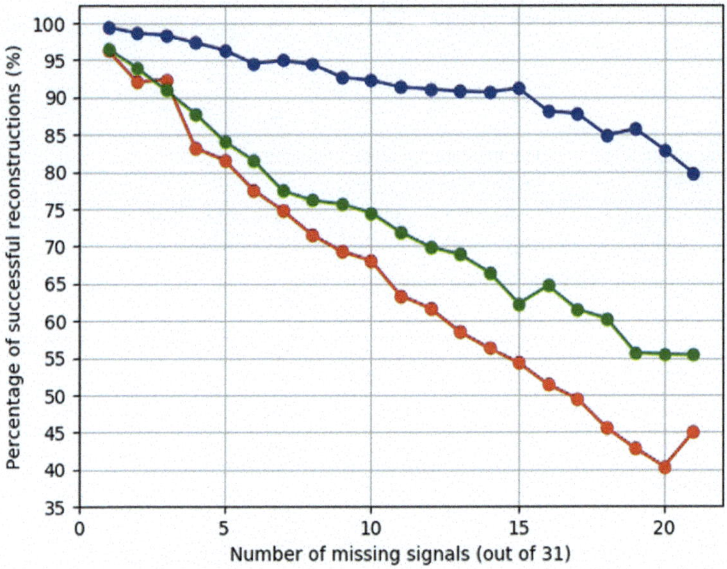

Fig. 3.29 Percentages of successful reconstructions. Red, green, and blue lines denote the results of Experiments I, II, and III, respectively. Reproduced with permission from Kim et al. (2020)

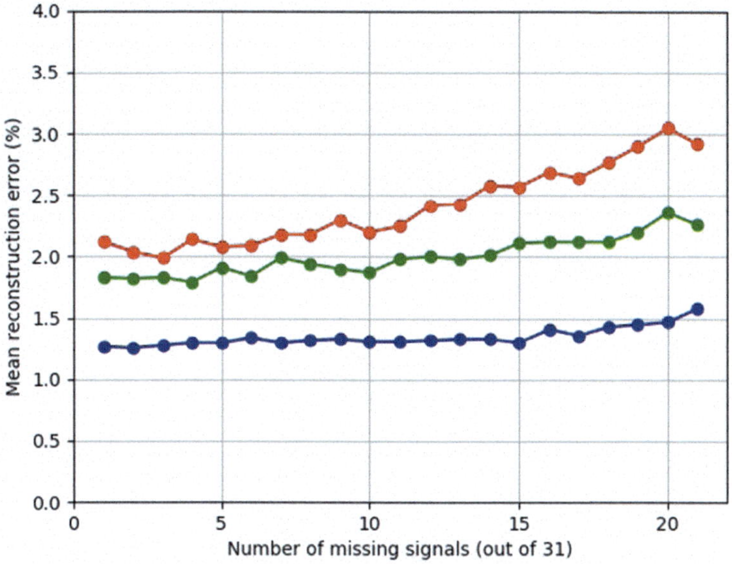

Fig. 3.30 Mean REs. Red, green, and blue lines denote the results of Experiments I, II, and III, respectively. Reproduced with permission from Kim et al. (2020)

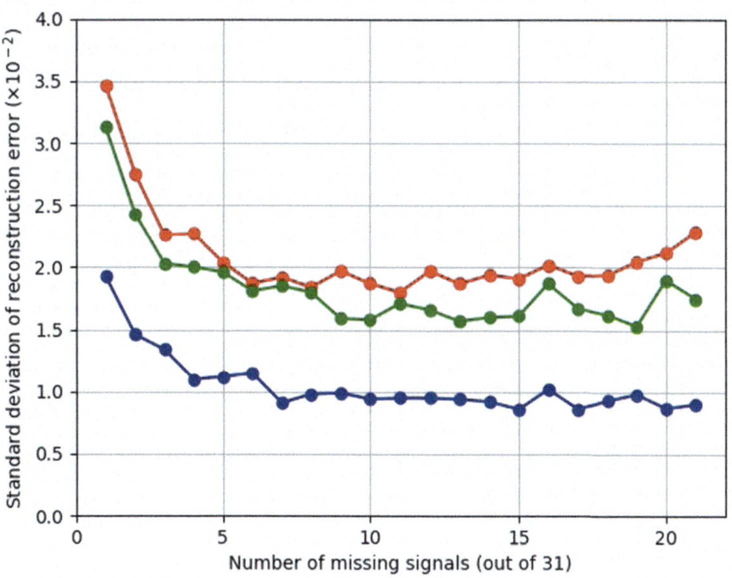

Fig. 3.31 Standard deviations of the RE. Red, green, and blue lines denote the results of Experiments I, II, and III, respectively. Reproduced with permission from Kim et al. (2020)

References

Albazzaz H, Wang XZ (2004) Statistical process control charts for batch operations based on independent component analysis. Ind Eng Chem Res 43(21):6731–6741

An J, Cho S (2015) Variational autoencoder based anomaly detection using reconstruction probability. Special Lect IE 2(1):1–18

Bae J, Ahn J, Lee SJ (2020) Comparison of multilayer perceptron and long short-term memory for plant parameter trend prediction. Nucl Technol 206(7):951–961

Bang D, Shim H (2021) MGGAN: solving mode collapse using manifold-guided training. In: Proceedings of the IEEE/CVF international conference on computer vision, pp 2347–2356

Baraldi P, Cammi A, Mangili F, Zio E (2010) An ensemble approach to sensor fault detection and signal reconstruction for nuclear system control. Ann Nucl Energy 37(6):778–790

Baraldi P, Di Maio F, Genini D, Zio E (2015) Comparison of data-driven reconstruction methods for fault detection. IEEE Trans Reliab 64(3):852–860

Boden M (2002) A guide to recurrent neural networks and backpropagation. The Dallas Project, 2(2):1–10.

Chen Y, Chen R, Pei L, Kröger T, Kuusniemi H, Liu J, Chen W (2010) Knowledge-based error detection and correction method of a multi-sensor multi-network positioning platform for pedestrian indoor navigation. In: IEEE/ION position, location and navigation symposium. IEEE, pp 873–879

Choi Y, Yoon G, Kim J (2021) Unsupervised learning algorithm for signal validation in emergency situations at nuclear power plants. Nucl. Eng. Technol

Choi J, Lee SJ (2020) Consistency index-based sensor fault detection system for nuclear power plant emergency situations using an LSTM network. Sensors 20(6):1651

Di Maio F, Baraldi P, Zio E, Seraoui R (2013) Fault detection in nuclear power plants components by a combination of statistical methods. IEEE Trans Reliab 62(4):833–845

Esteban C, Hyland SL, Rätsch G (2017) Real-valued (medical) time series generation with recurrent conditional gans. arXiv preprint arXiv:1706.02633

Fantoni PF, Hoffmann M, Htnes W, Rasmussen B, Kirschner A (2004) The use of non linear partial least square methods for on-line process monitoring as an alternative to artificial neural networks. Machine Intelligence: Quo Vadis? World Scientific

Fantoni PF, Mazzola A (1996) A pattern recognition-artificial neural networks based model for signal validation in nuclear power plants. Ann Nucl Energy 23(13):1069–1076

Gers FA, Schmidhuber J, Cummins F (2000) Learning to forget: continual prediction with LSTM. Neural Comput 12(10):2451–2471

Gertler J (1997) Fault detection and isolation using parity relations. Control Eng Pract 5(5):653–661

Hashemian H (2010) Aging management of instrumentation & control sensors in nuclear power plants. Nucl Eng Des 240(11):3781–3790

Hines J, Uhrig RE, Wrest DJ (1998) Use of autoassociative neural networks for signal validation. J Intell Rob Syst 21(2):143–154

Hines J (2009) On-line monitoring for calibration extension: an overview and introduction. US Nuclear Regulatory Commission

Hochreiter S, Schmidhuber J (1997) Long short-term memory. Neural Comput 9(8):1735–1780

Hwang I, Kim S, Kim Y, Seah CE (2009) A survey of fault detection, isolation, and reconfiguration methods. IEEE Trans Control Syst Technol 18(3):636–653

IAEA (2002) Accident analysis for nuclear power plants. International Atomic Energy Agency

IAEA (2006) Development and review of plant specific emergency operating procedures. International Atomic Energy Agency

Ito K, Xiong K (2000) Gaussian filters for nonlinear filtering problems. IEEE Trans Autom Control 45(5):910–927

KAERI (1990) Advanced compact nuclear simulator textbook. Nuclear Training Center in Korea Atomic Energy Research Institute

Kaistha N, Upadhyaya BR (2001) Incipient fault detection and isolation of field devices in nuclear power systems using principal component analysis. Nucl Technol 136(2):221–230

Kim SG, Chae YH, Seong PH (2020) Development of a generative-adversarial-network-based signal reconstruction method for nuclear power plants. Ann Nucl Energy 142:107410

Kim H, Arigi AM, Kim J (2021) Development of a diagnostic algorithm for abnormal situations using long short-term memory and variational autoencoder. Ann Nucl Energy 153:108077

Kim SG, Chae YH, Seong PH (2019a) Signal fault identification in nuclear power plants based on deep neural networks. Annals DAAAM Proc. 846–853

Kim YG, Choi SM, Moon JS (2019b) Development of convolutional neural networks diagnose abnormal status in nuclear power plant operation. KNS2019a, Korean Nuclear Society

Kingma DP, Ba J (2014) Adam: a method for stochastic optimization. arXiv preprint arXiv:1412.6980.

Kurki J, Seppälä M (2009) Thermal hydraulic transient analysis of the high performance light water reactor using apros and smabre.

Lee SJ, Kim J, Jang S-C, Shin YC (2009) Modeling of a dependence between human operators in advanced main control rooms. J Nucl Sci Technol 46(5):424–435

Li W, Peng M, Liu Y, Jiang N, Wang H, Duan Z (2018) Fault detection, identification and reconstruction of sensors in nuclear power plant with optimized PCA method. Ann Nucl Energy 113:105–117

Lin T-H, Wu S-C (2019) Sensor fault detection, isolation and reconstruction in nuclear power plants. Ann Nucl Energy 126:398–409

Lin T-H, Wang T-C, Wu S-C (2021) Deep learning schemes for event identification and signal reconstruction in nuclear power plants with sensor faults. Ann Nucl Energy 154:108113–108113

Mirza M, Osindero S (2014) Conditional generative adversarial nets. arXiv preprint arXiv:1411.1784

Nazir HZ, Schoonhoven M, Riaz M, Does RJ (2014) Quality quandaries: a stepwise approach for setting up a robust Shewhart location control chart. Qual Eng 26(2):246–252

Press O, Bar A, Bogin B, Berant J, Wolf L (2017) Language generation with recurrent generative adversarial networks without pre-training. arXiv preprint arXiv:1706.01399

Rabinovich SG, Rabinovich M (2010) Evaluating measurement accuracy. Springer

Shaheryar A, Yin X-C, Hao H-W, Ali H, Iqbal K (2016) A denoising based autoassociative model for robust sensor monitoring in nuclear power plants. Science and Technology of Nuclear Installations, 2016.

Simani S, Marangon F, Fantuzzi C (1999) Fault diagnosis in a power plant using artificial neural networks: analysis and comparison. In: 1999 European control conference (ECC). IEEE, pp 2270–2275

Xu H, Deng Y (2017) Dependent evidence combination based on Shearman coefficient and Pearson coefficient. IEEE Access 6:11634–11640

Xu X, Hines JW, Uhrig RE (1999) Sensor validation and fault detection using neural networks. In Proceedings of maintenance and reliability conference (MARCON 99), pp 10–12

Yang J, Kim J (2018) An accident diagnosis algorithm using long short-term memory. Nucl Eng Technol 50(4):582–588

Yang J, Kim J (2020) Accident diagnosis algorithm with untrained accident identification during power-increasing operation. Reliab Eng Syst Saf 202:107032

Yang Z, Xu P, Zhang B, Xu C, Zhang L, Xie H, Duan Q (2022) Nuclear power plant sensor signal reconstruction based on deep learning methods. Ann Nucl Energy 167:108765–108765

Yeh RA, Chen C, Yian Lim T, Schwing AG, Hasegawa-Johnson M, Do MN (2017) Semantic image inpainting with deep generative models. In: Proceedings of the IEEE conference on computer vision and pattern recognition, pp. 5485–5493

Yoo CK, Villez K, Lee IB, Van Hulle S, Vanrolleghem PA (2006) Sensor validation and reconciliation for a partial nitrification process. Water Sci Technol 53(4–5):513–521

Zavaljevski N, Gross KC (2000) Sensor fault detection in nuclear power plants using multivariate state estimation technique and support vector machines. Argonne National Laboratory, Argonne, IL (US)

Zio E, Di Maio F (2010) A data-driven fuzzy approach for predicting the remaining useful life in dynamic failure scenarios of a nuclear system. Reliab Eng Syst Saf 95(1):49–57

Zúñiga AA, Baleia A, Fernandes J, Branco PJDC (2020) Classical failure modes and effects analysis in the context of smart grid cyber-physical systems. Energies 13(5):1215–1215

Chapter 4
Diagnosis

The main objective of diagnosis in a general sense is to detect a failure or failures and identify the related cause. In a previous work, fault detection and diagnosis in NPPs were categorized into six applications: (1) instrument calibration monitoring, (2) dynamic performance monitoring of instrumentation channels, (3) equipment monitoring, (4) reactor core monitoring, (5) loose part monitoring, and (6) transient identification (Ma and Jiang 2011). Among them, this chapter is focused on transient identification, which is to identify the occurrence and cause of transients initiated by system failure(s) and external disturbances. It is critical that operators are able to mitigate transients in order to minimize their effects on the safety and efficiency of NPPs.

In current commercial NPPs, transients include both abnormal and emergency situations. When such a situation occurs, the operators need to specifically identify the current situation and also what caused it based on the plant symptoms, which are gathered from alarms, plant parameters, and system statuses. Once they diagnose the situation, the particular procedure relevant to the situation is selected that provides the proper mitigating actions.

This process, diagnosis, is regarded as one of the most demanding tasks for NPP operators. In an emergency situation, the plant changes dynamically, and the fast-changing parameters make it difficult for operators to diagnose the situation correctly. Additional factors that can lead to operator error in the diagnosis of an emergency include the presence of numerous alarms, time pressure, and the need to continuously monitor the plant status. On the other hand, for an abnormal situation, hundreds of potential cases exist in NPPs. For instance, typical NPPs in the Republic of Korea have about 100 different abnormal operating procedures, or AOPs, that are used to mitigate abnormal events, where each procedure covers one or more abnormal cases. Therefore, the selection of the relevant AOP, which is a critical diagnosis task, is highly challenging even for well-trained operators considering the hundreds of possible related events that can occur and the thousands of plant parameters that require monitoring (Park and Jung 2015).

J. Kim et al., *Autonomous Nuclear Power Plants with Artificial Intelligence*, Lecture Notes in Energy 94, https://doi.org/10.1007/978-3-031-22386-0_4

Owing to this difficulty, operator support during diagnosis has been the most popular application of AI techniques in this field. For instance, a fault diagnosis system was suggested for an NPP based on a state information-imaging method by applying kernel PCA using full-scope simulator data (Yao et al. 2020). Another group proposed a fault diagnosis technique with regard to tiny leakages in pipelines in an NPP using an integrated method of both knowledge-based and data-driven methods (Wang et al. 2019). An online diagnosis tool has also been developed to identify the severity of a LOCA in NPPs using ANNs to enhance robustness and quantify the confidence bounds associated with the prediction (Tolo et al. 2019). In addition, an approach to provide the "don't know" response capability to a DL-based system has been suggested for the NPP accident identification problem, which employs several data-driven methods such as particle swarm optimization, AE, and deep one-class AE (Pinheiro et al. 2020).

The function of diagnosis itself remains still a key element in related autonomous operation systems because the correct diagnosis or situation awareness is a prerequisite for the correct response of the autonomous system. This chapter introduces six examples of diagnostic algorithms using AI techniques.

4.1 Diagnosis of Abnormal Situations with a CNN

The number of variables that can be monitored in an NPP is enormous considering the number of components and the physical characteristics of the many NPP systems. If each of these variables is analyzed by applying a weight to it, then computational efficiency will decrease and data interpretation will be difficult. In this regard, DL technology is able to distinguish the useful features of input data and learn how much weight to assign them. As a DL method, the CNN provides a great advantage in terms of computational efficiency through its features of extracting local information from input data as opposed to analyzing all the information.

The CNN is inspired by the workings of the visual processing system of the human brain, which only responds to its local receptive field (Krizhevsky et al. 2017). This network type has shown remarkable success in image analysis tasks including face recognition (Li et al. 2015), handwritten character recognition (Ciresan et al. 2011), and medical image classification (Li et al. 2014). With its ability to deal with image data, the CNN can be effective in handling massive amounts of data assuming that the data is properly converted into an image format. From this point of view, the concept of using of a convolutional layer to identify the key characteristics of raw data can also apply to the analysis of NPP data accumulated in a time-series manner. With existing approaches, the computational complexity and costs will escalate if the dynamic aspects of individual systems at the plant-level are incorporated into the diagnosis of abnormal situations.

The operational data of NPPs have a form in which information on each variable is accumulated over time. To utilize the CNN's ability to identify spatial features, data preprocessing is performed on the NPP status at single moments in the shape of a square. In this way, as the processed data are transformed into snapshots every second, it may be difficult to detect a change in the NPP state from the normal state following the occurrence of an abnormal event. To overcome this problem, this section introduces a solution that adopts two-channel 2D images to describe the NPP state values. One channel represents the current NPP state values, while the other channel represents the extent of the changes in the NPP state values during a prescribed time period in the past. With this setup, the model is designed to deal with all the information that individual systems generate every second, as well as the dynamics of the states of individual systems.

The overall process of this methodology consists of the following four stages.

1. Raw data generation from an NPP simulator
2. Data transformation from raw data into two-channel 2D image data
3. Configuration of the CNN model with a description of each layer in the CNN
4. Performance evaluation with four typical evaluation metrics measuring the performance, reliability, and practicality of the model.

4.1.1 Raw Data Generation

AS a characteristic of supervised learning, the performance of CNNs depends on the number of data and their diversity. But in actual NPPs, the number of abnormal operation data generated to date is insufficient and has low accessibility for use in AI learning. Therefore, research related to abnormal situations normally produces training data through NPP simulators.

Abnormal operation data is labeled by dividing the normal states and the abnormal states from the moment when the abnormal event occurs. In the case of an abnormal event, multiple types of events requiring the implementation of various sub-procedures in the relevant AOP can be simulated, which gives the opportunity to create diverse abnormal operation scenarios. In addition, the diversity of the scenarios can be further expanded by adjusting the severity of the injected abnormal event.

The target systems for AOPs are widely located across the NPP. For example, about 80 AOPs exist for the Advanced Power Reactor (APR)-1400 reactor, and these procedures are relevant to almost all the systems in the NPP. To judge whether an NPP is in abnormal state through the plant parameters, it is necessary to select as many detailed variables as possible and analyze them all. Accordingly, the data obtained through abnormal operation should contain information on all observable variables.

4.1.2 Data Transformation

The data obtained in the previous step includes more than 1000 variables corresponding to NPP state values. But using such a large number of variables to build a classification model results in long training times and requires a huge number of training samples (Van Niel et al. 2005). A few studies have attempted to address this issue by representing the high dimensional data as images to be used as CNN inputs (Wen et al. 2017; Chang and Park 2018). The structure of a CNN reduces the number of parameters to be updated by performing the operations of convolution and pooling, thereby granting it the ability to effectively deal with high dimensional data while saving computational time and cost. Based on this, the developed two-channel 2D CNN in this section employs image data that (1) describe the NPP state values at a certain point in time, and (2) describe the changing patterns of the NPP state values during a certain time period in the past.

Data is transformed from raw data into two-channel 2D images with the following process. First, the raw data needs to be normalized because the ranges of the variables in the data vary from one to another. For data normalization, the min–max feature scaling method is used, which is based on the minimum and maximum values of each variable obtained from the NPP simulator.

$$X_{normalized} = \frac{X - X_{\min}}{X_{\max} - X_{\min}} \quad (4.1)$$

Second, the normalized data is transformed into a square 2D image. Generally, square 2D images are more readily handled by CNNs than rectangular 2D images. In cases in which the number of variables in the data is not a square number, adding zeros enables the data to be converted into a square, as shown in Fig. 4.1. It is known that zero-padding schemes such as this do not affect the performance of the classification model, as the zeros are merely dummy information in the data (Clark and Storkey 2015). Third, the extent of the changes in the NPP state values during a prescribed time period in the past is also converted into a 2D image in the same manner by assigning some value as the time-lag between the two images. Figure 4.2a, b show examples of these converted 2D images. Lastly, by overlapping the two 2D images generated in the second and third steps, two-channel 2D image data is constructed, as illustrated in Fig. 4.2c. These images are presented in grayscale, with black and white pixels indicating zero and one, respectively.

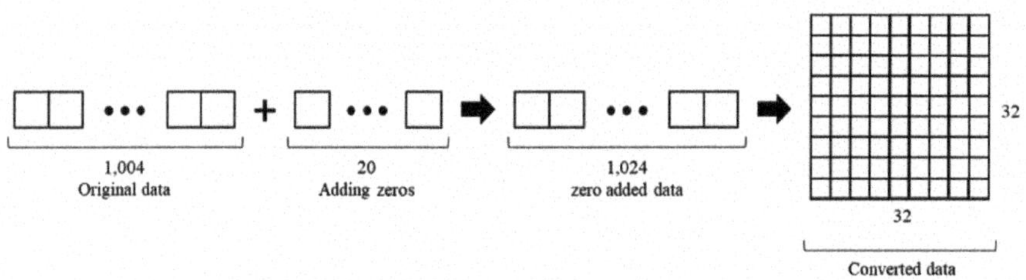

Fig. 4.1 Example of adding zeros. Reproduced with permission from Lee et al. (2021)

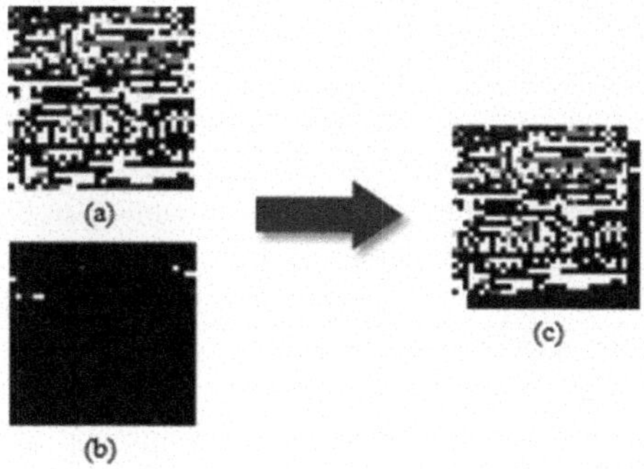

Fig. 4.2 Example of the result of data transformation. **a** Image for the NPP state values at a certain point in time. **b** Image for the changing patterns of NPP state values. **c** Overlapped two-channel 2D image. Reproduced with permission from. Reproduced with permission from Lee et al. (2021)

4.1.3 Structure of the CNN Model

As shown in Fig. 4.3, the two-channel 2D CNN model consists of two parts for feature extraction and classification. When the two-channel 2D image data from the previous data transformation step is fed as an input, a series of layers including convolution layers and pooling layers construct feature maps from the input data. Then a FC layer flattens the constructed feature maps and calculates classification scores for the different types of abnormalities.

4.1.3.1 Convolution Layer

In the convolution layer, images are taken from the previous layer, called input images, and feature maps are constructed through a convolution operation applying trainable filters to the transferred images. These convolution layer filters are a square matrix of a certain size made up of weights to be updated during training. The filters

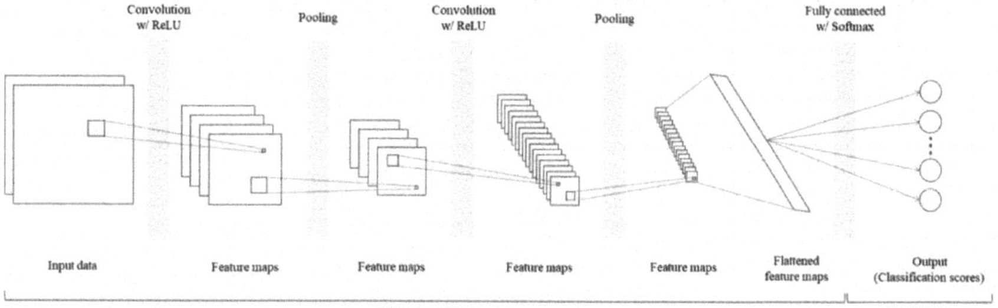

Fig. 4.3 Overall structure of the CNN model for abnormality diagnosis. Reproduced with permission from. Reproduced with permission from Lee et al. (2021)

each share weights and perform the convolution operation by passing over the image transferred from the previous layer, multiplying the matrix corresponding to each element of the image, and adding them together. In this convolution operation, the size of the filter determines the size of the feature maps to be constructed, and the number of filters corresponding to the number of channels in the convolution layer determine the depth or number of channels of the feature maps.

Figure 4.4 shows an example of the convolution operation process comprising three components assumed to be a square matrix: the input image, filter, and output image. In the example, the input image is represented as $I \in \mathbb{R}^{M_i \times N_i}$, where M_i and N_i indicate the size of the input image as the number of rows and columns, respectively. The output image, which is the feature map, is likewise represented as $O \in \mathbb{R}^{M_o \times N_o}$ with M_o and N_o indicating the output image size. The filter is expressed as $F \in \mathbb{R}^{M_f \times N_f}$, where F indicates the size of the filter. With these components, the output value is calculated as in Eq. (4.2).

$$
\begin{aligned}
(m_o, n_o) = \sum_{m_f=0}^{M_f} \sum_{n_f=0}^{N_f} f\left(m_f, n_f\right) \cdot i\left(m_i + m_f, n_i + n_f\right),\\
\mathit{where}\ m_o = 0, 1, \ldots, M_o;\ n_o\\
= 0, 1, \ldots, N_o;\ m_i = 0, 1, \ldots, M_i;\ n_i\\
= 0, 1, \ldots, N_i,\ \mathit{as\ indexes}
\end{aligned}
\tag{4.2}
$$

In this equation, $i(\cdot, \cdot)$ is each value of the input image, $f(\cdot, \cdot)$ is each weight of the filter, and $o(\cdot, \cdot)$ is each value of the output image. Each filter sweeps the input image at a certain stride, and therefore the size of the output image can be determined as follows.

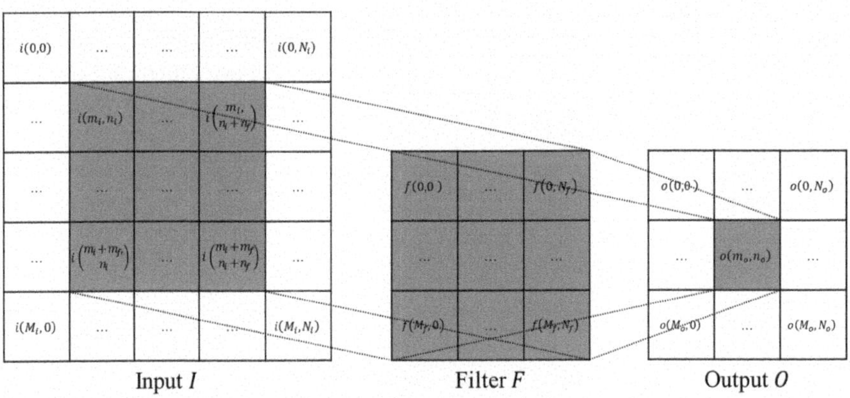

Fig. 4.4 Example of the convolution process. Reproduced with permission from Lee et al. (2021)

$$M_o = \frac{(M_i - M_f)}{S} + 1; N_o = \frac{(N_i - N_f)}{S} + 1, \; where\ S\ is\ stride \tag{4.3}$$

The output of each convolution layer generally passes through an activation function to introduce nonlinearity in the network before entering the next layer. As the nonlinear activation function, a ReLU is employed, which is formulated in Eq. (4.4).

$$f(x) = \max(0, x) \tag{4.4}$$

4.1.3.2 Pooling Layer

The purpose of the pooling layer is to conduct down-sampling in order to reduce the spatial size of the output feature maps, which decreases the number of parameters and prevents overfitting. The pooling layer adopts a filter similarly to the convolution layer, but with a difference in that the pooling layer filter spatially resizes the image with no weights to be trained. Two common types of pooling operation are max pooling and average pooling. The former selects the maximum value in a sliding window, while the latter finds the average value over the sliding window. As mentioned in Sect. 4.1.2, the second channel of the two-channel 2D image represents the extent that the NPP state values changed during a certain time period. In reality though, NPP state values seldom change while the plant is in a normal state, and only a handful of values may show fluctuation even in an abnormal state. In other words, most values of the images here are close to zero except for a small number of large values. Based on this, max pooling is employed to better capture the changes in the state values, as max pooling can lead to substantial information loss. For example, if a 2×2 filter with a stride of 2 is utilized in the pooling layer, then 75% of the values in the input image will be discarded through the max pooling operation. The remaining values, i.e., the maximum values in each sliding window, are delivered to the next layer.

4.1.3.3 FC Layer

In a typical CNN, a FC layer usually serves as the last hidden layer. It carries out the classification task based on the feature maps constructed in the previous layers. After several convolution and pooling layers, each 2D feature map is flattened into a 1D vector as an input to the last layer. The FC layer then calculates a classification score for the outputs, which in this case are predicted probabilities for the NPP abnormal events and normal state, through a matrix multiplication of the feature maps and weights in the layer, similar to the hidden layers of a traditional ANN. In this final layer, the softmax function is used as the activation function to convert the output values into a probability distribution over the predicted labels, as shown by Eq. (4.5).

$$f(x_i) = \frac{\exp(x_i)}{\sum_{i=0}^{k} \exp(x_i)}, \quad where\ i = 0, 1, \ldots, k \tag{4.5}$$

4.1.4 Performance Evaluation Metrics

Abnormality diagnosis in an NPP presents a multi-class classification problem, and therefore numerous performance evaluation metrics are considered to assess the performance of the approach after constructing a confusion matrix. We first measure the accuracy of each class and the overall accuracy of the approach as defined in Eqs. (4.6) and (4.7).

$$Accuracy_i = \frac{tp_i + tn_i}{fp_i + fn_i + tp_i + fn_i} \tag{4.6}$$

$$Overall\ accuracy = \frac{\left(\sum_{i=1}^{l} \frac{tp_i + tn_i}{fp_i + fn_i + tp_i + fn_i}\right)}{l} \tag{4.7}$$

The true positive (tp_i), true negative (tn_i), false positive (fp_i), and false negative (fn_i) for class i respectively represent the number of positive examples that are correctly classified, number of negative examples correctly classified, number of negative examples wrongly classified as positive, and number of positive examples wrongly classified as negative. l is the number of classes.

Next, we calculate the precision, recall, and F1 score to check reliability and practicality. Precision reflects the fraction of true positive results from among all results classified as positive, as in Eq. (4.8), and recall reflects the fraction of true positive results from among the results that should have been returned, as in Eq. (4.9). The F1 score in Eq. (4.10) measures the overall effectiveness of the classification model, which is defined as the harmonic average of the precision and recall. The values in this measure range between 0 and 1, where 1 represents perfect classification.

$$Precision_i = \frac{tp_i}{tp_i + fp_i} \tag{4.8}$$

$$Recall_i = \frac{tp_i}{tp_i + fn_i} \tag{4.9}$$

$$F_1\ score = 2 * \frac{precision * recall}{precision + recall} \tag{4.10}$$

4.1.5 Experimental Settings

The raw data for this experiment are produced with the 3KEYMASTER full-scope simulator (Western Service Corporation 2017). This simulator is designed for a generic 1400 MWe PWR that implements power generation and air circulation systems in addition to the main systems. A total of 10 abnormal events plus the normal state are selected to cover a portion of the AOPs. The abnormal events considered here, as summarized in the below descriptions, are judged to be more significant than others in terms of occurrence probability or consequence.

- POSRV: Leakage of the pilot-operated safety relief valve that depressurizes the reactor coolant system (RCS)
- LTDN: Abnormality in the letdown water system that controls the RCS inventory
- CHRG: Abnormality in the charging water system that controls the RCS inventory
- RCP: Abnormality in the reactor coolant pump (RCP)s that circulate coolant in the primary system
- SGTL: Leakage of tubes inside SGs
- CDS: Abnormality in the condenser vacuum for the cooling steam transferred from the SGs
- MSS: Abnormality in the main steam system that provides steam to the turbines
- CWS: Abnormality in the circulating water system that filters water before it is pumped to and through the condenser
- RMW: Valve abnormality in the reactor makeup water tank that provides coolant to the volume control tank in emergency situations
- MSIV: Abnormality in the main steam isolation valve that isolates main steam in emergency situations.

The simulator generated 300 scenarios as the causes of each abnormality, and thus the dataset comprises 3300 total scenarios. The same number of scenarios for each abnormal event and the normal state is used to prevent the imbalanced dataset problem known to cause poor classification performance (Peng et al. 2018). Each scenario includes NPP state data spanning 30 s after the simulation starts with 1004 numerical variables. This generated data can be divided into several subsamples; for instance, with a time-lag set to 5 s, the number of subsamples totals 82,500 (= 3300 × (30–5)). We add 20 zeros to the 1004 variables to give the data a total of 1024 values and create 32 × 32 square two-channel 2D images that describe the NPP state values at a certain point in time as well as the extent of the changes in the NPP state values during a prescribed time period in the past.

These subsamples are employed as the inputs of the developed CNN model. For a thorough performance evaluation, we apply a fivefold random sampling technique. A total of 80% of the samples are used as the training dataset, while the remaining 20% of the samples are utilized as the test dataset.

4.1.5.1 Hyperparameter Selection

As detailed previously, the CNN model consists of two alternating convolution and pooling layers and one FC layer (Fig. 4.3) that conduct the tasks of feature extraction and classification. Both pooling layers have a 2 × 2 pooling filter with a stride of 2. This model architecture follows the standard CNN setup that is known to work well on diverse classification problems using data-driven approaches. To find the optimal CNN model settings, the hyperparameters other than the pooling layer filter size and stride, specifically the number of filters in each convolution layer, the size of each convolution filter, and the number of hidden neurons in the FC layer, are determined in a grid search that inspects the classification performance with different hyperparameter values. Candidates for the number of filters in the two convolution layers are first set as three pairs: 8 and 16, 16 and 32, and 32 and 64 in the first and second convolution layers, respectively. Second, candidates for the size of the convolution filters are set to two square fields: 2 × 2 and 3 × 3 for each convolution layer. Candidates for the number of neurons in the FC layer are then set to 50 and 100. Several CNN structures are built by applying all combinations of the three hyperparameters, training them over 50 epochs with the collected and transformed data, and obtaining multiple classification performance results for the test dataset. Results of the hyperparameter selection process are presented in Table 4.1. While no notable differences among the 12 total combinations of hyperparameters are found, the CNN structure applying 32 and 64 filters in the two convolution layers, the 2 × 2 size of the convolution filters, and 100 neurons in the FC layer shows the highest classification accuracy for the test dataset, which is highlighted with bold in Table 4.1. This setting is accordingly This setting is accordingly chosen for the following comparative analysis and robustness test.

Table 4.1 Results of hyperparameter selection

Filters in the first layer	Filters in the second layer	Filter size	Neurons in the FC layer	Accuracy	Precision	Recall	F1 score
8	16	2	50	0.9992	0.9960	0.9959	0.9959
8	16	2	100	0.9987	0.9933	0.9931	0.9931
8	16	3	50	0.9982	0.9906	0.9900	0.9900
8	16	3	100	0.9977	0.9882	0.9874	0.9875
16	32	2	50	0.9991	0.9949	0.9948	0.9948
16	32	2	100	0.9986	0.9926	0.9923	0.9924
16	32	3	50	0.9991	0.9952	0.9950	0.9950
16	32	3	100	0.9987	0.9929	0.9927	0.9927
32	64	2	50	0.9992	0.9959	0.9958	0.9958
32	**64**	**2**	**100**	**0.9993**	**0.9961**	**0.9960**	**0.9960**
32	64	3	50	0.9987	0.9930	0.9927	0.9927
32	64	3	100	0.9989	0.9940	0.9938	0.9938

4.1.5.2 Model Training Method

For model training, the weights to be updated in each layer are typically estimated with backpropagation algorithms following a gradient descent method, where the weights are updated by calculating the derivatives of a loss function with respect to the weights of the network. In the current case, categorical cross entropy is used as the loss function and the Adam algorithm, known as an extension of the stochastic gradient descent method, is used as the optimizer to update the weights (Kingma and Ba 2014).

With the hyperparameter settings chosen in the previous section, we now conduct a comparative analysis between the two-channel CNN model and a one-channel CNN, an ANN (Ince et al. 2016; LeCun et al. 2015), and an SVM (Na et al. 2008; Zio 2007) representing other major classification models to evaluate the performance of the developed approach. For a fair comparison, the one-channel CNN is constructed in the same way as the developed CNN model with the only difference being that it employs one-channel 2D images that describe the current NPP state values as its inputs. The ANN takes the network structure of Embrechts and Benedek (2004) as its results show good performance (Embrechts and Benedek 2004). The SVM employed in this study is based on the work of Na et al. (2008), with the exception that we apply the one-against-one method in this case for multi-class classification (Zio 2007). Prior to the comparative analysis, we explored various configurations of the latter two models based on their references and empirically selected the configurations that showed the best performance, as follows: five hidden layers of 128, 128, 64, 32, and 16 neurons in the ANN, and the radial basis function kernel in the SVM. All classification models including the two-channel 2D image transformation method are implemented in a Python 3.6 environment with Keras and scikit-learn modules. We used four GPUs for multi-GPU computing in the training process and set the batch size to 128 for the training epoch in each neural network model.

4.1.6 Results

Tables 4.2, 4.3, 4.4 and 4.5 list the performance evaluation results comparing the developed approach and the three major models, respectively. The results demonstrate far superior performance of the developed two-channel CNN approach across all performance evaluation metrics compared to the other classification models. The ANN and SVM models achieve an average of nearly 90% accuracy in the results, but their precision, recall, and F1 scores are seen to be relatively low and have large variances compared to those of the two CNN-based models. This shows that the ANN and SVM can generate classification results that may be volatile and biased. The one-channel CNN shows a lower performance in the classification of the normal state than the abnormal states, which would likely result in a type 1 error in NPP abnormality diagnosis. Given that the F1 scores of the developed approach for every

Table 4.2 Performance evaluation metrics for the developed two-channel CNN

Class	Accuracy	Precision	Recall	F1 score
Normal	0.9987	0.9863	1.0000	0.9931
SGTL	0.9999	0.9993	1.0000	0.9997
CHRG	0.9999	1.0000	0.9991	0.9995
LTDN	1.0000	0.9996	1.0000	0.9998
CDS	1.0000	1.0000	1.0000	1.0000
POSRV	1.0000	1.0000	1.0000	1.0000
RMW	0.9997	1.0000	0.9965	0.9983
CWS	0.9993	1.0000	0.9921	0.9961
MSIV	1.0000	1.0000	1.0000	1.0000
RCP	0.9998	1.0000	0.9976	0.9988
MSS	1.0000	1.0000	0.9997	0.9999
Overall	0.9998	0.9987	0.9986	0.9986

Table 4.3 Performance evaluation metrics for the one-channel CNN

Class	Accuracy	Precision	Recall	F1 score
Normal	0.9452	0.7212	0.6471	0.6822
SGTL	0.9599	0.7902	0.7616	0.7756
CHRG	0.9477	0.6714	0.8317	0.7430
LTDN	0.9846	0.8707	0.9760	0.9203
CDS	0.9877	0.9914	0.8719	0.9278
POSRV	0.9887	0.9223	0.9558	0.9387
RMW	0.9914	0.9253	0.9848	0.9541
CWS	0.9660	0.8015	0.8317	0.8163
MSIV	0.9979	0.9932	0.9834	0.9883
RCP	0.9864	0.9946	0.8547	0.9193
MSS	0.9655	0.8404	0.7666	0.8018
Overall	0.9746	0.8656	0.8605	0.8607

abnormal event are close to 1 or at 1 with little variation, the two-channel CNN model achieves not only the best performance but also unbiased classification results.

From the results, it can be understood that the model performance may vary with the context of analysis. Two additional tests are therefore conducted applying different time-lag values and different output variables to assess the robustness of the developed classification model. First, the time-lag values are increased from 5 to 10 s and 20 s. The performance evaluation results are shown in Table 4.6. The two-channel CNN model exhibits the best performance at a time-lag of 20 s because longer time-lag values can represent richer information about the changes in the NPP state values. The performance results of the developed approach with the 5 s time-lag

Table 4.4 Performance evaluation metrics for the ANN

Class	Accuracy	Precision	Recall	F1 score
Normal	0.8469	0.3177	0.5960	0.4145
SGTL	0.9152	0.5653	0.2906	0.3838
CHRG	0.9520	0.9854	0.4789	0.6445
LTDN	0.9455	0.7020	0.6956	0.6988
CDS	0.9737	0.9911	0.7168	0.8319
POSRV	0.9576	0.9352	0.5737	0.7111
RMW	0.9678	0.9770	0.6609	0.7884
CWS	0.9247	0.5928	0.5488	0.5700
MSIV	0.9543	0.7832	0.6882	0.7326
RCP	0.9708	0.9655	0.7037	0.8141
MSS	0.7704	0.2051	0.5306	0.2958
Overall	0.9253	0.7291	0.5894	0.6260

Table 4.5 Performance evaluation metrics for the SVM (10% sample)

Class	Accuracy	Precision	Recall	F1 score
Normal	0.8920	0.4563	0.9800	0.6227
SGTL	0.9509	0.7078	0.7833	0.7437
CHRG	0.9363	0.6696	0.5900	0.6273
LTDN	0.9777	0.9871	0.7644	0.8616
CDS	0.9833	0.9455	0.8667	0.9043
POSRV	0.9951	0.9988	0.9467	0.9720
RMW	0.9966	1.0000	0.9622	0.9807
CWS	0.9674	1.0000	0.6411	0.7813
MSIV	0.9819	0.9865	0.8122	0.8909
RCP	0.9826	0.9223	0.8833	0.9024
MSS	0.9811	0.9709	0.8167	0.8871
Overall	0.9677	0.8768	0.8224	0.8340

are seen to be slightly lower than those with the longer time-lag values, but even with the shortest time-lag, the results demonstrate that almost every sample is classified into the correct abnormal state. In other words, the first test shows that the developed approach is robust against different time-lag values.

In the second test, the two-channel CNN conducts classification using a total of 19 sub-procedures as its output values based on the fact that each abnormal event includes multiple sub-procedures dedicated to different causes of the given abnormality. Table 4.7 lists the performance evaluation results of this test. Despite the precision showing a slight deterioration in the normal state classification, the results in general demonstrate an adequate performance of the developed approach

Table 4.6 Performance evaluation metrics for the developed two-channel CNN approach with different time-lag values

	5 s	10 s	20 s
Accuracy	0.9998	0.9999	1.0000
Precision	0.9987	0.9995	0.9999
Recall	0.9986	0.9995	0.9999
F1 score	0.9986	0.9995	0.9999

Table 4.7 Performance evaluation metrics for the developed two-channel CNN approach with 19 sub-procedures (time-lag = 5 s)

Class	Accuracy	Precision	Recall	F1 score
Normal	0.9987	0.9767	1.0000	0.9882
SGTL	0.9999	0.9989	1.0000	0.9995
CHRG [LN]	0.9999	1.0000	0.9975	0.9987
CHRG [PM]	0.9999	0.9985	1.0000	0.9993
CHRG [VV]	1.0000	1.0000	0.9999	0.9999
LTDN [LN]	1.0000	0.9999	1.0000	0.9999
LTDN [VV]	1.0000	1.0000	1.0000	1.0000
CDS	1.0000	1.0000	1.0000	1.0000
POSRV	1.0000	1.0000	1.0000	1.0000
RMW [LL]	0.9998	1.0000	0.9960	0.9980
RMW [LH]	1.0000	1.0000	1.0000	1.0000
CWS [LN]	1.0000	1.0000	1.0000	1.0000
CWS [VV]	1.0000	1.0000	1.0000	1.0000
CWS [PM]	1.0000	1.0000	1.0000	1.0000
MSIV	1.0000	1.0000	1.0000	1.0000
RCP [LC]	0.9990	1.0000	0.9801	0.9900
RCP [SD]	1.0000	1.0000	1.0000	1.0000
MSS [VV]	1.0000	1.0000	1.0000	1.0000
MSS [LN]	1.0000	1.0000	1.0000	1.0000
Overall	0.9999	0.9986	0.9986	0.9986

in achieving the classification of the 19 sub-procedures. In summary, the results of the two robustness tests indicate that the developed two-channel CNN model is robust across different contexts of analysis, namely time-efficient abnormality diagnosis and handling up to dozens of abnormal events.

For real-world application in an NPP to provide practical assistance to operators, it is essential that any abnormality diagnosis model have a high reliability of the classification of abnormal events and the normal state in real time. As the results shown here are limited to classification with a predetermined dataset, a question remains regarding whether the performance of the developed two-channel CNN approach can be consistent during real-time diagnosis or not. The current experiments examine classification performance without a measure of the reliability of the results, so it remains to be seen how confident the model is in providing the classification results.

Along these lines, a second set of additional experiments are conducted to more closely examine the practicality of the developed approach, in which the performance of the pre-trained two-channel CNN model and the reliability of its classification results for longer NPP simulator operating data are obtained. For this, extra operating data are collected for about 3 min of operation time simulating abnormal events and the normal state that were not previously included in the training or test datasets, and the developed CNN model attempts to diagnose the abnormal events from the extra operating data. In this test, reliability is defined as the highest probability that the model predicts among the outputs. Similar to the previous results, the model succeeds in classifying all the cases of extra operating data correctly; Fig. 4.5 plots a number of the results in graph form of the model reliability for the extra operating data. For instance, in Fig. 4.5a, the model predicts the normal state correctly with a high reliability on average. On the other hand, while Fig. 4.5b representing the abnormal event CHRG demonstrates that the model generally classifies the event correctly with a high reliability, in this case there exists a small area of fluctuation among the CHRG, SGTR, and POSRVL abnormal events. In the remaining graphs for LTDN, POSRVL, and RMW events, the model diagnoses each abnormal event perfectly

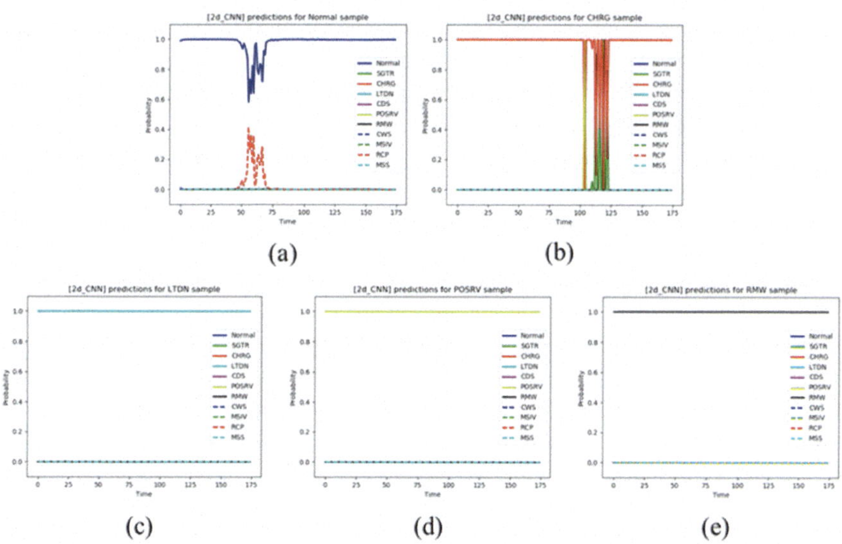

Fig. 4.5 Graphs of reliability for each predicted abnormal event or normal state along time. **a** Normal state, **b** abnormal event CHRG, **c** abnormal event LTDN, **d** abnormal event POSRVL, and **e** abnormal event RMW

with 100% reliability over the entire time period. Taken together, it can be said that in every case, the developed two-channel CNN classifies the abnormal events and normal state perfectly, with the reliability of the classification result reaching near 100% after a certain period of time. Accordingly, the additional reliability results confirm that the developed two-channel CNN model has the potential to be adopted as an operator support system for real-time abnormality diagnosis in actual NPP systems.

Despite these promising results though, some uncertainty remains as represented by the fluctuating region in some of the graphs in Fig. 4.5. One possible cause may be the scope of the training data, being limited to 30 s from the start of the simulation. In some abnormal events, characteristic changes in the input variables may appear after 30 s from the start of the event, and thus the performance of the developed model can be improved if data observed over longer time periods is utilized. To further improve the stability of the developed model for real-time diagnosis, additional research is needed to fine-tune the developed two-channel CNN structure with additional model training involving longer data periods.

4.2 Diagnosis of Abnormal Situations with a GRU

The method introduced in Sect. 4.1 converted NPP data into images for use in a CNN. Over the past few decades, various studies have applied ML algorithms for NPP state diagnosis through data-driven analysis. Anomaly detection technology is being researched using increasingly complex DL algorithms to detect NPP states out of normal situations by learning patterns when failures occur. Horiguchi et al. (1991) used the patterns of 49 normalized plant parameters during an abnormal situation for state detection with an ANN, while Santosh et al. (2007) implemented a resilient back propagation algorithm to solve the pattern recognition problem for the diagnosis of NPP transients. Serker et al. (2003) applied Elman's RNN to predict plant signals and detect damage on bearings, and Zhao et al. (2018) employed a local feature-based GRU for the prediction of tool wear, the diagnosis of gearbox faults, and the detection of bearing faults.

Prior research such as the above generally encounters two limitations regarding the selection of the input variables and the number of cases to be handled. When selecting the input variables, domain knowledge of NPP accidents has been considered by experts. For example, Ayodeji et al. applied PCA during preprocessing to reduce the dimensions of the input dataset and to filter noise (Ayodeji et al. 2018). Although PCA successfully achieved the anticipated results, this work covered only 43 variables involving only some of the primary systems.

Referring to these previous studies, the abnormality diagnosis model introduced in this section considers the full range of data through preprocessing to reduce dimensionality while not missing key features. Training of the preprocessed data is conducted through a GRU, and the model structure is divided into two stages to prepare for an increase in the number of labels in the future. The next section first introduces an analysis of the relationship between the variables constituting NPP operational data to find the basis for applying the preprocessing method in the subsequent section.

4.2.1 Characteristics of Abnormal Operation Data

Figure 4.6 is a heatmap of the correlations among a portion of the data. With the number of variables exceeding 1000, correlations are not apparent in the full heatmap on the left. The extracted heatmap on the right is therefore shown to indicate the high correlations among just 20 variables out of the entire dataset. Operational data with such a linear relationship in an abnormal state as well as in the steady state can greatly reduce the main features during preprocessing through feature extraction. Therefore, it is possible to use a small-dimensional dataset while considering all data and also minimizing information loss through feature extraction. Here, feature extraction refers to a method of acquiring a feature or features that best represent the information possessed by the data. In some cases, the features of a dataset are arranged according to their importance as a ranking, or they are excluded one by one to find the most influential features. The methodology presented in this chapter applies PCA to find a new axis that best represents the information in the data. PCA can significantly reduce the thousands of variables recorded during operation in detecting anomalies.

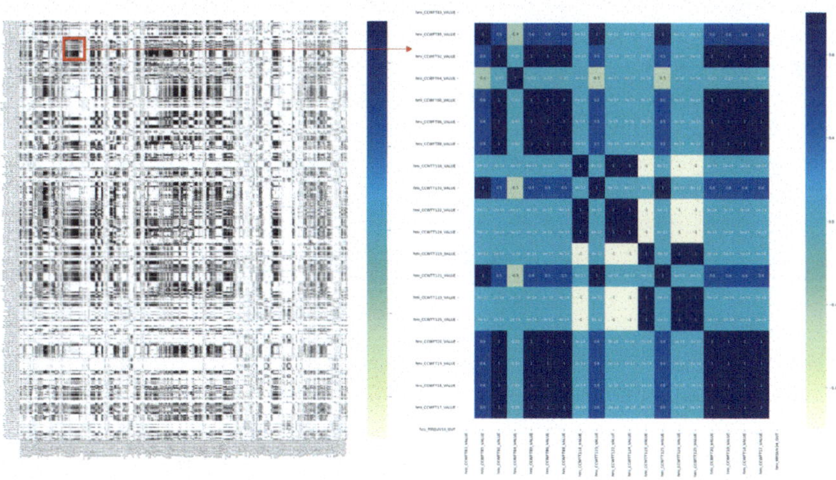

Fig. 4.6 Heatmap of the variances of sample data (bright: 1, dark: –1). Reproduced with permission from Kim et al. (2020)

The data with reduced dimensionality is used to train an AI model that can classify abnormalities following the developed methodology. Considering that NPP operational data are recorded over time, GRU algorithms are chosen among RNNs having advantages in the analysis of time-series data. The GRU algorithm is known as a solution to overcome the vanishing gradient problem through the use of more sophisticated activations than those in traditional RNNs (Bengio et al. 1994; Cho et al. 2014; Chung et al. 2014). Even in the case that symptoms of an abnormal situation may not appear at the beginning of the event, failures can still be identified by recognizing any change from the point of observation. Compared to a CNN, which learns by extracting local information from instantaneous data, RNNs learn the trend of variable change over time.

4.2.2 PCA

The CNN model used in Sect. 4.1 had an additional channel to learn the changing trends of the variables, but the RNN series introduced in this section can use the data as it is. Since the training of an AI network can be difficult when the dimensions increase, there are many methods to decrease data dimensionality. As a typical method to reduce data dimensionality, feature extraction works by extracting the latent features from the original variables. This is advantageous in the development of a real-time fault diagnosis system through extracting only the key characteristics from the entire dataset. As NPP simulator data cover a wide range of properties, e.g., pressure, temperature, and flow rate, PCA is chosen as the preprocessing method. PCA is a method to convert some original data into a dimensionally reduced dataset (Jolliffe 2002); to do so, PCA calculates a linear transformation matrix in the following steps.

First, a correlation matrix or covariance matrix of the original data is obtained during the process of normalizing the original data matrix. The unique features of data can be expressed mathematically as a total variance. The maximum-minimum normalization method is applied as follows.

$$x^*_{t,i} = \left\{x_{t,i} - \min(x_i)\right\} \\ /\{\max(x_i) - \min(x_i)\}(i = 1, 2, \ldots, p) \quad (4.11)$$

Here, $x_{t,i}$ is the i-th input variable at time t and $x^*_{t,i}$ is the normalized variable. The maximum and minimum values are extracted from the i-th dataset.

Second, the correlation matrix is broken into eigenvectors and eigenvalues, where the former represents the principal components (PCs) and the latter represent the percentage of variance given by the corresponding eigenvectors. This can be expressed with Eq. (4.12).

$$\begin{aligned} \vec{z_1} &= a_{11}\vec{x_1} + a_{12}\vec{x_2} + \ldots + a_{1p}\vec{x_p} = \vec{a_1}^T X \\ \vec{z_2} &= a_{21}\vec{x_1} + a_{22}\vec{x_2} + \ldots + a_{2p}\vec{x_p} = \vec{a_2}^T X \\ &\cdots \\ \vec{z_p} &= a_{p1}\vec{x_1} + a_{p2}\vec{x_2} + \ldots + a_{pp}\vec{x_p} = \vec{a_p}^T X \end{aligned} \tag{4.12}$$

$$\mathbf{z} = \begin{bmatrix} \vec{z_1} \\ \vec{z_2} \\ \cdots \\ \vec{z_p} \end{bmatrix} = \begin{bmatrix} \vec{a_1}^T X \\ \vec{a_2}^T X \\ \cdots \\ \vec{a_p}^T X \end{bmatrix} = \begin{bmatrix} \vec{a_1}^T \\ \vec{a_2}^T \\ \cdots \\ \vec{a_p}^T \end{bmatrix} X = A^T X$$

Among the terms, $\vec{z_k}$ is the k-th vector of the PCs ($k = 1, 2, \ldots, p$), A^T is the orthogonal matrix in which the k-th column, $\vec{a_k}$, is the k-th eigenvector of the covariance matrix, and X is the normalized dataset.

In the third step, the number of PCs is determined from the cumulative percentage of the described variance. For instance, a sum of the top 10 PCs of 0.90 means that these 10 PCs contain 90% of the information in the original data.

4.2.3 GRU

The vanishing gradient problem, as mentioned numerous times in Chap. 2, is that past data tends to be ignored for long data observation times. The GRU has a solution to this problem by applying gates to the information moved between cells, as follows. GRU uses two gates, namely update and reset. The activation h_t, which is a linear interpolation between the previous activation h_{t-1} and the candidate activation $\tilde{h}_t$, is computed as follows.

$$h_t = z_t h_{t-1} + (1 - z_t)\tilde{h}_t \tag{4.13}$$

Update gate z_t determines the extent that the unit updates its activation, or content. The updated gate is computed with Eq. (4.14).

$$update\ gate : z_t = \sigma\left(W^{(z)} x_t + U^{(z)} h_{t-1}\right) \tag{4.14}$$

Here, σ is the logistic sigmoid function, and x_t and h_{t-1} are the input at time t and the previous hidden state, respectively. $W^{(r)}$ and $U^{(r)}$ are trained weight matrices. The candidate activation $\tilde{h}_t$ is obtained as follows.

$$\tilde{h}_t = \tanh(W x_t + U(r_t \odot h_{t-1})) \tag{4.15}$$

In this expression, r_t is a set of reset gates, and $\odot$ is an element-wise multiplication. Similar to the update gate, the reset gate is computed as below.

$$\text{reset gate} : r_t = \sigma\left(W^{(r)}x_t + U^{(r)}h_{t-1}\right) \tag{4.16}$$

Figure 4.7 depicts a GRU algorithm with these two gates controlling how much of the information passed between cells is reflected in the current state. This enables the GRU to reduce the complexity of the computations, which increases the computational efficiency. In this way, the GRU can handle data with many variables, the dimension of which is reduced through PCA.

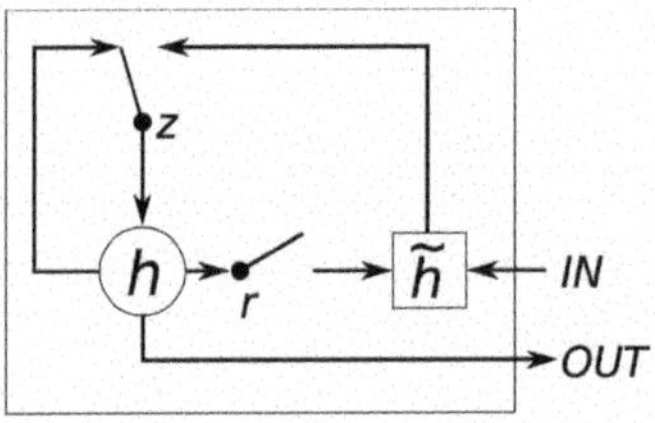

Fig. 4.7 Schematic diagram of the gated recurrent unit algorithm Chung et al. (2014)

4.2.4 Two-Stage Model Using GRU

As discussed above, NPP operational data is recorded as time-series data of two dimensions, having a time axis and a characteristic axis. RNNs have a structure that makes complex judgments on the input data every second and the information processed in the previous step.

Figure 4.8 shows the flow of the overall model development process. Data generation with an NPP simulator is followed by data preprocessing and then the training of the GRU algorithms in both main and substages.

In more detail, this model is designed to handle all monitorable parameters with two judgment processes employing the GRU algorithm for its strength in time-series data analysis. The PCA method is first used to extract the features and reduce the

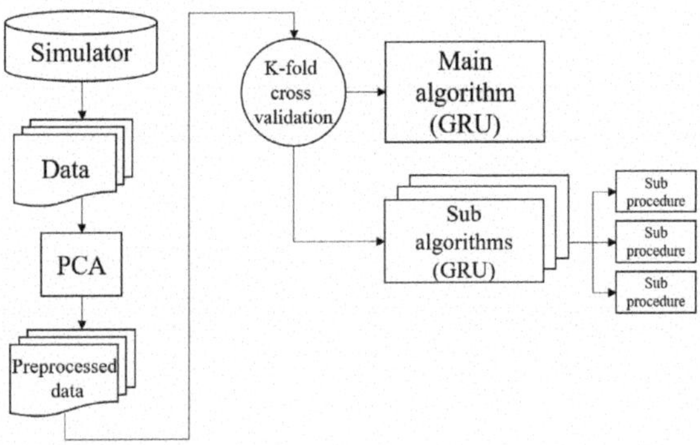

Fig. 4.8 Process to develop the GRU-based abnormal state diagnosis model. Reproduced with permission from Kim et al. (2020)

dimensionality of the abnormal operation data from the simulator. These preprocessed data are then used for the training and testing of the diagnosis model. The model structure is divided into two levels or stages to consider the possible cases that may occur. The main level determines the appropriate AOP corresponding to the current event, and the sub-level determines the detailed cause of the event. In both levels, GRU algorithms are used to handle the preprocessed datasets.

Since the amount of information to be dealt with at this time may become larger than necessary, the number of features is reduced to 10 by preprocessing the input data with PCA. The operational data of an NPP typically shows a linear characteristic with strong correlation between data, and thus even if the number of new axes through PCA is small, the amount of information stored does not change.

Through this, the GRU model receives data that has undergone PCA as inputs and is then trained, while a model structure that can reduce the number of classes requiring prediction is applied. This results in high accuracy.

When diagnosing an abnormal situation in an NPP, the large number of target systems and corresponding outputs create difficulty. Figure 4.9 shows the determination process according to the two-stage model (TSM) utilizing GRU algorithms introduced above. New input data is preprocessed and then judged in two-stage.

In the model structure, processing all cases with one algorithm is avoided; here, the first diagnosis or main stage predicts the relevant AOP, and the substage predicts the sub-procedures of the individual AOPs with dedicated algorithms. In other words, following the prediction results of the algorithm in the main stage, the corresponding substage shows the correct sub-procedure of the AOP. Dividing the decision process into two such stages increases the opportunity to detect errors. Another advantage is that each algorithm in the TSM model considers a smaller number of labels compared to employing a single algorithm to cover the entire process.

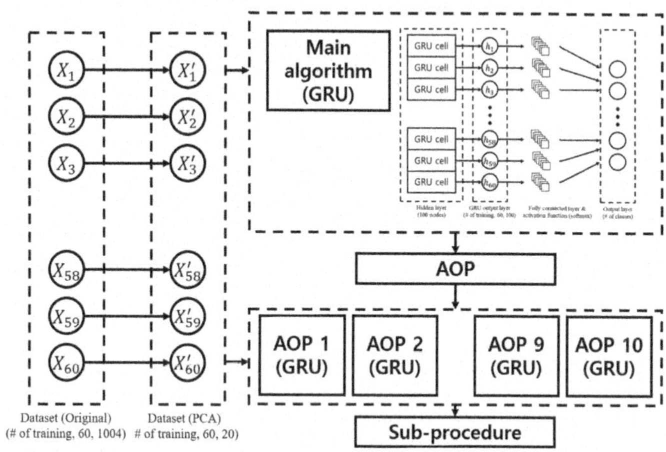

Fig. 4.9 Determination process of the abnormal state diagnosis model using GRU algorithms in two stages. Reproduced with permission from Kim et al. (2020)

4.2.5 Experimental Settings

The data used in the experiment through the 3KEYMASTER simulator and the performance evaluation metrics are the same as those in Sect. 4.1. All the variables that can be monitored in the NPP simulator are used as the model inputs. A total of 10 AOPs are selected that include 18 sub-procedures, as listed in Table 4.8. The related systems include both primary and secondary side systems to evenly represent an NPP.

4.2.6 Results

Data preprocessing and model development are implemented in a Python 3.6 environment with Keras and scikit-learn modules. Based on the selected AOPs, the TSM has 1 algorithm in the main level and 10 algorithms in the sub-level, one for each

Table 4.8 List of selected abnormal operation scenarios

#	Title of abnormal operating procedure	#	Sub-procedure
1	SGTL	1–1	SG 1,2 tube leakage
2	CHRG	2–1	Normal charging pump trip (PM)
		2–2	Charging valve abnormality (VV)
		2–3	Water line leakage (LN)
3	LTDN	3–1	Water line leakage (LN)
		3–2	Letdown valve abnormality (VV)
4	CDS	4–1	CDS vacuum release
5	POSRVL	5–1	POSRV leakage (VV)
6	RMW	6–1	VCT low level (LL)
		6–2	VCT high level (LH)
7	CWS	7–1	LP condenser tube leakage (LN)
		7–1	IP condenser tube leakage (LN)
		7–1	HP condenser tube leakage (LN)
		7–2	Valve abnormality (VV)
		7–3	Pump trip (PM)
8	MSIV	8–1	MSIV abnormality
9	RCP	9–1	RCP CCW loss (LC)
		9–2	RCP seal damage (SD)
10	MSS	10–1	SBCS valve abnormality (VV)
		10–2	Main steam leakage (LN)

VCT volume control tank; *LP* low pressure; *IP* intermediate pressure; *HP* high pressure; *CCW* component cooling water; *SBCS* steam bypass control system

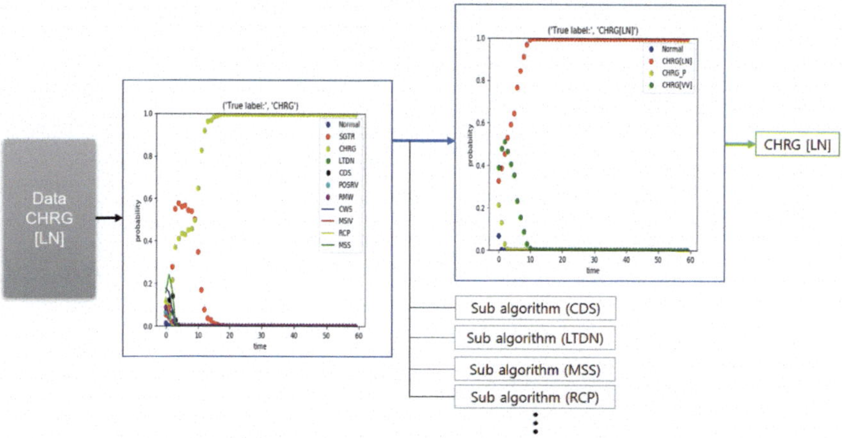

Fig. 4.10 Prediction process in the two-stage model for a charging water system abnormality with water line leakage. Reproduced with permission from Kim et al. (2020)

abnormal event. For a given abnormal event prediction, the related sub-level algorithm determines the sub-procedure based on the results by the main level algorithm. An example of this process is illustrated in Fig. 4.10. The main level algorithm determines the title of the AOP, which is CHRG in this example, and then the sub-level algorithm for CHRG determines the specific sub-procedure, which in this case is water line leakage, or sub-procedure 2–3 in Table 4.8. As the graphs show, the probabilities of the predicted labels are marked each second.

To increase the accuracy, 20 PCs are conservatively chosen with more than 99.99% of the information conserved. We note that each algorithm in the TSM is trained in the same training environment. The performance evaluation metrics of the main algorithm of the TSM are given in Table 4.9 along with results from ANN and SVM models for comparison. The results indicate that the TSM achieves the best performance among the models. Table 4.10 lists the performance evaluation metrics of the TSM sub-algorithms, with the results showing that the sub-layer algorithms achieve almost 100% accuracy for the given data. Therefore, securing accurate prediction in the main stage is the key for the TSM.

Table 4.9 Performance evaluation metrics in the model performance comparison

Model	TSM (main)	ANN	SVM
Accuracy	0.9982	0.9590	0.7563
Precision	0.9982	0.9715	0.7610
Recall	0.9983	0.9590	0.7563
F1 score	0.9982	0.9621	0.7464

Table 4.10 Performance evaluation metrics for the 10 sub-algorithms of the TSM

Model	CDS	CHRG	CWS	LTDN	MSIV
Accuracy	1.0000	0.9917	1.0000	1.0000	0.9983
Precision	1.0000	0.9917	1.0000	1.0000	0.9983
Recall	1.0000	0.9919	1.0000	1.0000	0.9983
F1 score	1.0000	0.9917	1.0000	1.0000	0.9983
Model	MSS	POSRV	RCP	RMW	SGTL
Accuracy	1.0000	0.9983	0.9944	0.9978	1.0000
Precision	1.0000	0.9983	0.9945	0.9978	1.0000
Recall	1.0000	0.9983	0.9944	0.9978	1.0000
F1 score	1.0000	0.9983	0.9944	0.9978	1.0000

The analysis of AOPs involves great amounts of time and effort from untrained NPP operators. Fortunately, by applying PCA, the data dimensionality can be efficiently reduced while maintaining as much information as required. To first validate the PCA method for NPP diagnosis, a test is conducted for comparison with a model considering parameters selected by experts based on an AOP analysis (knowledge-based parameters). A total of 62 parameters are chosen that relate to symptoms and alarms in the AOP, including critical parameters of the primary system such as the RCS pressure. Figure 4.11 shows that the algorithm applying 20 PCs reached convergence in terms of accuracy within a smaller number of epochs compared to the one using knowledge-based parameters.

Considering the high correlation among plant parameters, a relatively small number of PCs can sufficiently preserve the information of the original data. In the first test, for example, only 10 PCs are selected but more than 98% of the information is preserved. By selecting 20 PCs, the data contains more than 99% of the original information.

Under the same training environment, an ANN can show strong performance with an accuracy of more than 95%. But without preprocessing of the input dataset, the

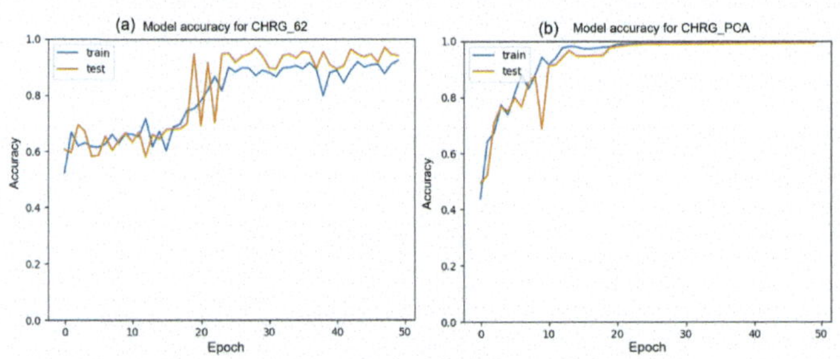

Fig. 4.11 Model accuracy graphs of **a** knowledge-based parameters and **b** PCA (epoch: number of training cycles). Reproduced with permission from Kim et al. (2020)

more than 1000 variables would complicate the algorithm and increase the computation time for diagnosis. It should be noted here that in this section, 10 out of the 82 total AOPs for the APR-1400 are considered; therefore, the performances of the tested models might show greater diversity when handling entire sets of AOPs. The purpose of this section focused on identifying the feasibility of the GRU-based abnormality diagnosis model; future work needs to consider more complex situations by increasing the number of scenarios as much as possible.

One particular advantage of the TSM is its ability to achieve higher efficiency in terms of training. When new data is generated, only the main algorithm needs to be updated to account for the AOPs, while the individual sub-procedures are determined by the algorithms in the sub-level. This is in contrast to a model with only one algorithm, which would require new training to update the whole model, including all AOPs and corresponding sub-procedures, for any new data.

As for decision time, directly diagnosing any changes in NPP status in a direct manner is advantageous, even including cases where no alarms are actuated. As operators typically check symptoms only after being alerted, the detection of a plant status change may be delayed if the severity of the abnormal event is low. It was found in preliminary tests here that, even in cases with low severity, the TSM is able to complete the prediction in under 1 min.

The TSM also grants benefits in solving given problems in a top-down manner, which is similar to the judgment process of human operators. If a complex accident situation occurs or if a judgment by an operator is wrong, it is believed that the multiple steps in the TSM will allow for review and mistake identification mid-process.

4.3 Diagnosis of Abnormal Situations with an LSTM and VAE

Numerous diagnostic algorithms using ANNs have been shown to perform well in trained situations, but they have some weaknesses. First, existing diagnostic algorithms cannot correctly evaluate anonymous cases as an unknown situation if such an unknown situation occurs. As some abnormal situations are not known and are therefore by definition unpredictable in actual NPPs, defining all possible abnormal situations is infeasible. Wrong diagnostic results from an algorithm for an unknown situation could harm the safety of NPPs.

In addition, current algorithms typically produce diagnosis results in the form of probabilities, leaving operators to make the final decision based on the probability. In some situations, the diagnostic algorithms may provide ambiguous results with competing probability values. For example, consider that a diagnostic algorithm produces an output such that Event A has a probability of 0.6 and Event B has a probability of 0.4. In such cases with competing diagnosis results, operators might make errors due to confusion and uncertainty, especially during the initial period of an abnormal situation when the probabilities of the potential events may not be clearly distinguished. In some abnormal situations, fast decision making is critical. Hence, confirming the diagnostic result of the algorithm during the initial period of an abnormal situation is necessary.

To this end, this section introduces a diagnostic algorithm for abnormal situations: unknown events are identified, the diagnostic results are confirmed, and the final diagnosis through the algorithm is conducted, along with efforts to increase the reliability of the diagnostic results. The algorithm combines LSTM and VAE networks for identifying unknown situations and confirms the diagnosis with an LSTM. First, the methodologies are briefly introduced in Sect. 4.3.1, after which the development of the algorithm's functional architecture consisting of several functions is discussed. The various functions and interactions between the functions are covered in Sect. 4.3.3, followed by detailed descriptions of the implementation, training, and testing of the diagnostic algorithm. The training and testing data include Gaussian noise to reflect actual situations in NPPs. The algorithm is trained and implemented using a CNS, with the reference plant based on the three-loop Westinghouse 990 MWe PWR.

4.3.1 Methods

This section briefly explains the LSTM and VAE that are adopted for developing the diagnosis algorithm. An LSTM is applied as the primary network for diagnosing abnormal situations, while VAE-based assistance networks are incorporated to ensure the credibility of the diagnosis result from the LSTM network.

4.3.1.1 LSTM

The LSTM, as introduced in Sect. 2.4.2.2, is an RNN-based neural network architecture capable of dealing with the vanishing gradient problem that stems from processing data of long temporal sequences. Although the LSTM has a similar structure as other RNNs, it employs a different equation to calculate the hidden states. In this network, a structure called a memory cell replaces the typical RNN neuron. This allows the network to combine fast training with efficient learning of the tasks, which requires sequential short-term memory storage over many time steps per trial. The LSTM can learn to bridge small time lags over 1000 discrete time steps through these special memory cell units. It works by determining whether the previous memory value should be updated and calculating the value to be stored in the current memory based on the current state and the memory cell input value. This makes the structure highly effective in storing and processing long sequences (Hochreiter and Schmidhuber 1997; Yang and Kim 2020).

Figure 4.12 illustrates the architecture of the LSTM cell considered in this section. The input x passes through every gate like on a conveyor belt. In an LSTM model,

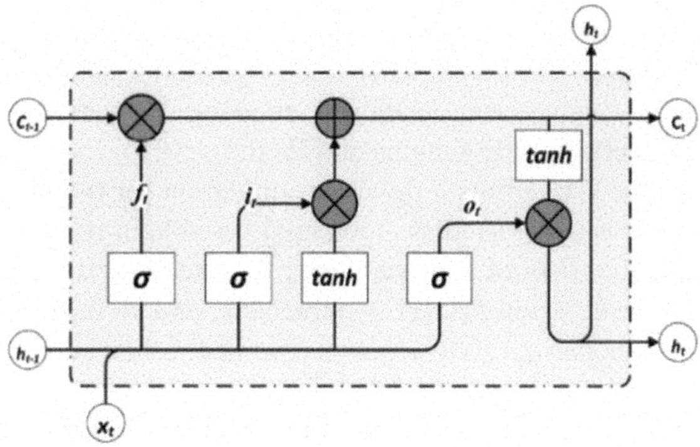

Fig. 4.12 Architecture of the LSTM. Reproduced with permission from Kim et al. (2021)

each cell adjusts the output value based on the input gate, forget gate, and output gate while the structure of the cell is maintained. As the operation of each gate includes additional operations attached to the cell state, the vanishing gradient problem is avoided.

The three gates work as follows. The input gate determines the level of the input value, the forget gate determines the degree to which the previous cell state is forgotten, and the output gate determines the value of the output. Equations (4.17) to (4.19) show the input gate i, forget gate f, and output gate o calculations, where σ and t represent the sigmoid function and time state, respectively. The cell state C is derived in Eq. (4.20). Lastly, the LSTM produces the output h of the network via Eq. (4.21) (Hochreiter and Schmidhuber 1997).

$$f_t = \sigma(W_f \cdot [C_{t-1}, h_{t-1}, x_t] + b_f) \tag{4.17}$$

$$i_t = \sigma(W_i \cdot [C_{t-1}, h_{t-1}, x_t] + b_i) \tag{4.18}$$

$$o_t = \sigma(W_o \cdot [C_t, h_{t-1}, x_t] + b_o) \tag{4.19}$$

$$C_t = i_t * \tanh(W_c \cdot [h_{t-1}, x_t] + b_c) + f_t * C_{t-1} \tag{4.20}$$

$$h_t = o_t * \tanh(C_t) \tag{4.21}$$

4.3.1.2 VAE

The VAE (see Sect. 2.4.3) is an unsupervised DL generative model that can represent distributions of training data (Kingma and Ba 2014). When the input data and training data are similar, the output appears to be mostly similar to the input. In other cases, a probabilistic measure considering the variability in the distribution of variables decreases. Several fault detection algorithms using the reconstruction log-likelihood of VAE have been developed, showing compatibility between a VAE and LSTM (Kim et al. 2021; Kingma and Welling 2013).

The VAE offers a flexible formulation for the interpretation and encoding of z, which is considered as a potential variable in probabilistic generation models. The VAE consists of a probabilistic encoder $q_\phi(z|x)$ and decoder $p_\theta(x|z)$; the posterior distribution $p_\theta(z|x)$ is intractable, and so the VAE approximates $p_\theta(z|x)$ using the encoder $q_\phi(z|x)$, which is assumed to be Gaussian and parameterized by Ø. In Fig. 4.13 showing this VAE architecture, the input x goes through the encoder, and the parameters of the latent space distribution are obtained. The latent variable z is first acquired from sampling in the current distribution and then used to generate a reconstructed sample through the decoder (Kingma and Welling 2013). Thus, the encoder is able to learn and predict the latent variable z, making it possible to draw samples from the distribution.

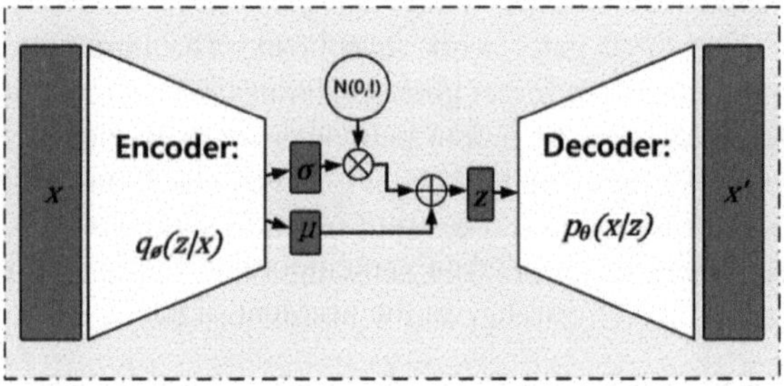

Fig. 4.13 Architecture of the VAE. Reproduced with permission from Kim et al. (2021)

To decode a sample z drawn from $q_\phi(z|x)$ to the input x, the reconstruction loss via Eq. (4.22) should be minimized. The first term in this equation is the KL divergence between the approximate posterior and the prior latent variable z. This term aligns the posterior distribution with the prior distribution by operating as a term for regularization. The second term in Eq. (4.22) represents the reconstruction of x through the posterior distribution $q_\phi(z|x)$ and the likelihood $p_\theta(x|z)$.

$$L(\theta, \phi; x) = -D_{KL}(q_\phi(z|x)||p_\theta(z)) + \mathbb{E}_{q_\phi(z|x)}[logp_\theta(x|z)] \tag{4.22}$$

Here, choosing the appropriate distribution type is important since the VAE models the approximated posterior distribution $q_\phi(z|x)$ from the prior $p_\theta(x)$ and likelihood $p_\theta(x|z)$. A Gaussian distribution is one typical choice for the posterior, where the standard normal distribution $N(0, 1)$ is used for the prior $p_\theta(x)$.

4.3.1.3 Softmax

For the post-processing of the LSTM output, the softmax function is used, which is an activation function commonly applied in the output layer of DL models as it can categorize more than three classes of output. Softmax works by significantly separating the values and then normalizing the output, as shown in Eq. (4.23). In $y \in \mathbb{R}^k$, which is the input vector of the softmax function, k is the number of classes of output. For $S(y)_1 \ldots S(y)_k$, the normalized output values are between 0 and 1, and the sum of the output values is always 1.

$$S(y)_i = \frac{e^{y_i}}{\sum_{j=1}^{k} e^{y_j}} \; (for\, i = 1, \ldots, k) \tag{4.23}$$

4.3.2 Diagnostic Algorithm for Abnormal Situations with LSTM and VAE

The overall structure of the LSTM-VAE diagnostic algorithm for abnormal situations can now be described. Figure 4.14 depicts the designed functional architecture of the algorithm, comprising four functions as follows: an input preprocessing function, an unknown event identification function, an event diagnosis function, and a confirmation of diagnosis result function, which are detailed in the next subsections. Table 4.11 provides descriptions as well as the inputs and outputs of each function.

4.3.2.1 Preprocessing Function

The first function of the algorithm is to preprocess the plant parameters in order to put them in suitable form as network inputs. The network inputs are selected from operating procedures considering their importance and their ability to influence the NPP state or the availability of systems. While network inputs need to have values ranging between 0 to 1, plant parameters have various value ranges and states, e.g., 38% PZR level, or on/off for an alarm. Variables with higher values generally have a greater influence on the network results; however, high values are not necessarily more meaningful for prediction, a consequence of which is the appearance of local minima. To prevent this, the input preprocessing function takes the regular plant parameters as inputs and outputs normalized plant parameters for use in the following networks.

Specifically, min–max normalization is applied to prevent local minima. The input of the network is calculated via Eq. (4.24), where X is the current value of the plant parameter, and $X_{\min}$ and $X_{\max}$ are the minimum and maximum values of the collected data, respectively. As a normalized value, X_{input} has a range of 0–1.

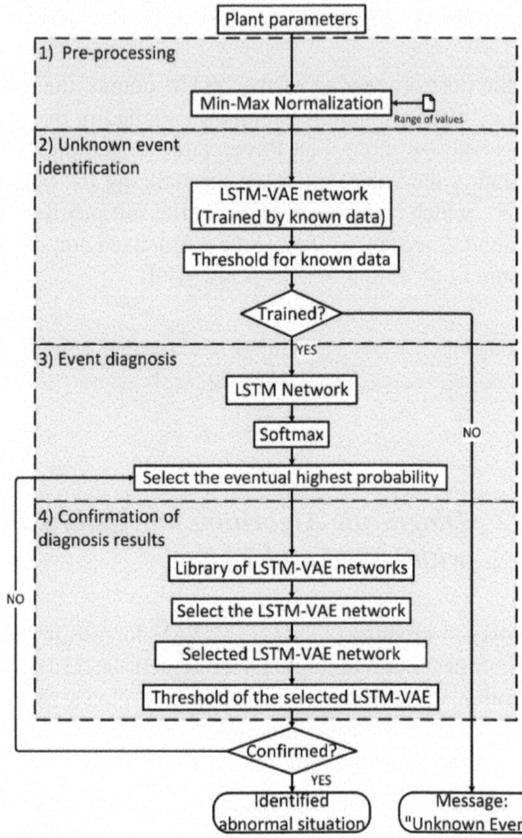

Fig. 4.14 Functional architecture of the LSTM and VAE-based diagnostic algorithm design. Reproduced with permission from Kim et al. (2021)

$$X_{input} = \frac{(X - X_{\min})}{(X_{max} - X_{min})} \tag{4.24}$$

4.3.2.2 Unknown Event Identification Function

The second function, which is for unknown event identification, is used to identify unknown events through a combination of LSTM and VAE networks. This combination not only supports the sequence of the input data but also captures the complex temporal dependences in the time series (Park et al. 2018). As inputs, this function receives normalized NPP parameters from the preprocessing function and identifies unknown events in real time. An anomaly score is assigned to indicate any discrepancies between the actual data and the trained data used for this function. For anomaly scores below a pre-set threshold, the event is identified as a known event that the diagnosis network in the next function has been trained for. But for anomaly scores

Table 4.11 Summary of diagnostic algorithm functions

No.	Function	Input	Output
1	Preprocessing function	– Plant parameters (from the NPP)	– Normalized plant parameters (to networks)
2	Identification of unknown event function	– Normalized plant parameters (from the preprocessing function)	– Identification as a known event (to the event diagnosis function) – Identification as an unknown event (to operators)
3	Event diagnosis function	– Identification of known event (from the identification of unknown event function) – Normalized plant parameters (from the preprocessing function)	– Eventual highest probability event (to the confirmation of diagnosis result function)
4	Confirmation of diagnosis result function	– Highest probability event (from the event diagnosis function) – Normalized plant parameters (from the preprocessing function)	– Identified event (to operators) – Incorrect diagnosis results (to the event diagnosis function)

above the threshold, the event is classified as unknown, with the message "Unknown event occurrence" provided as the output to operators.

The process of the unknown event identification function is illustrated in Fig. 4.15. To account for the temporal dependency of time-series data in a typical VAE, the VAE in this architecture is combined with LSTMs by replacing the feed-forward network in the VAE. For the multi-parameter input $x_t \in \mathbb{R}^D$, where D is the number of input parameters, made up of $x_{1,t}, \ldots x_{D,t}$ at time t which are normalized plant parameters from the preprocessing function, the encoder approximates the posterior $p(z_t|x_t)$ by the LSTM in the encoder and estimates the mean $\mu_{z_t} \in \mathbb{R}^M$ and the variance $\sigma_{z_t} \in \mathbb{R}^M$ of the latent variable $z_t \in \mathbb{R}^M$. A randomly sampled z_t from the posterior $p(z_t|x_t)$ is then fed to the LSTM in the decoder. The mean $\mu_{x_t} \in \mathbb{R}^D$ $(\mu_{x_{1,t}}, \ldots \mu_{x_{D,t}})$ and variance $\sigma_{x_t} \in \mathbb{R}^D\left(\sigma_{x_{1,t}}, \ldots \sigma_{x_{D,t}}\right)$ of the reconstruction are computed as the final output.

In order to identify an unknown event, the unknown event identification function recognizes an anomalous execution when the current anomaly score exceeds the threshold α′, as shown in Eq. (4.25). The anomaly score calculator is represented by the term $f_s(x_t, \emptyset, \theta)$. This score is defined as the negative log-likelihood of x_t, as in Eq. (4.26), with respect to the reconstructed distribution of x_t from the LSTM-VAE network. In the equation, μ_{x_t} and σ_{x_t} are the mean and the variance of the reconstructed distribution from the LSTM-VAE network with parameters $\emptyset$ and θ, respectively. Figure 4.16 depicts the calculation of an anomaly score as an example. The normalized input parameters are $x_t \in \mathbb{R}^3$ (for $x_{1,t}, x_{2,t}, x_{3,t}$), the reconstruction

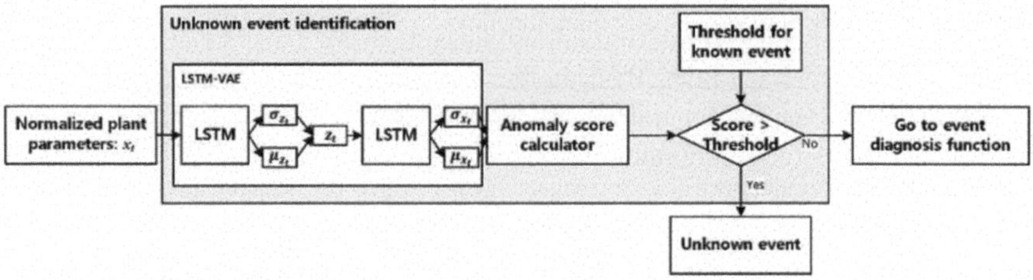

Fig. 4.15 Various processes in the unknown even identification function. Reproduced with permission from Kim et al. (2021)

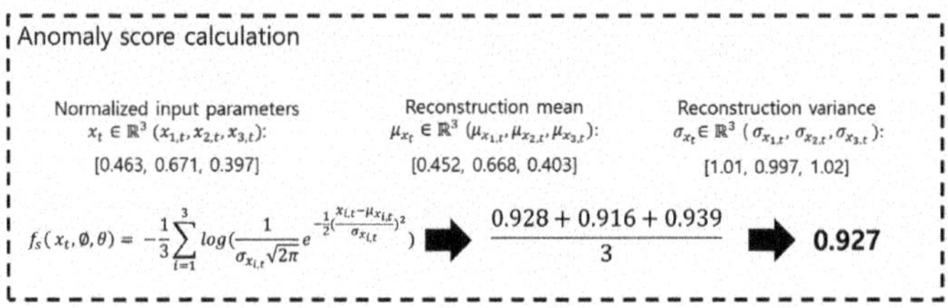

Fig. 4.16 Example of anomaly score calculation. Reproduced with permission from Kim et al. (2021)

mean is $\mu_{x_t} \in \mathbb{R}^3$ (for $\mu_{x_{1,t}}, \mu_{x_{2,t}}, \mu_{x_{3t}}$), and the reconstruction variance is $\sigma_{x_t} \in \mathbb{R}^3$(for $\sigma_{x_{1,t}}, \sigma_{x_{2,t}}, \sigma_{x_{3t}}$). Each element is processed via Eq. (4.26) and the anomaly score is calculated as a result. In the example shown in the figure, there are three input parameters (D), and the anomaly score is calculated to be 0.927.

$$\begin{cases} Unknown\, event, & if\, f_s(x_t, \emptyset, \theta) > \alpha' \\ Known\, event, & otherwise, \end{cases} \tag{4.25}$$

$$f_s(x_t, \emptyset, \theta) = -\frac{1}{D}\sum_{i=1}^{D} logp\left(x_{i,t};\, \mu_{x_{i,t}}, \sigma_{x_{i,t}}\right) \tag{4.26}$$

To set the threshold α', a three-sigma limit is applied following the consideration of the anomaly score distribution of the training data. Anomaly scores achieving small values indicates that the output data, in other words the reconstructed data from the LSTM-VAE, is similar to the training data, with the threshold reflecting the upper control limit. In the calculation shown in Eq. (4.27), S_{mean} and S_{std} are the mean and standard deviation of the anomaly scores of the training data, respectively. This sets a range for the process parameter at a control limit of 0.27%, which corresponds to the three-sigma in the normal distribution, in addition to minimizing the cost associated with preventing errors in terms of classifying a known event as unknown. Section 4.3.3.2 describes how the threshold and hyperparameters are determined for this function.

$$\alpha' = S_{mean} + 3 * S_{std} \tag{4.27}$$

4.3.2.3 Event Diagnosis Function

The third function of the algorithm uses an LSTM to provide the diagnosis results of the plant situation; Fig. 4.17 shows the flow of the event diagnosis function. The LSTM receives normalized plant parameters from the preprocessing function and outputs the identified abnormal event along with its probability, which represents the confidence level of the identified event. The output is post-processed using the softmax function. The function selects the event with the highest probability among the diagnosis results and passes it to the confirmation function. Here, multiple events can be identified with varying probabilities in the event diagnosis function. After receiving the selected event, the confirmation function may return information indicating that the current situation is not consistent with the diagnosis result, in which case the event diagnosis function selects the event with the next highest probability and sends it to the confirmation function until the current situation is consistent with the diagnosis result. The process to determine the model hyperparameters including the number of layers and nodes is discussed in Sect. 4.3.3.3.

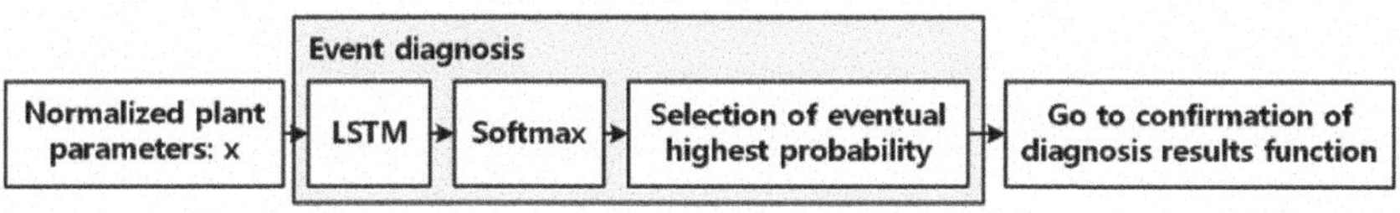

Fig. 4.17 Processes in the event diagnosis function. Reproduced with permission from Kim et al. (2021)

4.3.2.4 Confirmation of Diagnosis Result Function

The last function is to confirm whether the current abnormal situation matches the event selected in the previous event diagnosis function. The confirmation function has a library composed of LSTM-VAE networks for trained events. For confirmation, the particular LSTM-VAE network for the selected event is used to check that the selected event is the same as the trained event.

Figure 4.18 shows the process of confirming the selected event. First, after receiving the event selected by the previous event diagnosis function, the confirmation function selects the LSTM-VAE network from the library that corresponds to the identified event. Next, the function verifies that the current situation is identical to the selected event through the LSTM-VAE network. To determine the anomaly score, the confirmation function employs negative log-likelihood similarly as the unknown event identification function. For negative log-likelihood values below the threshold,

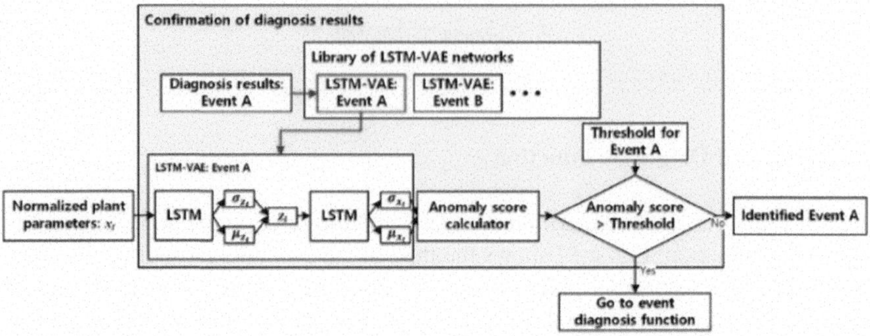

Fig. 4.18 Processes in the confirmation of diagnosis results function. Reproduced with permission from Kim et al. (2021)

the algorithm affirms that the diagnosis result from the event diagnosis function is correct. For negative log-likelihood values above the threshold, the algorithm returns to the previous event diagnosis function to select another event. The thresholds of the LSTM-VAE networks in this function are determined in a manner similar to that discussed in Sect. 4.3.2.2.

4.3.3 Implementation

A CNS is used as a real-time testbed to implement, train, and validate the algorithm. In the data collection, a total of 20 abnormal situations and 558 cases are simulated. Table 4.12 lists the abnormal scenarios along with the number of simulations for each event. The scenarios include representative abnormal situations in actual NPPs, namely instrument failures (Nos. 1–6), component failures (Nos. 7–16), and leakages (Nos. 17–20).

From the AOPs of the reference plant, 139 plant parameters are chosen for the inputs, including plant variables like pressure or temperature and component states like the status of valves or pumps. The parameters are collected every 1 s over the simulations.

For training, 15 out of the selected 20 scenarios containing 409 cases are used. The remaining 5 scenarios, i.e., scenario Nos. 3, 13, 14, 15, and 20, are used for testing untrained events with a total of 149 cases. Among them, 115 cases are used for determining the two thresholds, namely those of the unknown event identification and confirmation functions.

As the CNS produces data without noise, as shown in Fig. 4.19a, ± 5% Gaussian noise is added to the collected data to more realistically reflect actual signals from NPPs. The noise intentionally added to the CNS data is shown in Fig. 4.19b.

Table 4.12 Abnormal scenarios and number of simulations Kim et al. (2021)

No.	Scenario	Training cases	Test cases	Total cases
1	Failure of PZR pressure channel (High)	14	4	18
2	Failure of PZR pressure channel (Low)	20	6	26
3	Failure of PZR water level channel (High)	–	6	6
4	Failure of PZR water level channel (Low)	11	4	15
5	Failure of SG water level channel (Low)	32	8	40
6	Failure of SG water level channel (High)	35	6	41
7	Control rod drop	38	10	48
8	Continuous insertion of control rod	7	1	8
9	Continuous withdrawal of control rod	6	2	8
10	Opening of PZR power-operated relief valve	42	10	52
11	Failure of PZR safety valve	35	8	43
12	Opening of PZR spray valve	41	9	50
13	Stopping of charging pump	–	1	1
14	Stopping of two main feedwater pumps	–	3	3
15	Main steam line isolation	–	3	3
16	Rupture at the inlet of the regenerative heat exchanger	40	10	50
17	Leakage from chemical volume and control system to component coolant water (CCW)	40	10	50
18	Leakage at the outlet of charging control flow valve	24	6	30
19	Leakage into the CCW system from the RCS	24	6	30
20	Leakage from SG tube	–	36	36
Total		409	149	558

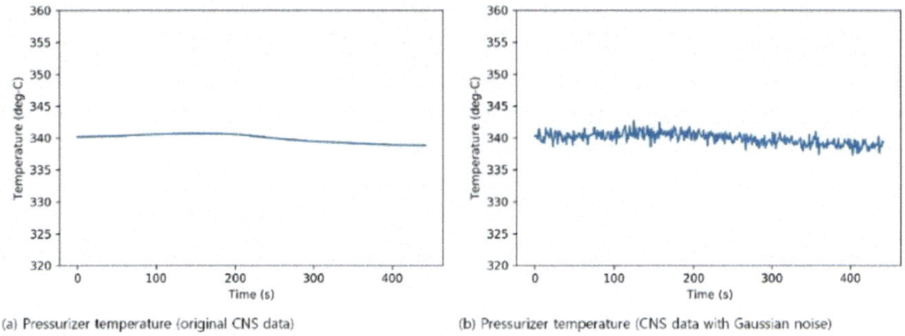

Fig. 4.19 Examples of PZR temperature data. **a** Original CNS data and **b** data with ± 5% added Gaussian noise. Reproduced with permission from Kim et al. (2021)

4.3.3.1 Preprocessing

For data preprocessing, the maximum and minimum values of the parameters are determined for all of the collected data (558 cases) to implement the min–max normalization in the preprocessing function. However, adding ± 5% Gaussian noise to the simulator data to better reflect an actual NPP environment can result in data that is higher than the maximum values or lower than the minimum values, in which case the data would be outside the 0 to 1 range when normalized. To prevent this issue, 10% and –10% margins, respectively, are added to the maximum and minimum values of each parameter. An example of a normalized PZR temperature is shown in Fig. 4.20.

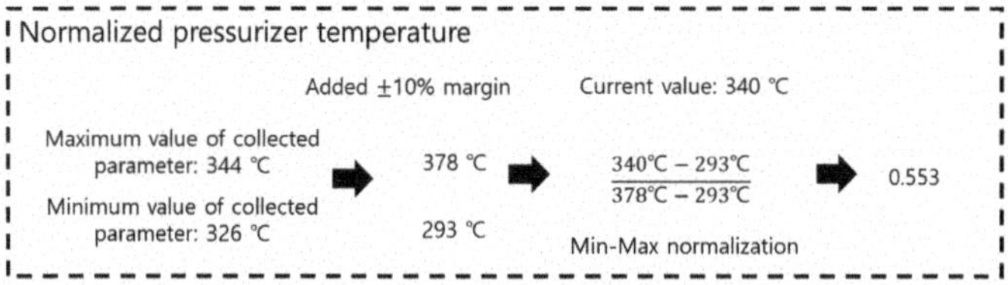

Fig. 4.20 Example of a normalized PZR temperature. Reproduced with permission from Kim et al. (2021)

4.3.3.2 Unknown Event Identification

The unknown event identification function is implemented with the LSTM-VAE network. In this network, the datasets have a sequence of 10 s and 139 input values. As mentioned in Sect. 4.3.3, 409 scenarios are trained in this implementation, resulting in 192,637 datasets for the 139 parameters. Figure 4.21 illustrates how the unknown event function works within the LSTM-VAE network. The VAE does not necessarily tune the hyperparameters; rather, it creates a variational information bottleneck to prevent overfitting, which consequently has the effect of providing an optimal bottleneck size for each hyperparameter (Tishby and Zaslavsky 2015; Alemi et al. 2016; Ruder 2016). Otherwise, the LSTM-VAE network uses the Adam optimizer (Kingma and Ba 2014) with a learning rate of 0.0001 and runs for 100 epochs.

To evaluate the performance of the LSTM-VAE network, the receiver operating characteristic (ROC) curve is considered. An ROC curve is made by plotting the true positive rate, which is the ratio of the correctly predicted positive observations to all observations in the actual class, against the false positive rate, which is the ratio of the incorrectly predicted negative observations to all observations in the actual class. It offers a useful method to interpret the performance of binary classifiers. The area under the ROC curve (AUC) is an effective measure to determine the overall accuracy of the diagnosis, which can be interpreted as the average value of sensitivity for all possible values of specificity. This measure takes any value between 0 and 1,

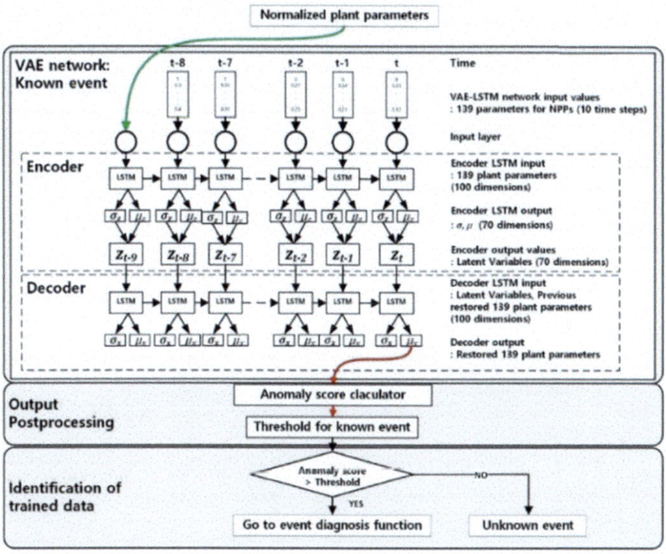

Fig. 4.21 Identification of unknown event function using the LSTM-VAE network. Reproduced with permission from Kim et al. (2021)

where 0 indicates a wholly inaccurate test. The closer the AUC is to 1, the stronger the overall diagnostic performance of the model. An AUC value of 0.5 typically suggests no discrimination, 0.7–0.8 is considered acceptable, 0.8–0.9 is considered excellent, and over 0.9 is considered outstanding (Park et al. 2004; Bergstra and Bengio 2012; Mandrekar 2012). The threshold value is calculated to be 0.923 using Eq. (4.27). Figure 4.22 plots the result of the ROC and AUC of the identification of unknown event function.

Fig. 4.22 ROC and AUC of the identification of unknown event function. Reproduced with permission from Kim et al. (2021)

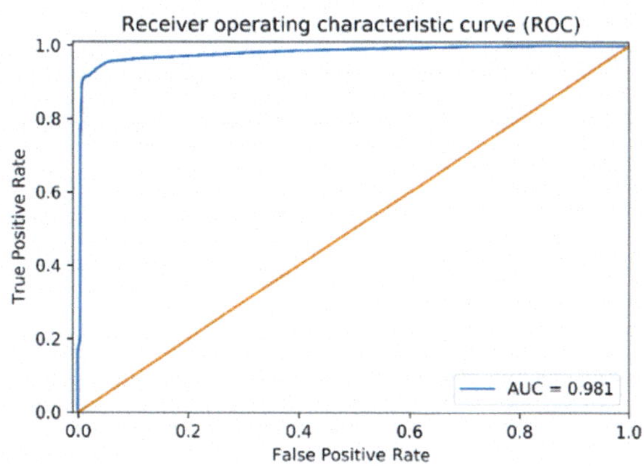

Table 4.13 Accuracy comparison results of various configured networks

No.	Time step	Batch size	Layers	Val_Accuracy
1	5	32	2	0.9668
2	5	32	3	0.9638
3	5	64	2	0.9634
4	5	64	3	0.9650
5	**10**	**32**	**2**	**0.9768**
6	10	32	3	0.9746
7	10	64	2	0.9764
8	10	64	3	0.9741
9	15	32	2	0.9767
10	15	32	3	0.9762
11	15	64	2	0.9764
12	15	64	3	0.9766

4.3.3.3 Event Diagnosis

The event diagnosis function adopts the LSTM and softmax function. A total of 409 scenarios producing 192,637 datasets for 139 parameters are used for training. For validation to prevent overfitting, 20% of the training data (38,527 datasets) is randomly selected, and to optimize the LSTM in the event diagnosis function, a manual search method is used in which the hyperparameters such as the input sequence length, batch size, and number of layers are individually adjusted. In general, no single standard exists for hyperparameter determination in the optimization of a network (Ruder 2016; Bergstra and Bengio 2012; Kim et al. 2021).

Table 4.13 lists the results of an accuracy comparison between different configured networks. Accuracy is defined here as the ratio of correctly predicted data to the total validation data. According to the results, the optimal LSTM network has 2 layers, a time step of 10 s, and a batch size of 32. This network also applies the Adam optimizer (Kingma and Ba 2014) with a learning rate of 0.0001 and runs for 100 epochs. Figure 4.23 depicts the event diagnosis function with the LSTM and softmax function.

4.3.3.4 Confirmation of the Diagnosis Result

To implement the confirmation function, an LSTM-VAE library is developed that contains 16 LSTM-VAE networks trained for each known event, in other words 15 abnormal events and 1 normal state. An illustration of the confirmation of diagnosis result function is shown in Fig. 4.24 for the diagnosed event Ab 01. All LSTM-VAE networks and thresholds are developed in a similar manner as those in the unknown

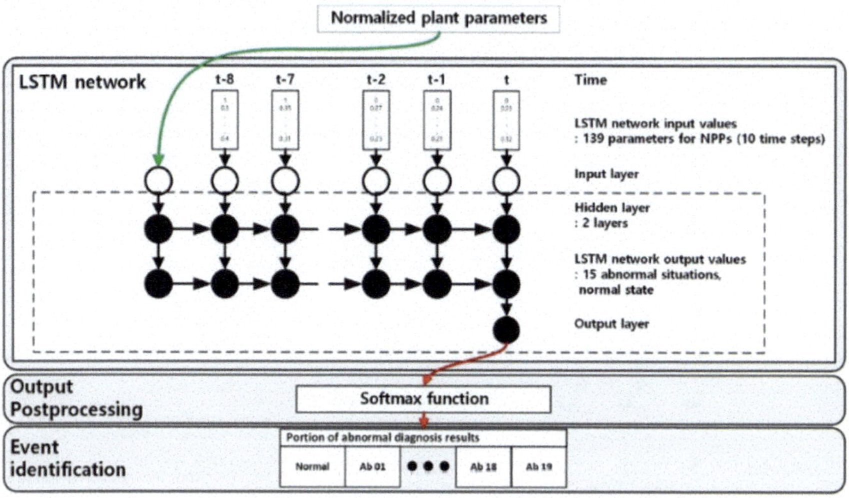

Fig. 4.23 Event diagnosis function with an LSTM and softmax function. Reproduced with permission from Kim et al. (2021)

event identification function. Table 4.14 lists the number of training datasets, AUC, and threshold of each LSTM-VAE network in the confirmation function.

4.3.3.5 Results

A test of the network is performed for 149 scenarios with 57,309 datasets. Among them, 47,987 datasets from 100 cases are used as the trained events, while 9322 datasets from 49 cases are used as untrained events. In the test results, the accuracy of the network for unknown event identification, which respectively predicts trained and untrained events as known and unknown events, is 96.72%. The accuracy of the network for confirmation of the diagnosis result is 98.44%. With these results, the test demonstrates that the algorithm can successfully diagnose the trained events and identify the untrained events. An example of the diagnosis of an untrained event, in this case Ab 20 (leakage from SG tubes), is shown in Fig. 4.25. Here, the identification of unknown event function classifies Ab 20 as an untrained event because its anomaly score exceeds the threshold, and the message "Unknown Event" is provided.

Figure 4.26 illustrates the diagnosis process of the algorithm for a trained event, in this case Ab 08 (continuous insertion of control rods). First, the preprocessing function normalizes the plant parameters, and the identification of unknown event function highlights the event as a trained event. The event diagnosis function then determines the event to be Ab 08, and the diagnosis result confirmation function affirms that the result is correct. The message "Ab 08: continuous insertion of control rods" is finally presented as the diagnosis of the current situation.

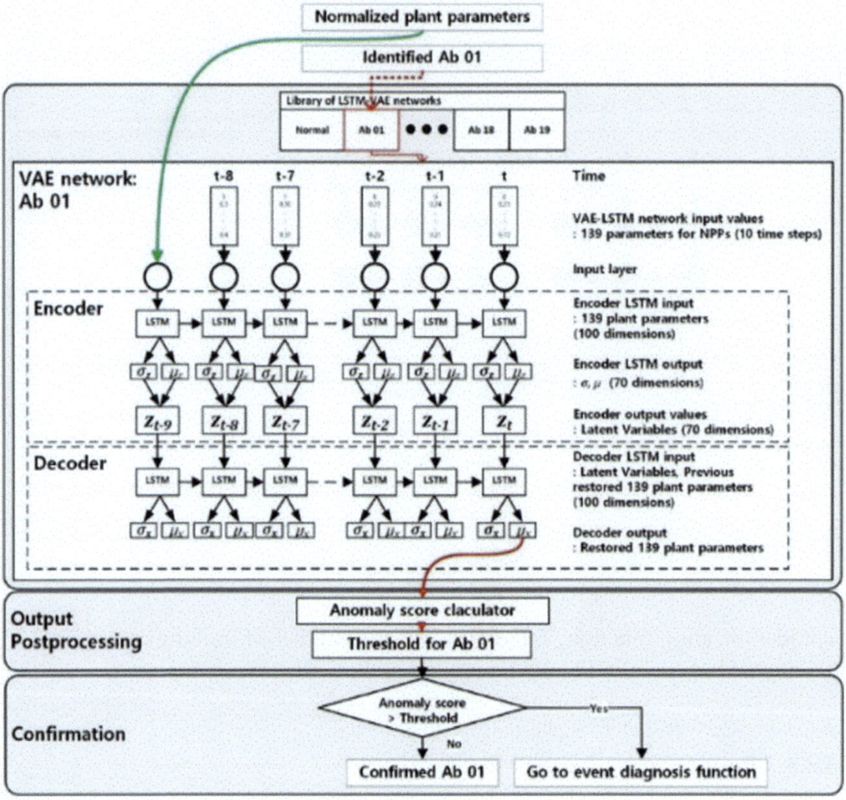

Fig. 4.24 Illustration of the confirmation of diagnosis result function. Reproduced with permission from Kim et al. (2021)

4.4 Sensor Fault-Tolerant Accident Diagnosis

Prompt reactions to anomalies are essential to minimize the potential consequences, especially in safety–critical systems. As NPPs may threaten public safety in case of a large release of radioactive materials as a result of an accident, accurate diagnosis is crucial to determine the ORPs that contain the essential mitigation tasks for the particular accident (USNRC 1982). Diagnosis procedures follow intuitive logics to identify accidents based on a series of symptom checks, which as previous discussed can be a highly demanding task for plant operators because the early phases of an emergency can have a great effect on the consequences of the accident. In this light, an inaccurate diagnosis leading to the selection of the wrong ORP could result in numerous human errors and impair mitigation.

One factor in the misdiagnosis of severe accidents such as the TMI and Fukushima accidents is sensor faults (Mizokami and Kumagai 2015; Norman 1980). In particular, the TMI accident represents a case where sensor faults led to a critical misdiagnosis

Table 4.14 Training data, performance, and thresholds of the LSTM-VAE networks in the confirmation function

No.	Name	Trained data	AUC	Threshold
1	VAE Ab 01	3459 datasets (Ab 01)	0.995	0.923
2	VAE Ab 02	3785 datasets (Ab 02)	0.934	0.923
3	VAE Ab 04	5658 datasets (Ab 04)	0.993	0.922
4	VAE Ab 05	5730 datasets (Ab 05)	0.998	0.921
5	VAE Ab 06	8312 datasets (Ab 06)	0.974	0.923
6	VAE Ab 07	34,735 datasets (Ab 07)	0.999	0.920
7	VAE Ab 08	2831 datasets (Ab 08)	0.958	0.924
8	VAE Ab 09	2070 datasets (Ab 09)	0.955	0.927
9	VAE Ab 10	8394 datasets (Ab 10)	0.985	0.921
10	VAE Ab 11	8004 datasets (Ab 11)	0.938	0.921
11	VAE Ab 12	23,885 datasets (Ab 12)	0.996	0.920
12	VAE Ab 16	24,600 datasets (Ab 16)	0.965	0.920
13	VAE Ab 17	30,904 datasets (Ab 17)	0.953	0.934
14	VAE Ab 18	14,805 datasets (Ab 18)	0.984	0.922
15	VAE Ab 19	1546 datasets (Ab 19)	0.967	0.920
16	VAE Normal	7602 datasets (Normal state)	0.995	0.923

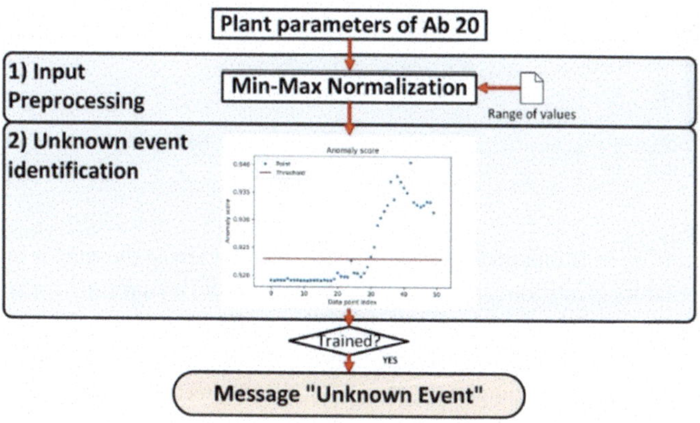

Fig. 4.25 Process of diagnosing an untrained event. Reproduced with permission from Kim et al. (2021)

error. In short, electromagnetic relief valves, also called power-operated relief valves (PORVs), were accidentally stuck open, but the related indicator showed that the valves were closed. From this error, the operators mistakenly understood the situation and turned off the safety-injection system, which had automatically actuated to cool down the reactor core. The well-documented accident progressed from there.

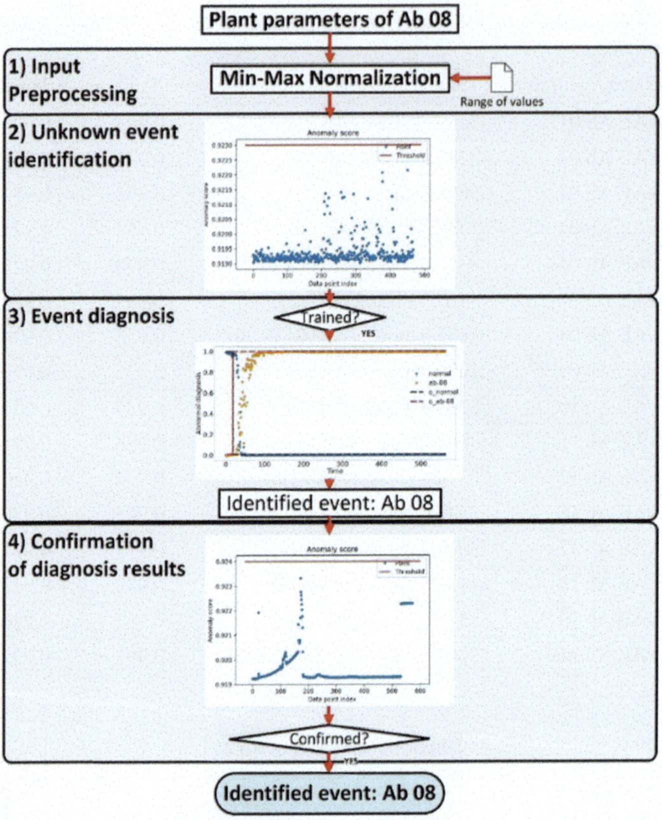

Fig. 4.26 Process of diagnosing a trained event. Reproduced with permission from Kim et al. (2021)

In this section, an accident diagnosis algorithm considering the sensor fault state to aid plant operators is introduced. Numerous models based on data-driven approaches, which typically require longitudinal multivariate plant parameters that contain accident symptoms, have been suggested to reduce the number of required tasks while securing accurate accident diagnosis. However, the means to account for sensor faults during an accident sequence have not been discussed in existing diagnosis procedures or the more recently developed diagnosis models. This seems to imply that current diagnosis models are vulnerable to sensor errors, based on their lack of consideration of sensor fault-induced diagnosis failure. The diagnosis model in this section adopts the sensor fault monitoring system described in Sect. 3.1 and adds a sensor fault mitigation system that conditionally isolates any faulty sensor data to achieve sensor fault-tolerant diagnosis.

4.4.1 Sensor Fault-Tolerant Diagnosis System Framework

The sensor fault-tolerant diagnosis system is made up of two subsystems: one for sensor fault detection, and one for sensor fault mitigation. Figure 4.27 shows a basic schematic. Following a reactor trip, sensor fault monitoring is initiated as the first function of the system. If no sensor faults are detected, the accident diagnosis system generates an output that identifies the accident. If a sensor fault is detected by the monitoring system, the faulty information is transferred to the sensor fault mitigation system, which either substitutes the faulty sensor data via data imputation or weakens its influence. This way, accident diagnosis results that are robust against sensor error are generated.

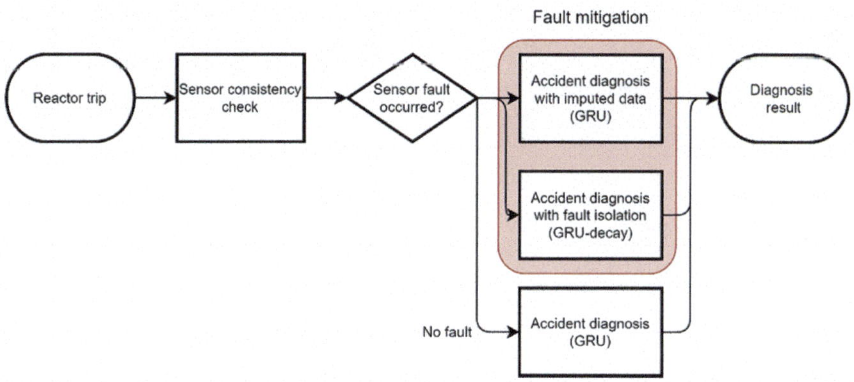

Fig. 4.27 Overall framework for fault detection and fault-tolerant accident diagnosis. Reproduced with permission from Choi and Lee (2020)

4.4.1.1 Accident Diagnosis Algorithm Using GRU

Correctly identifying the accident type is crucial to mitigate an emergency. Emergency situations create great pressure and stress for operators that invite potential human errors in following the complicated diagnosis procedures (Park et al. 2005). For operator support, ML-based and statistical model-based accident identification algorithms have been proposed to secure accurate diagnosis outputs quickly. The tasks included in diagnosis involve checking multiple process parameters, namely their trends or if any value exceeds its threshold, which requires knowledge of multivariate parameters. Accordingly, accident diagnosis algorithms are also required to classify multivariate data via temporal analysis.

In the current system, the accident diagnosis algorithm is constructed based on a GRU to monitor diagnosis performance through tests for any degradation from faulty inputs. It has one hidden layer with 64 nodes and uses the Adam optimizer for training. The hyperparameters including the number of hidden layers and nodes were determined from a pilot study for satisfactory diagnosis performance. Outputs of the algorithm are generated with the softmax function, which normalizes the output with a probability distribution (Fig. 4.28).

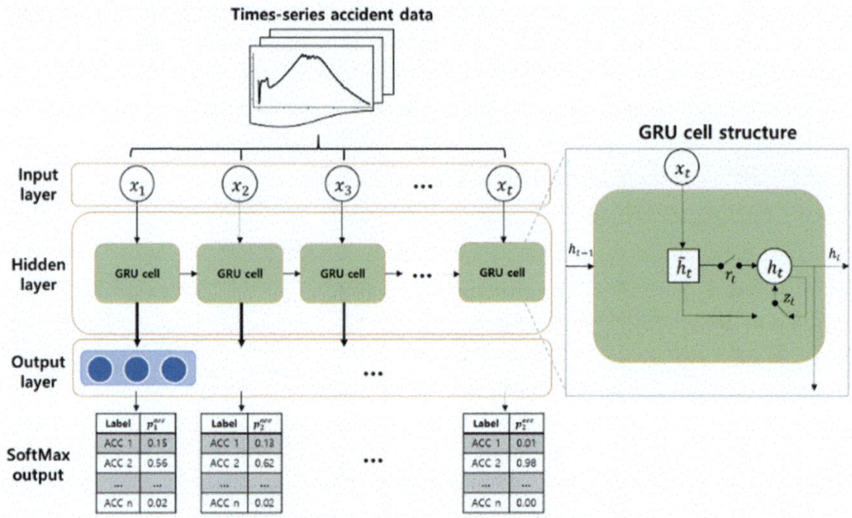

Fig. 4.28 GRU-based accident diagnosis algorithm with multivariate time-series input and softmax output. Reproduced with permission from Choi and Lee (2020)

The characteristics of GRU are favorable for accident diagnosis in terms of the connections between the cells and its forward propagation aspect transferring contextual information. Issues with the robustness of neural network models have been raised though, and thus measures to strengthen network robustness receive continued attention. One work was able to achieve higher performance of an RNN in the reconstruction of energy production data but it was accompanied by much lower robustness compared with other data-driven approaches (Baraldi et al. 2015). Another work compared the performance of various neural network models and found a lower robustness of the RNN, with missing values in some cases (Kim et al. 2018). As the robustness of RNN models can be questioned, measures to counteract the threat of possible sensor errors in the current case need to be prepared to prevent the failure of the model from wrong inputs.

4.4.1.2 Sensor Fault Detection System

The RNN-based sensor error detection system detailed in Sect. 3.1 that treats several time-series multivariate data to account for sensor faults in an emergency situation acts as the base of the system in the present section. As discussed in that section, the

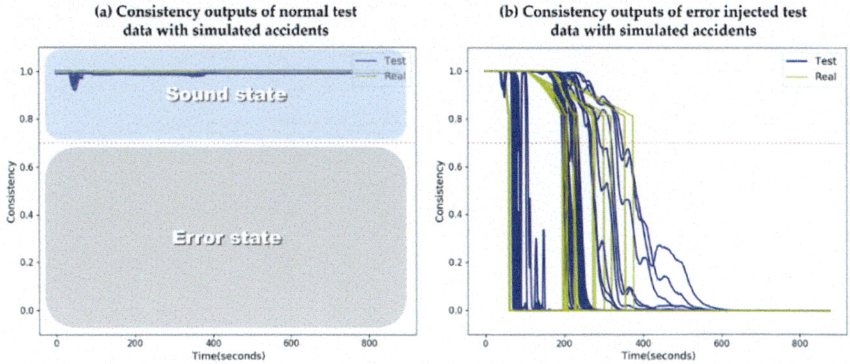

Fig. 4.29 Examples of sensor state identification with **a** normal data test results and **b** error injected data test results from the sensor fault detection model using supervised learning (Sect. 3.1). Reproduced with permission from Choi and Lee (2020)

output of the detection system takes the form of a consistency index, representing numerically normalized scores for sensor health. Figure 4.29 shows an example of the consistency index hovering around 1 in normal states but decreasing along the degree of deviations in the sensor signal. The criteria for the consistency index as a fault threshold was previously determined from empirical test results to be 0.7 considering detection speed as well as uncertainties. The consistency index allows sensor fault information to be derived with masking inputs, which have the same data structure as time-series inputs. Masking inputs indicate the absence of data in a binary manner, where 1 reflects a normal sensor value and 0 reflects missing data. The masking inputs are obtained as in Eq. (4.28).

$$m_t^i = \begin{cases} 1, & C_t^i \geq 0.7 \\ 0, & C_t^i < 0.7 \end{cases} \tag{4.28}$$

The sensor error modes previously selected consider relations to human error and commonness. Accident data were injected with drift and stuck errors and the consistency was evaluated. In the present section, the same sensor errors are implemented to check for any degradation in the performance of the diagnosis algorithm. In this case, drift error with rates of change of 2 and 10 times in both upward and downward directions as well as stuck at zero error are added to the accident data. Examples of the drift and stuck errors occurring at 100 s are shown in Fig. 4.30. All error injection is added at 1 s as the pilot study found that error injection close to 0 s resulted in large deviations of the diagnosis results.

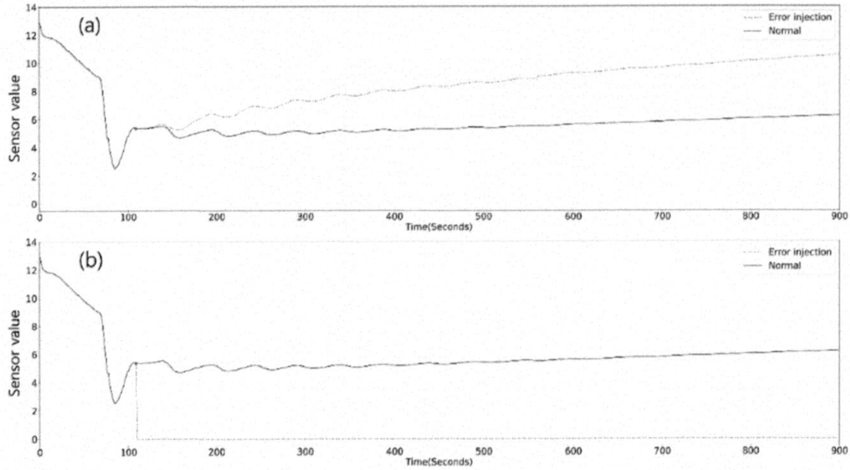

Fig. 4.30 Example trends of **a** drift and **b** stuck at zero sensor errors. Reproduced with permission from Choi and Lee (2020)

4.4.1.3 Fault-Tolerant Accident Diagnosis System

With successful results of the sensor fault detection system, continuous information of the states of the sensors in an accident can be monitored with confidence. But in the event that a sensor fault is detected, the inherently unreliable incoming data from the faulty sensor should be removed. This presents a problem as GRU-based accident diagnosis algorithms cannot accept an empty input based on the GRU structure where each input influences all the functions and the outputs via interconnections. It is therefore necessary to estimate the missing values or employ a modified RNN structure in cases with missing data. In this section, various imputation approaches that substitute missing data with estimations are applied to the GRU model and compared to construct the fault-tolerant accident diagnosis system.

Simple imputation methods for time-series data include moving window imputation and last observed carried forward imputation. Such methods are unsuitable in the present case because they cannot reflect the characteristics of multivariate time-series data, as a faulty sensor will generate inaccurate data continuously following the fault occurrence. Moreover, a statistical model or models should be included in the imputation to cover the diverse plant symptoms based on the accident type. Three imputation methods to substitute the faulty sensor data are compared below.

K-Nearest Neighbors (KNN)

K-nearest neighbors (KNN) is a common single imputation method performing a single calculation only. The basic principle is to make an estimation based on the average of multiple neighbors. In the K parameter setting, data in the same vicinity are grouped via distance calculations such as Manhattan distance, Euclidian distance, and correlation distance (Zhang 2012). Figure 4.31 shows the basic premise.

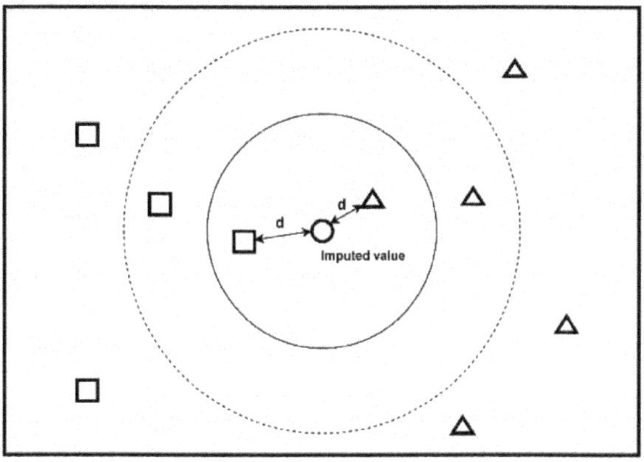

Fig. 4.31 KNN imputation

In case of missing data, the missing values are substituted for with the weighted average of the nearest neighbors. The present model applies a Euclidian distance-based KNN method for data imputation.

Multivariate Imputation with Chained Equations (MICE)
Single imputation methods have limitations that can be addressed with the emergence of multiple imputation methods. Basic multiple imputation such as in the multivariate imputation with chained equations (MICE) approach which depicts the multiple implementation of single imputation to make up the random losses in multivariate data is as follows.

(1) A simple imputation, such as mean, is conducted for every missing data as a place holder
(2) The variable with the largest portion missing is returned to the missing data
(3) The variable is regressed from other variables
(4) The regression repeats until convergence is obtained.

Typical regression models employ linear, logistic, or Poisson regression (White et al. 2011). In the different MICE software packages, regression models and convergence criteria differ. In the present case, linear regression and a convergence criteria of $\Delta < 0.1$ is applied; in other words, the relative change in a new imputation value compared to the old imputation value is under 0.1.

Missforest
Originally proposed in the medical industry to handle big data containing missing sections, the Missforest method is a means of multiple imputation that can work with either categorical or numerical data. It follows a similar imputation process as MICE except for the regression method. Missforest adopts iterative random forest regression, which is a popular ML method in which numerous decision trees are constructed following training and a mean regression is generated by the ensembling of multiple decision trees (Liaw and Wiener 2002). Computation time for this method

is adjusted by the inputs, number of trees, and number of iterations. Figure 4.32 depicts the Random forest regression process which is basis of Missforest.

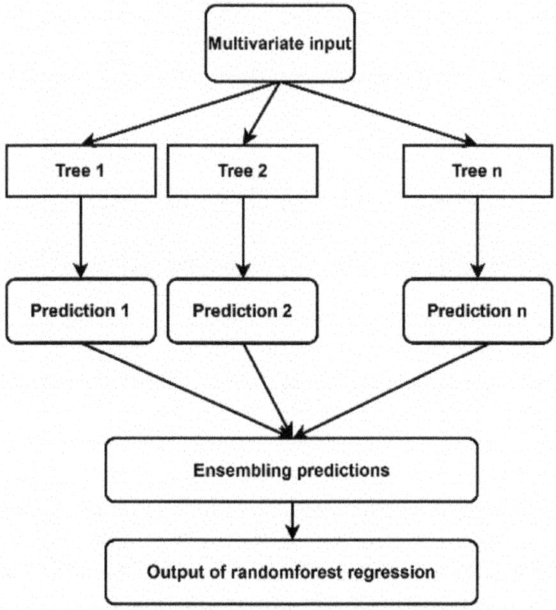

Fig. 4.32 Random forest regression

The above imputation methods work by replacing missing data. In addition to this, the RNN structure can also be modified to utilize missing data. The RNN structure has the same parameters throughout all time-series data, which means that the input data must match the data dimensions of the trained model. But it is known that data loss may occur for any variable and scale, in which case a typical RNN model cannot produce outputs when inputs are missing. To address this problem, the GRU-decay (GRUD) model has been developed for multivariate time-series data having missing sections (Che et al. 2018).

GRU-decay

The GRUD model complements the basic GRU structure with a simple imputation and weight decay mechanism to minimize the effect of any missing data; Fig. 4.33 shows its cell structure. A decay term, γ, is determined from training and used to represent the decrease in the missing data. The decay mechanism with decay term γ is shown in the following equations.

$$\gamma_t = \exp\left(-\max\left(0, W_\gamma \sigma_t + b_\gamma\right)\right) \tag{4.29}$$

$$\hat{x}_t = m_t x_t + (1 - m_t)\left(\gamma_{x_t} x_t + \left(1 - \gamma_{x_t}\right) \breve{x}_t\right) \tag{4.30}$$

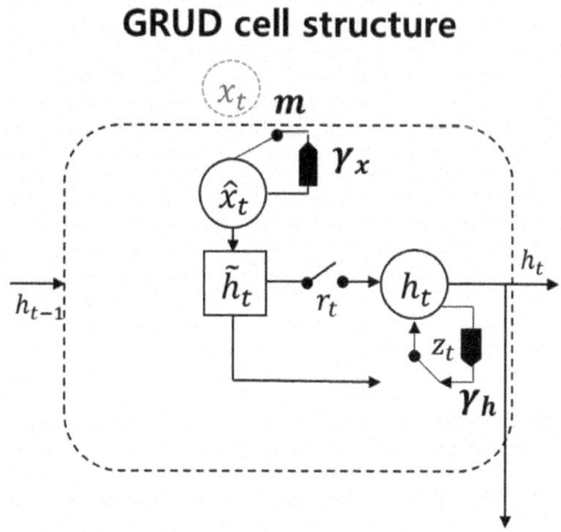

Fig. 4.33 GRUD scheme. Reproduced with permission from Choi and Lee (2020)

$$\hat{h}_{t-1} = \gamma_{h_t} \odot h_{t-1} \tag{4.31}$$

Here, γ_t is the decay term and m is the masking, which gives the missing features of the data. The decay term affects the input and hidden state, which is advantageous in unified designs combining imputation and RNN models performing actual work, such as the accident diagnosis classification of the present model. Based on the trained RNN, imputation manipulates the decays of the missing variables with trained decay rates. Hence, the GRUD model is suitable to generate an RNN-based diagnosis output when specific variables are missing by sensor failure. The masking inputs can mirror the sensor fault monitoring results from the front system.

With these four approaches, two fault-tolerant diagnosis structures are constructed. The first replaces the missing data by performing regression-based imputation via KNN, MICE, or Missforest, which are compared in Sect. 4.4.2.3, and then makes a diagnosis with GRU. The second structure inputs the data with missing sections directly to GRUD for diagnosis. Their diagnosis performance as strategies to mitigate sensor error in simulated NPP accidents is compared in the next section.

4.4.2 *Comparison Results*

4.4.2.1 Data Descriptions

For a proper accident diagnosis, proper situational awareness is required to check the various accident symptoms. Here, the diagnosis procedures guide operators in a standardized way by providing conditional logics as shown in Fig. 4.34, which

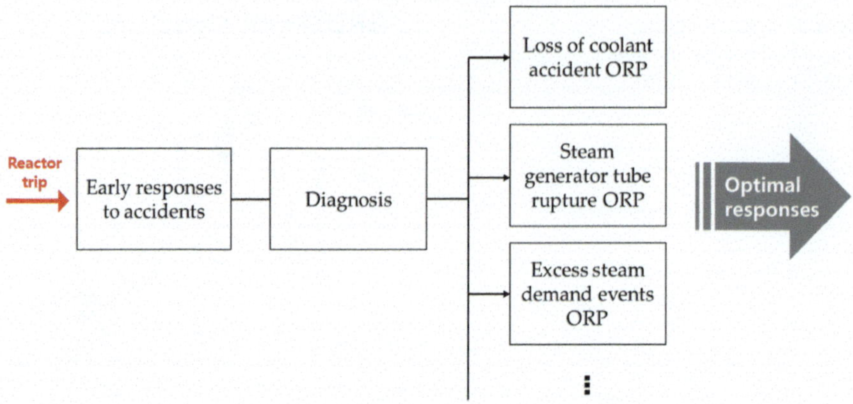

Fig. 4.34 Symptom-based EOP package. Reproduced with permission from Choi and Lee (2020)

includes the early responses to an accident, the diagnosis procedure, and the appropriate ORPs depending on the particular accident (IAEA 2006). It is noted here that this flow is vulnerable to inaccurate process parameters from sensor faults. Even one wrong transition in a procedure could result in a diagnosis failure, which would prevent the optimal response and lead to the commission of unnecessary tasks or potentially harmful actions. It is clear that securing robust and accurate diagnosis of an NPP accident is indispensable for safe operation.

A CNS developed by KAERI of the Westinghouse 990 MWe PWR is used to generate the nuclear accident data. Such simulators are commonly used as data sources for several data-driven ML applications in the nuclear field. The CNS can generate emergency or accident data with options for detailed malfunctions. It is a simplified 1D model with theoretical assumptions and cannot simulate all accident phenomena; however, it can generate large amounts of data in a short time. In this case, from the 2217 process parameters generated by the CNS, 41 parameters are chosen based on the diagnosis procedure in the EOPs. The parameters are also selected to include ones that reflect specific accident symptoms. Data acquisition is set to 900 s referring to accident diagnosis time limits recommended in IAEA safety reports.

All the selected parameters show nonlinear and variable changes in emergency situations following reactor trip and the actuation of various safety systems. In addition, the data includes unexpected phenomena such as parameter oscillation by vaporization, and the diverse accident symptoms differ from the detailed malfunction options. Table 4.15 details the simulated accidents and lists the numbers of datasets.

As shown in the table, five broad NPP accident categories are considered, for which data from nine detailed accident sequences are extracted. A total of 1850 data are divided by severity or break location to generate various accident features, as follows. The LOCA data is classified as small/medium LOCA and large LOCA by break size from a reference, and the PORV LOCA type is added for its distinctive symptoms compared to other LOCA types. The ESDE or MSLB is separated into in- and out-CTMT because each involves different symptoms. RCP failure and RPS

Table 4.15 Number of training and test sets for the nine simulated accidents

Accident type	Detailed accident type	Accident label	No. of training sets	No. of test sets
LOCA	Small/medium LOCA	S/MLOCA	228	72
	Large LOCA	LLOCA	390	126
	PORV LOCA	PORVLOCA	54	17
SGTR	SG tube rupture	SGTR	111	36
ESDE	In-CTMT ESDE	ESDE_IN_CTMT	216	72
	Out-CTMT ESDE	ESDE_OUT_CTMT	186	70
LOAF	LOAF	LOAF	112	38
Reactor trip	RCP failure	RCP fail	50	11
	RPS failure	RPS fail	50	11
Total			1397	453

failure are included as accidents with spurious reactor trips. Among the data, 1397 are randomly selected for training and validation, and 453 are used for testing. For an efficient training of the neural network model, min–max normalization of all the collected maximum and minimum variables in all datasets is performed for the training and test data.

4.4.2.2 Results

At the end of the GRU model, the softmax function generates a numerically normalized output, but in this case there are no exact criteria for identifying the states. Despite the fact that accidents have a wide range of symptoms according to the accident type and scale, RNNs have demonstrated successful accident identification in many studies. For both stable and robust diagnosis algorithm performance, consistently high outputs of the true labels should be generated. In the present case, simple criteria are set for a thorough evaluation of the diagnosis algorithm. It should be noted here that sufficient time for the accident symptoms to appear is needed following the accident occurrence. Previous research showed that sensor faults were detected around an average of 140 s. Based on this, the success criteria for accident diagnosis is assumed as when the true softmax output maintains its maximum value from 200 s to the end of the simulation, or 900 s.

First, the developed diagnosis algorithm with GRU is initially assessed prior to testing with injected sensor errors. The GRU model test results are given in Table 4.166 for classifying the accident type from among 453 test sets with fault-free data. Unstable trends were observed in S/MLOCA and LLOCA test data, which have break sizes near the boundary value. Despite this, the output maintained the true diagnosis across all time sequences. The performance degradation of the diagnosis algorithm is then analyzed for five sensor error modes, giving a total of 2265 test data for each sensor. The sensor error data are generated to target seven process parameters that

Table 4.16 Diagnosis accuracy of GRU with error-injected and no-error sensor data

Unit: %	PZR Pressure	Secondary RAD	CTMT Pressure	Cold Leg #1 Temp	Flow SG to RCP #1	RV Water Level	SG #3 Level
Normal data	100						
Faulty data	92.98	60.04	68.65	83.05	93.20	93.47	87.55

show diverse trends depending on the accident type and significantly influence the diagnosis procedures. In this section, a single sensor error is assumed. Table 4.16 shows the results, namely that sensor error deteriorates the diagnosis performance at varying degrees by the particular error-injected sensor. The largest drop is seen for the secondary radiation sensor error because this parameter is critical to discriminate SGTR from other accident types. Figure 4.35 shows the successful identification of SGTR from normal data along with its diagnosis failure as a consequence of the secondary radiation sensor error.

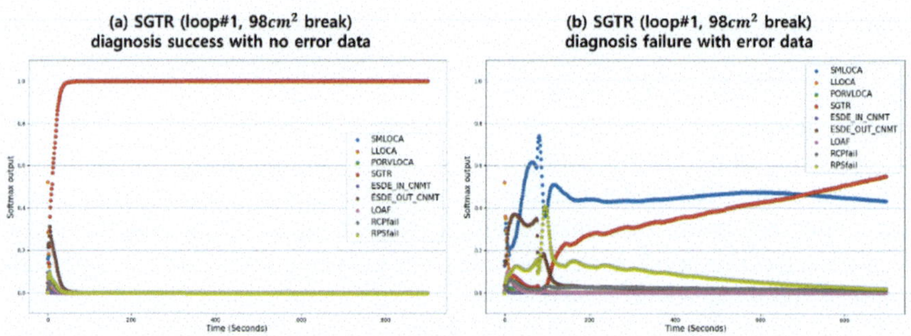

Fig. 4.35 Example trends of **a** accident diagnosis success in the normal data test, and **b** diagnosis failure from error-injected data for the secondary radiation sensor. Reproduced with permission from Choi and Lee (2020)

4.4.2.3 Performance Evaluation of Imputation Models

To select the best imputation model for the diagnosis algorithm, the accuracies of the imputation methods described in Sect. 4.4.1.3 are compared. Mean imputation is also included as a base method for KNN, MICE, and Missforest. Errors in the data reconstructed from the original values are collected in the form of average over time. Computation times are also included in the performance comparison as a crucial factor in accident situations that require rapid responses. While adjusting the model parameters may affect both imputation accuracy and computation time, such adjustments in this case are not a major determiner of model performance. The imputation model parameters, such as the number of nearest neighbors for KNN or the number of decision trees for Missforest, are fixed based on pilot tests that found the best performance in under 20 s of computation time. Assuming only single sensor faults as previously mentioned, the MICE and Missforest methods deal with single fittings from linear and random forest regression in the following tests.

Certain NPP process parameters present zero values, such as specific radiation alarms that are maintained at zero in the absence of a radioactive material leak. Actual values of the sensors should be inserted as the denominator in the error percentage metrics though, such as for the mean absolute percentage error. To cover this problem, the symmetric mean absolute percentage error (MAPE) metric (Armstrong 2010) is applied, which is defined as follows.

$$symmetic\ MAPE = \frac{1}{n}\sum_{t=1}^{n}\frac{2 \cdot |A_t^v - F_t^v|}{|A_t^v| + |F_t^v|} \tag{4.32}$$

In this expression, A_t denotes the actual measured value at time t and F_t denotes the imputed value. Asymmetric issues are known with symmetric MAPE (Goodwin and Lawton 1999), but these are not a concern with actual data of non-negative values. Evaluation with the symmetric MAPE metric is conducted to select the optimum imputation method among mean, KNN, MICE, and Missforest, where each involves a crucial parameter that determines the computation time and imputation accuracy. Imputation model comparisons are based on regressions of the missing sensor values from the same sample data as the training data. Tables 4.17 and 4.18 respectively list the percentage errors and computation times of the comparisons, where 5 variables are randomly selected from the 41 plant parameters (see Sect. 4.4.2.1) to compare the reconstruction performances independent of the features of the variables.

From the results, Missforest is the most stable with the lowest error. MICE demonstrates good performance with Var #4 having a simple pattern, regardless of the accident. This seems to be a characteristic of linear regression. But in the case of Var #2 and #3, which contain constant zero data, MICE performance rapidly declines, as

Table 4.17 Comparative evaluation of imputation accuracy with the symmetric MAPE metric

Unit: %	Accuracy	Variable #1	Variable #2	Variable #3	Variable #4	Variable #5
Mean	Max/Min	79.98/8.82	198.00/108.34	198.81/3.13	82.04/4.75	63.04/2.77
	Average	39.94	195.35	135.97	18.06	30.09
KNN	Max/Min	34.35/0.01	82.73/0.00	76.96/0.00	56.00/0.00	21.15/0.00
	Average	3.93	1.94	3.25	4.21	2.50
MICE	Max/Min	22.65/2.79	198.87/11.18	198.41/3.26	6.34/0.0042	51.80/0.86
	Average	7.63	177.95	95.91	0.21	6.80
Missforest	Max/Min	14.93/0.076	44.61/0.00	38.27/0.00	23.67/0.0008	17.99/0.00
	Average	1.68	1.41	0.27	0.77	1.48

Table 4.18 Total average sMAPE and computation time of the imputation methods

	Mean	KNN	MICE	Missforest
Symmetric MAPE	83.88	3.17	57.70	1.12
Computation time	0.0198 s (SD = 2.45 × −4)	17.31 s (SD = 1.24 × −1)	4.53 s (SD = 1.53 × −1)	9.06 s (SD = 2.33)

shown in Fig. 4.36d. While KNN shows good performance in min error, it is unstable with unusual peaks. Considering the computation time, which as previously stated is a critical factor for applicability to real NPP emergencies, MICE shows an average computation time of under 5 s, Missforest about 9 s, and KNN about 17 s. Missforest presents a range of computation times among the variables because the computation time in Missforest derives from the number of branches, and thus the regression of sensor values with diverse trends like Var #4 take longer to calculate. Taken together, as Missforest-based imputation presented the best performance among the tested methods considering mean error, maximum peak error, and affordable computation time, it was chosen as the imputation tool to replace the missing data from sensor faults for signal reconstruction in the diagnosis algorithm.

4.4.2.4 Fault Mitigation Results

The selected imputation method, Missforest, is now compared to GRUD by applying unreliable sensor data to test the sensor fault mitigation strategies. Each imputation method is tested with a test data set containing seven sensor errors. As shown in Table 4.19, GRUD achieves notable performance recovery from the error states. In this case, its recovered diagnosis accuracy is directly affected by the level of degraded accuracy. For instance, the lowest accuracy among the faulty data for 'Secondary radiation' is related to the lowest recovered accuracy in the result. As for Missforest, all sensor errors are recovered to 100% diagnosis accuracy. Figure 4.37 plots the performance degradations and recovered diagnosis accuracies of the two methods for the seven sensor errors.

To summarize the results shown in Table 4.19, we first confirm that the base GRU diagnosis algorithm could successfully diagnose the 453 test data at an assumed threshold. Injecting sensor errors results in a diagnosis accuracy drop to an average of 82.71%, i.e., the failure of 2742 cases out of the total 13,113 test data. The GRUD-based fault-tolerant strategy is then applied, which achieves a notable accuracy recovery to 96.75%, with 103 failure cases out of the total 3171 test data. As a second strategy, Missforest achieves a complete diagnosis accuracy recovery, or in other words 0 failures out of the 3171 test data.

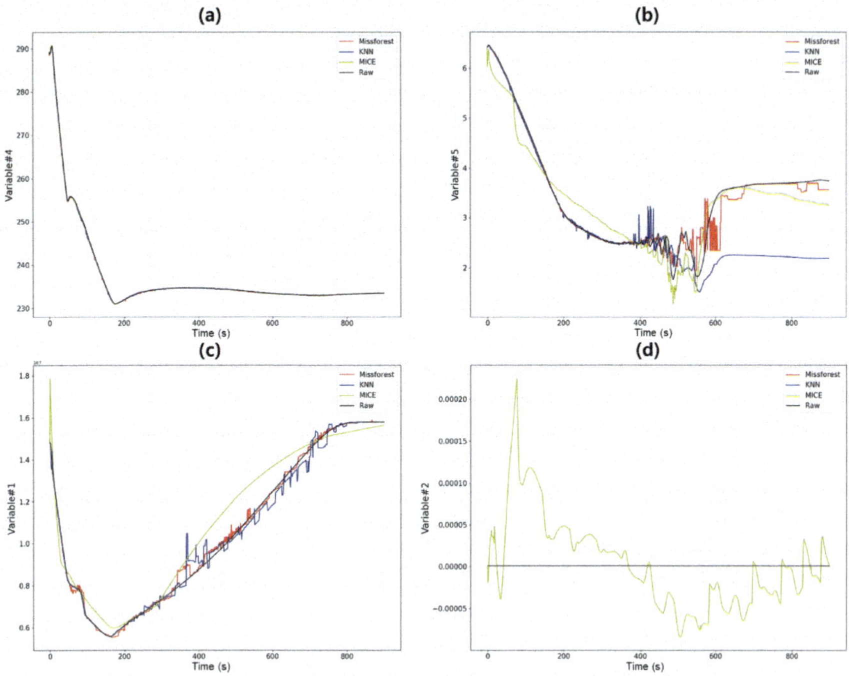

Fig. 4.36 Comparison of imputation results showing **a** fine imputation by all models for a variable #4 fault, **b** the maximum error of K-nearest neighbors (KNN) for a variable #5 fault, **c** accurate performance of Missforest for a variable #1 fault, and **d** the maximum error of multivariate imputation with chained equations (MICE) for a variable #2 fault. Reproduced with permission from Choi and Lee (2020)

Table 4.19 Total diagnosis accuracies

	Normal data	Faulty data	GRUD	Missforest
Total	100% (453/453)	82.71% (13,113/15,855)	96.75% (3068/3171)	100% (3171/3171)

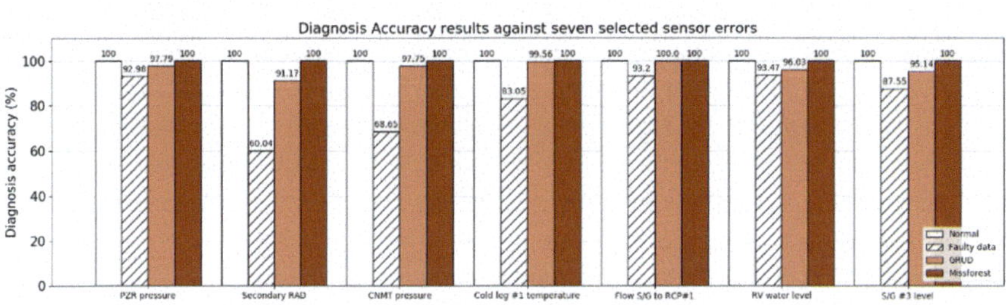

Fig. 4.37 Diagnosis accuracy results of seven sensor errors with error-free data (white), error-injected data (dashed), and the application of GRUD (orange) and Missforest (brown) to the error-injected data. Reproduced with permission from Choi and Lee (2020)

4.4.3 Considerations for Optimal Sensor Fault Mitigation

Several models for NPP accident identification are being actively researched, for which faulty input data from sensor networks should be considered. One notable finding in this section is that the GRU-based diagnosis algorithm is not robust against injected sensor faults, as its diagnosis accuracy dropped to about 80% from sensor errors in the performance test. Moreover, the sensor features largely determine both the performance degradation and the recovery. Table 4.16 highlights that one specific sensor parameter can play a crucial role in distinguishing two different accident types, with sensor errors of such types resulting in a significant degradation in the accuracy of the accident diagnosis system. More specifically, secondary radiation is a key factor for discriminating SGTR from LOCA, while CTMT pressure divides in/out CTMT ESDE accidents. Sensor error tests in these cases show performance degradations down to 60.05% and 68.65% accuracy. Diagnosis algorithm applications therefore require dedicated error mitigation strategies to compensate such cases.

The Missforest mitigation strategy achieves a complete recovery with 100% diagnosis accuracy, while GRUD reaches 96.75% accuracy. Despite this lack of complete recovery, GRUD offers advantages in computation time and code complexity. The GRUD structure includes a decay mechanism, meaning only simple imputations need to be computed. Conversely, Missforest takes an average of 9.06 s (Standard deviation = 2.552) of computation to generate the imputed data. In an emergency situation, this longer calculation time could potentially delay the proper mitigation measures. For application of the sensor fault-tolerant diagnosis system in real NPPs, the computational time and performance trade-off needs to be examined in more detail.

The developed fault-tolerant diagnosis system, as an advisory support system work, is designed to provide operators with accurately identified information during an accident situation as well as abnormal situations and start-up and shutdown operations. It also has potential for adoption in other industries requiring process parameter-based reactions that are sensitive to sensor faults.

For further improvement of the diagnosis performance of the GRUD-based system, the GRUD structure should be developed further. For example, multi-directional and bi-directional RNNs enable models to consider reverse directional or row-wise contextual situations with decay mechanisms for any missing data. In addition, the present system assumes error data from a single sensor fault only with no operator actions in the emergency situation. But in reality, simultaneous sensor errors may occur, and actions by human operators based on judgments at any given moment can influence the process parameters. A larger database therefore needs to be gathered for more comprehensive model training and testing.

4.5 Diagnosis of Multiple Accidents with a GNN

NPPs provide operators with detailed procedures to efficiently and accurately respond to accidents while reducing human error. There are two types of emergency situations that can occur during NPP operation. The first group represents single accidents, such as a reactor trip, LOCA, SGTR, ESDE, LOAF, LOOP, and SBO. In such single-accident cases, the ORP corresponding to the accident is performed. The second group represents multiple accidents, for which accident diagnosis becomes difficult because the plant behavior does not permit a definitive diagnosis. Rather than an ORP, a functional recovery procedure (FRP) that ensures the integrity of the CSFs is performed in this case, without directly responding to the accident.

With a quick and accurate determination of whether an emergency is a single or multiple accident situation, the FRP can be quickly selected in case of multiple accidents, thereby providing additional time for operator response. Accident management can likewise be improved if operators can clearly identify the particular combination of accidents, which are typically difficult to distinguish because of the similar effects they have on the plant behavior. To support operators in various accident situations, a diagnosis system or agent requires a high diagnosis resolution. And as the accident may distort the measurement values, performing the diagnosis with as little data as possible is recommended.

This section introduces a method achieving high diagnosis resolution with limited measurement variables. For increased diagnosis resolution and an optimized amount of required data for the neural network agent, the developed algorithm is based on a GNN.

4.5.1 GNN

The basic structure and algorithm of the GNN are described in Sect. 2.4.4. In this section, its major characteristics and the network structure employed in the current model development are mainly discussed.

4.5.1.1 GNN Details

A type of ANN structure, the GNN was first proposed by Scarselli (2008). While the CNNs and RNNs applied in various fields take vectors or matrices as inputs, the input for GNNs is a graph structure. With matrices used as inputs, CNNs are well suited for data in Euclidean space, like images or text, but conversely they are not well suited for data in non-Euclidean space. On the other hand, GNNs are applicable to data in a non-Euclidean space as local connections can be represented in a graph structure. Figure 4.38 illustrates the CNN and GNN structures for comparison.

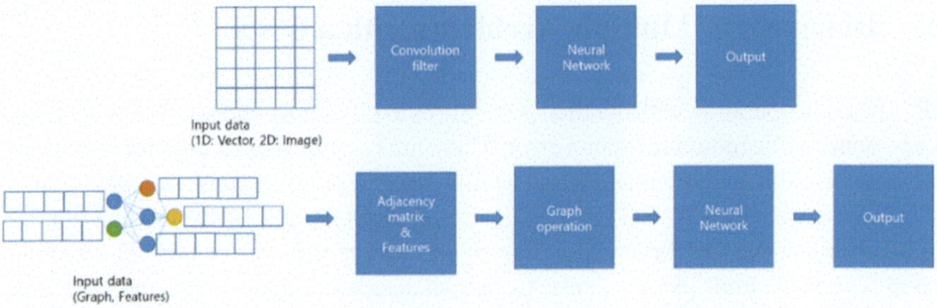

Fig. 4.38 Schematic diagrams of CNN and GNN structures. Reproduced with permission from Chae et al. (2022)

In a CNN, a convolution filter is used to extract data features in Euclidean space, and the output of this, called a feature map, is used to train the neural network. By modifying the size of the convolution filter, CNNs can be applied to various data types. In a GNN, both the graph structure and the features of the nodes are used. Here, the graph structure in the form of a matrix, called the adjacency matrix, is input to the neural network. The basic GNN algorithm can be described as follows. The state of a vertex ($\mathbf{h_v}$) is a function of the four following elements: the features of node v ($\mathbf{x_v}$), features of the edges ($\mathbf{x_{co[v]}}$), states of neighboring nodes ($\mathbf{h_{ne[V]}}$), and features of neighboring nodes ($\mathbf{x_{ne[v]}}$), as shown in Eq. (4.33).

$$\boldsymbol{h}_v = f\left(\boldsymbol{x}_v, \boldsymbol{X}_{co[v]}, \boldsymbol{h}_{ne[v]}, \boldsymbol{x}_{ne[v]}\right) \tag{4.33}$$

The output of vertex v is obtained using the output function g, the inputs for which are the node state ($\mathbf{h_v}$) and node features ($\mathbf{x_v}$). The output can be written as in Eq. (4.34).

$$\boldsymbol{o}_v = g(\boldsymbol{h}_v, \boldsymbol{x}_v) \tag{4.34}$$

Loss can then be defined as the difference between the target value t and the output of function g.

$$\text{loss} = \sum t_i - o_i$$
$$\text{where } i \text{ is the number of target data} \tag{4.35}$$

These f and g functions are modeled using a neural network. The training of a GNN is conducted by updating the weights of functions f and g according to the loss. Zhou et al. describes various related algorithms in detail (Zhou et al. 2020).

4.5.1.2 Spectral Graph Convolution Neural Network (Spectral GCN)

To improve the diagnosis accuracy of the agent with a small dataset, the spectral GCN model is adopted, which stands for spectral graph convolution neural network. First proposed by Bruna (Bruna et al. 2013), the spectral GCN applies a convolution layer before the neural network layers. While its overall shape is similar to that of a conventional CNN, a spectral GCN can receive inputs in the form of a graph, where the filter g is assumed as a learnable parameter. In addition, spectral GCNs can work with multi-channel inputs such as images. From Eq. (4.33), the operation in the convolution layer in this case can be defined as in Eq. (4.36).

$$H^{k-1}_{:,j}(\text{Output from k} - \text{th hidden layer}) = \sigma\left(\sum_{i=1}^{f_{k-1}} U\Theta^{\mathrm{k}}_{i,j}U^T H^{k-1}_{:,i}\right) \tag{4.36}$$

where, σ is activation function,

Θ is convolution filter which is form of diagonal matrix filled with learnable parameters.

H is input signal from k–1 th node $(\mathbf{H}^0 = \mathbf{X})$.

Employing a spectral GCN instead of a conventional CNN grants the following advantages.

1. Grid structures (i.e., non-Euclidean data) can be analyzed
2. More abundant relationships between different nodes can be expressed.
3. The entities properties in the graph can be considered differently

As shown by Eq. (4.36), not only the input signal (**X**) but also the matrix from the graph ($\mathbf{U}$, $\Theta^{\mathrm{k}}_{\mathrm{i,j}}$, U^{T}) are utilized in the calculation of the output signal with a GNN.

4.5.2 GNN-Based Diagnosis Algorithm Representing System Configuration

In the diagnosis algorithm, the structure of an NPP control system is graphed and provided to the neural network as an additional feature. The form of such a system in an NPP is presented in such a way that it does not structurally change during operation. Figure 4.39 shows a basic schematic of the shape of an NPP.

The typical power plant is separated into primary and secondary sides. The primary side is a closed loop in which the heat from the reactor increases the temperature of the coolant (water), the RCP circulates the water, and heat is exchanged in the SG. The secondary side also forms a closed loop, in which heat is exchanged in the SG, the feedwater pump circulates the water, and the steam is cooled in the condenser. Any problem with the PZR will be delivered to the reactor immediately, but related information will be delivered to the relatively distant condenser in a slow or attenuated manner. As such, the operating variables of actual NPPs are not

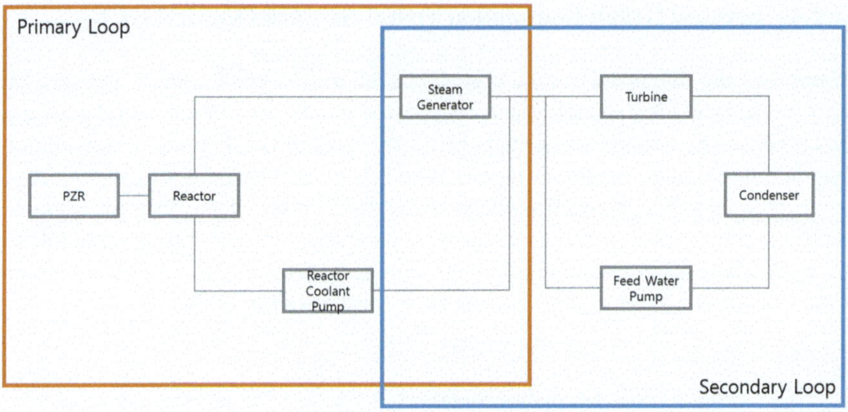

Fig. 4.39 Schematic diagram of an NPP. PZR: pressurizer. Reproduced with permission from Chae et al. (2022)

independent at each data point, but rather form a dataset in non-Euclidean space with relationships between data points. To represent these non-Euclidean characteristics, the operating variables of the power plant are expressed in graph form, as shown in Fig. 4.40, based on the shape of the plant.

The system configuration determines the graph, as follows.

1. All measured values of a given component are interconnected
2. When connecting components, measurements of the same type are connected
3. If it is not possible to connect measurements of the same type, all variables are connected.

For instance, the PZR has measured values of level, pressure, and temperature, which are all connected to each other by a first-order neighbor. A flow measurement value is connected with the other flow measurement values, and a pressure measurement value is connected with the other pressure measurement values. In the case of the RCP, because it only measures the flow rate, both the pressure and level of the

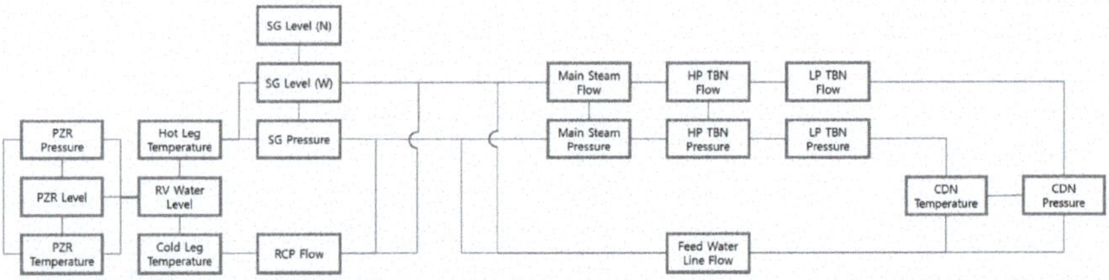

Fig. 4.40 Graph of relationships between operating variables. PZR: pressurizer, RCP: reactor coolant pump, SG: steam generator, HP: high pressure, LP: low pressure, TBN: turbine, CDN: condenser. Reproduced with permission from Chae et al. (2022)

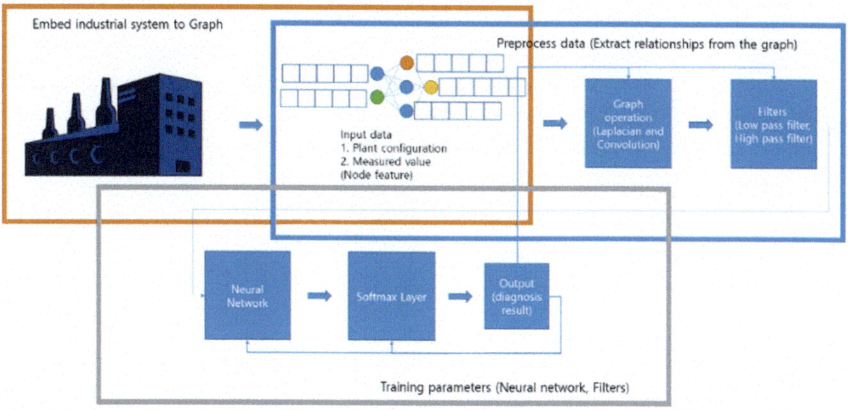

Fig. 4.41 Diagnosis agent. Reproduced with permission from Chae et al. (2022)

SG are connected to the RCP. Based on the graph in Fig. 4.40, an adjacency matrix is constructed for use as the GNN input.

As mentioned in the previous section, the current ANN structure is based on a spectral GCN, which receives graphs as inputs and passes them through various filters that convert the data into useful information for training and automatically process outliers. The filter variable is a training parameter of the network. Figure 4.41 shows a schematic of the developed diagnosis agent model.

The agent has three stages. The first stage is to transform the system configuration to graph form. The main output of the first stage is a node, edge relations in the form of an adjacency matrix, and measured value. The NPP is simplified as in Fig. 4.40 to test the accident diagnosis capability of the agent. The second stage is to preprocess the data, similar to a CNN, where Laplacian and convolution operations are conducted to extract more relation information from the graph. After this convolution operation, high pass and low pass filters are applied to remove signals with high and low frequency ranges that result from the Fourier transform. The preprocessed input then goes to the neural network, or third stage, which compares the network output to the actual diagnosis result and calculates the loss.

Among the two filters used in the model, the first is a low pass filter described by Eq. (4.37).

$$F_1(\lambda) = \left(1 - \frac{\boldsymbol{\lambda}}{\lambda_{max}}\right) \tag{4.37}$$

where λ is an eigenvalue and λ is a set of eigenvalues

The second filter is a high pass filter described as follows. The eigenvalue of each filter is calculated by graph convolution.

$$F_2(\lambda) = \left(\frac{\lambda}{\lambda_{max}}\right) \tag{4.38}$$

where λ is an eigenvalue and λ is a set of eigenvalues

The approach to graphing the system in this section is not only applicable to NPPs but other industrial plant systems such as chemical plants. The system in graph form is given as an input to the neural network, and GNN-based training is performed to provide the graph as a feature.

4.5.3 Experiments

For a performance assessment of the GNN-based accident diagnosis agent, we compared its performance with that of a CNN. To check for a sufficient diagnosis resolution of the methodology to identify multiple accidents, three different diagnosis tasks are tested. The first scenario attempts to diagnose two different single accidents, and the second scenario attempts to diagnose two types of single accidents in addition to a multiple accident combining the two single accidents. The third scenario attempts to diagnose six types of accidents, namely three different multiple accidents combining different sets of single accidents and three types of single accidents.

4.5.3.1 Data Acquisition and Preprocessing

While the use of real data from actual power plants would be preferable for verification of the agent, such data are limited considering the scarcity of NPP emergency situations. Instead, a KAERI-developed CNS is used to generate data. The back-end of the simulator is built on the SMABRE code, and the reference plant is a Westinghouse three-loop 900 MW PWR.

The DBAs included in the tests are a reactor trip, LOCA, SGTR, ESDE, LOAF, LOOP, and SBO. Data are collected for the pipe breakage accidents having similar behaviors (LOCA, SGTR, and ESDE). As multiple accidents, data on LOCA + SGTR, LOCA + ESDE, and SGTR + ESDE are also collected. For the LOCA, the situation is assumed to result from a stuck-open PZR safety valve (PSV), as a PSV-LOCA is more likely to be misdiagnosed than a LOCA caused by a general pipe breakage (Kim et al. 2003).

The 19 variables shown in Fig. 4.40 are collected as measured variables, which are also listed in Table 4.20. Descriptions of the data acquired for the different accident scenarios are as follows.

Single Accidents

- PSV-LOCA: valve position [1%, 2%, …, 100%]

  i.e., PSV-LOCA with 1% valve stuck open, PSV-LOCA with 2% valve stuck open, …

  → 100 scenarios.

Table 4.20 Measured variables

Component	Measured variable				
PZR	Pressure	Level	Temperature		
Hot leg	Temperature	–	–	–	–
Cold leg	Temperature	–	–	–	–
RV	Level	–	–	–	–
RCP	Flow	–	–	–	–
SG	Level (Narrow)	Level (Wide)	Pressure	Main steam flow	Main steam pressure
HP turbine	Flow	Pressure	–	–	–
LP turbine	Flow	Temperature	–	–	–
Condenser	Temperature	Pressure	–	–	–
Feed water pump	Flow	–	–	–	–

- SGTR: tube rupture size [10 cm^2, 20 cm^2, …, 2000 cm^2]

  i.e., SGTR with 10 cm^2 tube rupture, SGTR with 20 cm^2 tube rupture, …
  → 200 scenarios.

- ESDE: steam line break size [10 cm^2, 20 cm^2, …, 2000 cm^2]

  i.e., ESDE with 10 cm^2 steam line break, ESDE with 20 cm^2 steam line break, …
  → 200 scenarios.

Multiple Accidents

- PSV-LOCA + SGTR: [1%, 11%, …, 91%] + [10 cm^2, 110 cm^2, …, 1910 cm^2]

  i.e., 1% PSV-LOCA and 10 cm^2 SGTR, 1% PSV-LOCA and 110 cm^2 SGTR, …
  10 PSV-LOCA conditions and 20 SGTR conditions
  10 × 20 → 200 scenarios.

- PSV-LOCA + ESDE: [1%, 11%, …, 91%] + [10 cm^2, 110 cm^2, …, 1910 cm^2]

  i.e., 1% PSV-LOCA and 10 cm^2 ESDE, 1% PSV-LOCA and 110 cm^2 ESDE, …
  10 PSV-LOCA conditions and 20 ESDE conditions
  10 × 20 → 200 scenarios.

- SGTR + ESDE: [1%, 11%, …, 91%] + [10 cm^2, 110 cm^2, …, 1910 cm^2]

  i.e., 10 cm^2 SGTR and 10 cm^2 ESDE, 10 cm^2 SGTR and 110 cm^2 ESDE, …
  20 SGTR conditions and 20 ESDE conditions
  20 × 20 → 400 scenarios.

For equal amounts of data, augmentation is performed to create 400 scenarios per case. The raw data are normalized using the formula in Eq. (4.39), and 90%

(10%) of the data is used for training (testing). To prevent overfitting, the tenfold cross-validation technique is applied.

$$\mathbf{X}_{\mathbf{Normalized}} = \frac{\mathbf{X} - \mathrm{X}_{\min}}{\mathrm{X}_{\max} - \mathrm{X}_{\min}} \tag{4.39}$$

In the CNN data preprocessing, the data are arranged in 2D as $\mathbf{R}^{19\times300}$. First, the data extracted from each component at a given time is in the form $\mathbf{R}^{1\times300}$. Then attaching this data in parallel produces input data expressed in the form of a matrix, $\mathbf{R}^{19\times300}$. The vertical axis represents the measured variable, and the horizontal axis represents the time step. Figure 4.42 illustrates the form of the CNN input data, and Table 4.21 lists the model hyperparameters.

In the GNN data preprocessing, the data are separated into two different types. Datasets are constructed by dividing the adjacency matrix expressing the correlation between the measured variables and the node features associated with the measured variables. The adjacency matrix takes the form $\mathbf{R}^{19\times19}$, and the measured variables take the form $\mathbf{R}^{19\times300}$. Figure 4.43 illustrates the form of the GNN input data, and Table 4.21 lists the model hyperparameters. While the CNN only handles the measured variables, the GNN utilizes both adjacency matrix and measured variables (Table 4.22).

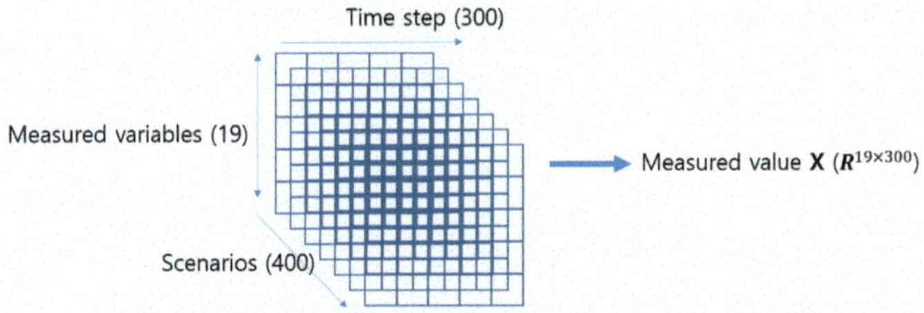

Fig. 4.42 Inputs for the CNN. Reproduced with permission from Chae et al. (2022)

Table 4.21 Hyperparameters of the CNN

Hyperparameter	Diagnosis agent
Number of convolution layers	2(5 × 5 × 32, 5 × 5 × 64)
Number of hidden layers	2
Activation function of convolution layers	ReLU
Activation function of hidden layers	ReLU (first layer), softmax (second layer)
Number of neurons in hidden layers	1024
Input dropout	10%
Optimizer	Adam (Kingma and Ba 2014)
Learning rate	5e-4

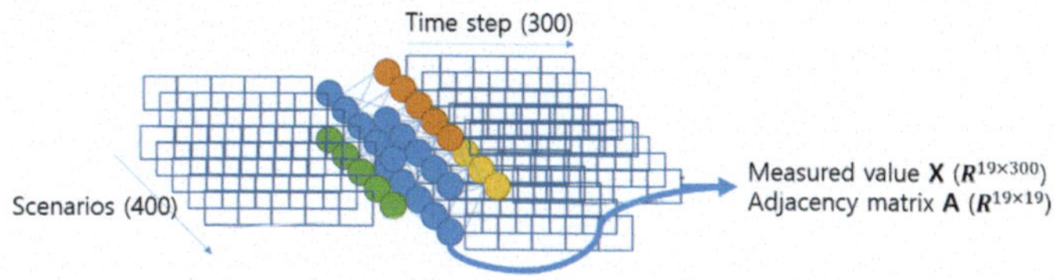

Fig. 4.43 Inputs for the GNN. Reproduced with permission from (Chae et al. 2022)

Table 4.22 Hyperparameters of the GNN

Hyperparameter	Diagnosis agent
Activation function of hidden layer	ReLU
Activation function of output layer	Softmax
Biases of hidden layer	False
Biases of output layer	True
Input dropout	10%
Weight decay	10%
Optimizer	Adam (Kingma and Ba 2014)
Learning rate	10e-5

4.5.3.2 Experiment Results

The performance of the two diagnosis agents is analyzed for three cases, as follows: case 1 is the diagnosis of two accidents, LOCA and SGTR; case 2 is the diagnosis of three accidents, LOCA, SGTR, and LOCA + SGTR; and case 3 is the diagnosis of six accidents, LOCA, SGTR, ESDE, LOCA + SGTR, LOCA + ESDE, and SGTR + ESDE.

In case 1, both the CNN and GNN show good diagnosis performance, as seen in Fig. 4.44. This is mainly because LOCA and SGTR have relatively clear characteristics, and thus it is relatively easy to distinguish them.

In case 2, Fig. 4.45 shows that while the CNN can identify the individual accidents with good performance, its diagnostic success probability for the multiple accidents (LOCA + SGTR) lowers to approximately 78%. Examining the successful and unsuccessful results reveals that most of the failures were diagnosed as SGTR when combining small-break LOCA and large-break SGTR, or as a single LOCA when combining small-break SGTR and large-break LOCA. In contrast, the GNN results confirm that the agent can identify the three accidents with nearly perfect accuracy from the correlations among the 19 limited variables.

As for case 3, Fig. 4.46 shows a diagnosis accuracy of approximately 40% for the CNN. With a more complex network and more than 19 measured variables, the CNN may achieve good performance, but the diagnosis of multiple accidents is almost impossible with this network in the current setup. On the other hand, the GNN exhibits an accuracy of 98% in the diagnoses of the single and multiple

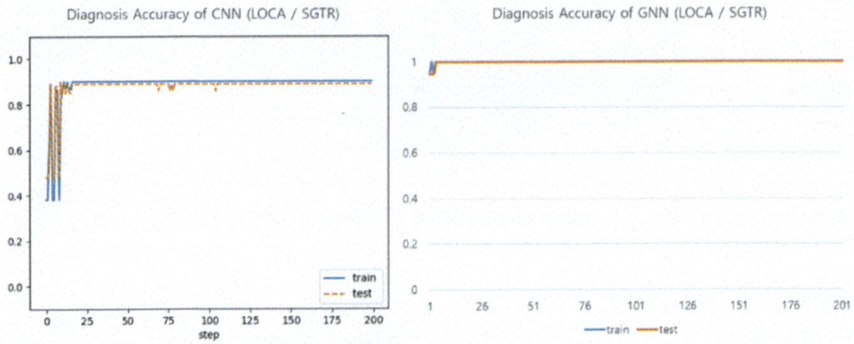

Fig. 4.44 Diagnosis accuracy of the CNN and GNN for case 1. Reproduced with permission from Chae et al. (2022)

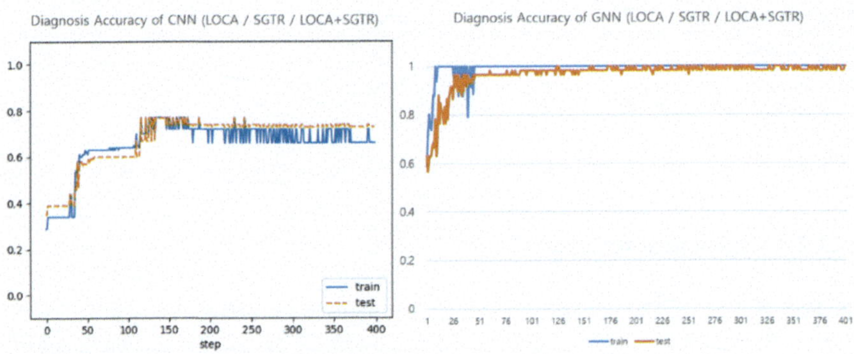

Fig. 4.45 Diagnosis accuracy of the CNN and GNN for case 2. Reproduced with permission from Chae et al. (2022)

accidents using only 19 measurement variables. According to Lee (Lee et al. 2021), reaching above 90% accuracy with a CNN requires around 1000 measured variables to train the neural network. The results of the current tests confirm that a significant improvement in performance can be achieved when the structure of the system is represented in the neural network training.

To support operators in various situations, the diagnosis agent should have a high diagnosis resolution. Moreover, as the accident progression may affect the measured values, it is desirable to consider as few measurements as possible in the diagnosis. The developed GNN-based diagnosis method addresses both of these points.

One novelty of this method is the efficient training of the neural network and the quick graphing of the form of the system, which has not been used before. To date, it has been difficult to know the importance of the data and what kinds of relationships between the data can be found when training a neural network. Such preprocessing is then completely delegated to the network, resulting in very low training speeds. To reduce the required variables, human intuition is also used,

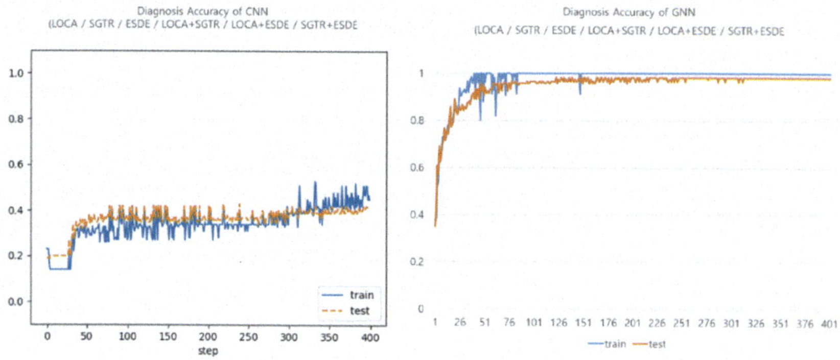

Fig. 4.46 Diagnosis accuracy of the CNN and GNN case 3. Reproduced with permission from Chae et al. (2022)

where experts with a high understanding of the system can efficiently select the data. However, non-experts with a lower understanding of the given system have difficulty in data selection. By providing the neural network with an intuitive physical form of the system as a hint, training that is faster and more efficient can be achieved. The physical system form used in this methodology maps only the physical connections between the components, and thus it can be easily modeled by not only experts but also beginners. With this connection data in addition to the measured variable data, diagnosis resolution dramatically improves. The successful diagnosis results of multiple complex accidents, which has traditionally been difficult to perform due to resolution issues, indicate that transforming physical systems into graph form for use in training can result in more efficient and high performance neural network analysis of industrial systems.

4.6 Interpretable Diagnosis with Explainable AI

4.6.1 Need for Interpretable Diagnosis

In the previous sections, several algorithms for diagnosing abnormal events and severe accidents in NPPs have been introduced. Many AI models studied in this way have the purpose to support the diagnostic tasks of operators to reduce human error and increase NPP safety. However, it is contradictory to apply technology to increase the safety of NPPs that has not been proven to be safe. Therefore, proper validation of a developed AI model is required to apply the model as an actual operator support system in NPPs, for which safety is the primary concern.

One area that validation needs to cover is the potential biases that may occur in the training of NPP accident diagnosis models. In addition, if a trained model incorrectly diagnoses the NPP state, it should be able to analyze the errors and reflect feedback.

Therefore, interpretation of the AI model forming the basis of the NPP state diagnosis system is essential.

Another area to consider for the application of such an AI-based NPP state diagnosis system for operator support is as follows. When an abnormal situation occurs in an NPP, operators in the MCR recognize the problem, make a diagnosis of the plant status, and carry out the appropriate operating procedures. Following all the diagnosis tasks, the subject who makes the final diagnosis is the operator. A support system providing operators with the diagnosis results of the model in the course of the diagnosis tasks can be confusing if the support information is unreliable. Another potential problem is operators relying on unproved results (Lee and Seong 2007). In other words, it is uncertain whether providing only the diagnosis results of the model directly to the operators can support their tasks and decisions. Additional ways to take these issues into account need to be explored.

An AI model that has a high level of interpretability can be used in various ways. For example, a close analysis of the causes of the model diagnosis results can provide additional information about the diagnosis to operators, which can increase the reliability of the information and therefore the model results. By internally determining the appropriateness of the provided model diagnosis and deciding whether to accept it, operators can increase their task efficiency while maintaining the diagnosis subject. In order to apply an NPP state diagnosis model as an actual operator support system, it is necessary to utilize interpretable AI.

4.6.2 Explainable AI

The developed AI models in this chapter have tried to solve difficult classification problems following various approaches. They represent algorithms that conduct classification through a simple tree structure as well as more complex algorithms that conduct classification through an ANN structure. As the classification performance of a model improves, so does its complexity; in other words, the complexity of a model is closely related to its performance. However, the more complex the model is, the more difficult it is for humans to analyze the model. This trade-off relationship between model performance (accuracy) and interpretability is shown in Fig. 4.47 (Duval 2019).

Since the ANNs being actively studied as classification models have complex structures, it is difficult to explain their results logically. This is because the logical development of the relationship between the input and output with all the various weights for the numerous nodes inside the ANN is unclear. For the same reason, when an ANN model gives an incorrect classification result, the cause of the error cannot immediately be found within the internal structure. This makes it challenging to correct the internal structure to improve the model. As such, ANN models present a type of black-box problem. To solve the black-box problem of various models, including ANNs, numerous model interpretation techniques are under research.

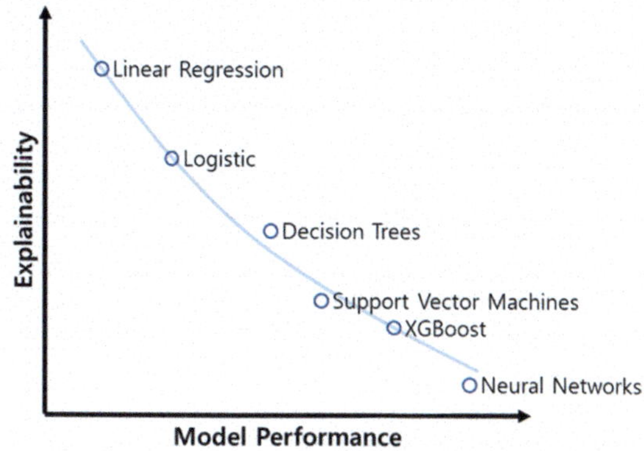

Fig. 4.47 Relation between model performance and model explainability

Explainable AI refers to AI that can convey information on its actions and judgments in a form that humans can understand. As stated above, any AI-based classification model applied to real industrial systems must be sufficiently safe and reliable, and if any problem occurs in the system, feedback on the system should be possible by identifying the cause of the problem. Explainable AI can facilitate these points by solving the black-box problem, thereby increasing the likelihood of adoption in real industries.

4.6.3 *Examples of Explanation Techniques*

Support systems such as those adopting NPP state diagnosis models are designed for human operators as the end-users. Therefore, when the model diagnoses the current situation as a specific NPP state, the model should give operators the highest level of understanding as possible. Techniques developed to explain the classification results of AI models work by calculating the relevance of the input values, such as pixels in images or parameters in table data, to the classification. The calculated relevance can be expressed as a heatmap or ranking; each technique takes a different approach to describing the AI model results. The following sections introduce some techniques developed to describe AI models.

4.6.3.1 DeepLIFT (DL Important FeaTures)

The DL important features or DeepLIFT technique defines a contribution score C as in Eq. (4.40) (Shrikumar et al. 2017). Here, t is the node to be interpreted and x_i is the node that affects the interpreting target node.

$$\sum_{i=1}^{n} C_{\Delta x_i \Delta t} = \Delta t \tag{4.40}$$

In this equation, Δt and Δx_i are the amount of change in t and x_i, respectively, compared to the reference state. The reference state is a domain-specific value that is defined to calculate the contribution score by comparing it with judgment.

4.6.3.2 SHAP (SHapley Additive exPlanations)

Shapley values are a means of interpreting the contributions of features that make up the inputs through the difference in values when the feature to be calculated is included and when it is predicted from all possible combinations of the other features (Strumbelj and Kononenko 2010; Kuhn and Tucker 1953). This method has limitations in application to ANNs with deep structures, as the calculations are performed for all possible combinations of features connected to the input of a specific node.

To address this, Deep SHAP refers to calculation using the Shapley values of a multiplier between the output layer and the previous hidden layer. The multiplier is the contribution defined above divided by Δx_i. This can be interpreted as a kind of gradient. In order to interpret the deep structure of an ANN, Deep SHAP calculates the effect of the input on the output through one-pass backpropagation up to the input layer after calculating the Shapley value only for the output layer and the previous layer, considering it as a contribution score.

4.6.3.3 Grad-CAM

Gradient-weighted Class Activation Mapping (Grad-CAM), one of the explanation methods for CNNs (Selvaraju et al. 2017), calculates the contribution score of each feature using information about the gradient of the last convolutional layer. Its related terms are as follows. A^k is the k-th feature map of the last convolutional layer, as in Eq. (4.41) to calculate a_k^c using the partial derivative of the A^k of the input value of the softmax layer.

$$\alpha_k^c = \frac{1}{Z}\sum_i\sum_j \frac{\partial y^c}{\partial A_{ij}^k} \tag{4.41}$$

The a_k^c term represents the importance of A^k. Using this, Grad-CAM can be calculated through Eq. (4.42).

$$L_{Grad-CAM}^c = ReLU\left(\sum_k \alpha_k^c A^k\right) \tag{4.42}$$

This equation, as the result of Grad-CAM, is derived by applying the ReLU activation function. In other words, in order to ignore the features necessary to determine negative labels, the negative contribution is set to zero and the contribution is evaluated only for positive values.

Grad-CAM can highlight which features contribute to the model classification. However, in the case of simulator data, this technique may place a parameter in a highly contributing location by the difference of just one pixel. Therefore, the resolution level of the class activation map from the explanation method is considered highly important. Prior research into interpreting an image classification model showed that it is possible to obtain a higher resolution heatmap by multiplying the Grad-CAM and the guided backpropagation results, referred to as guided Grad-CAM. A more accurate explanation is expected for a given model by confirming the result of guided Grad-CAM in this way.

4.6.4 Application to an Abnormal Event Diagnosis Model

This section demonstrates the application of explainable AI to a simple NPP abnormality diagnosis. As the diagnosis model, the two-channel CNN introduced in Sect. 4.1 is adopted for training simple NPP abnormal state datasets. The trained model diagnoses the test datasets representative of each abnormality, after which the diagnosis results are interpreted through two model explanation techniques: DeepLIFT + SHAP and guided Grad-CAM. The explanation results allow users to determine the contribution of all the NPP monitoring parameters included in the input data to the diagnosis of each state.

4.6.4.1 Datasets

The abnormal events considered in this application are shown in Table 4.23. The previously introduced 3KEYMASTER full-scope simulator produces scenarios consisting of 10 abnormal event situations (Western Service Corporation 2017). For the training dataset, 20 data files per abnormal event (200 data files total) are acquired. All data are obtained by injecting malfunctions specific to the abnormal event with varying intensities. The test dataset has a total of 10 data files with median values in each abnormal intensity range used to produce the training dataset. Each data file contains information about 744 human–machine interface parameters over 60 time steps (60 s).

4.6.4.2 Base Model

The inputs of the two-channel CNN model consist of a channel representing the current parameter values for every time step and a channel representing the extent

Table 4.23 Data description with abnormality label

Label	Abnormal event
CDS	Loss of condenser vacuum
CHRG	Charging line break upstream of FT-121
CWS	Circulating water tube leak in low-pressure condenser
LTDN	Letdown line leak inside CTMT
MFW	Main feedwater pump recirculation valve FV2B leak
MSIV	Main steam isolation valve HV14 leak
POSRV	PZR pilot-operated safety relief valve HV456A leak
RCP	Loss of seal injection water with valve HV8351A leak
SGTL	SG A tube leak
TCS	High-pressure turbine control valve CV1 leak

CDS condensate system abnormality; *CHRG* charging system abnormality; *CWS* circulating system abnormality; *LTDN* letdown system abnormality; *MFW* main feedwater system abnormality; *MSIV* main steam isolation valve abnormality; *POSRV* pressurizer pilot-operated safety relief valve abnormality; *RCP* reactor coolant pump abnormality; *SGTL* SG tube leakage abnormality; *TCS* turbine control system abnormality

of the changes from the parameter values 5 time steps prior to the current state. To compose each channel for model training, 744 parameter values points are arranged in a square shape of 28 * 28 with 40 padded zeros, as shown in Fig. 4.48. In other words, since the two channels for every time step constitute one input data, it can be seen that one data file contains 55 datasets. The two-channel CNN model trains with a two-channel image structure as the input.

The whole model consists of two convolutional layers and two batch normalization layers for simple training. The output of the model, which shows the classification results for the 10 abnormal events, is produced by the FC layer after the flattening

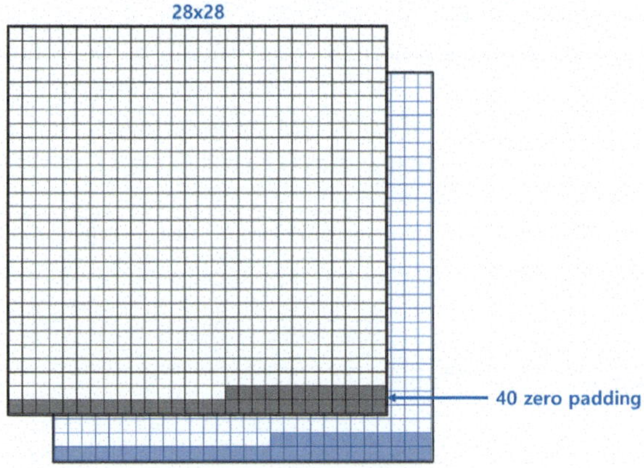

Fig. 4.48 Two-channel structure for the model inputs

layer. The structure of the two-channel CNN model is shown in Fig. 4.49, and the model hyperparameters are given in Table 4.24.

The model is trained for 100 epochs per dataset, and 30% of the training dataset is used as the validation dataset. The final information about the model for the epoch step with the lowest logloss is saved for the validation dataset. As a result, the final model has an accuracy of 1.0 on both the training and validation datasets. Figure 4.50 and Fig. 4.51 plot learning curves showing the accuracy and the logloss, respectively, for the training and validation datasets at every epoch step. It can be confirmed that the training progression reaches the optimum point for the model through the convergence of both the accuracy and logloss for the validation dataset. The trained model achieved an appropriate diagnosis for all 10 test scenarios.

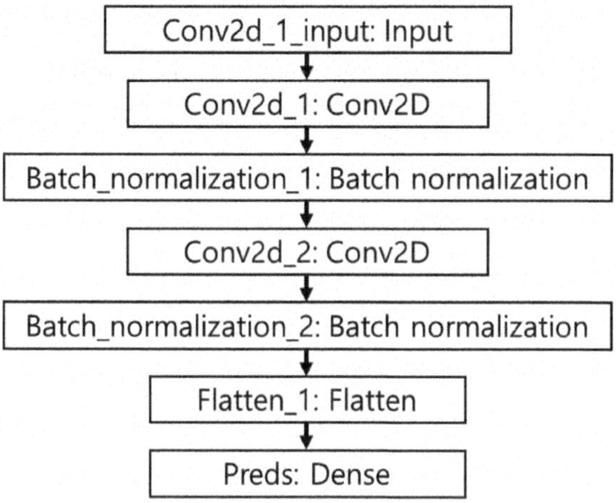

Fig. 4.49 Model structure

Table 4.24 Model hyperparameters

Filter number of convolution layer	32
Kernel size of convolution layer	3 * 3
Activation function of convolution layer	ReLU (Goodfellow et al. 2016)
Activation function of dense layer	Softmax (Nair and Hinton 2010)
Loss function	Categorical cross-entropy
Optimizer	Adam (Kingma and Ba 2014)

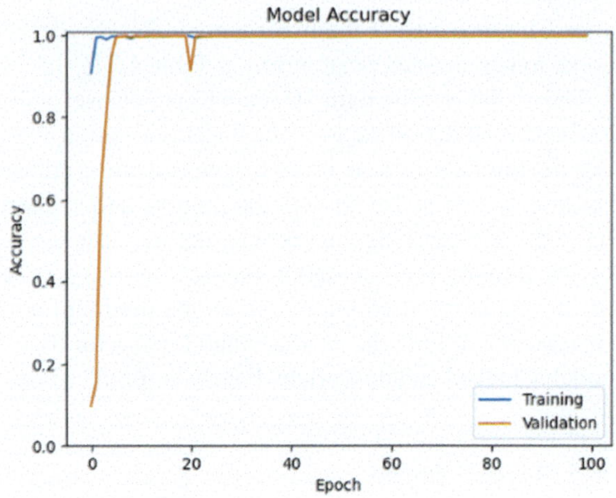

Fig. 4.50 Learning curve with accuracy

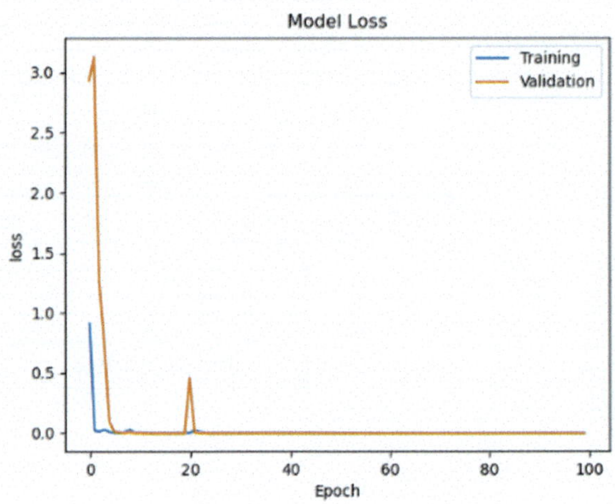

Fig. 4.51 Learning curve with logloss

4.6.4.3 Deep SHAP Results

As above, the trained model performed a successful diagnosis of the appropriate abnormal event for all 10 test scenarios. The model's diagnosis is now interpreted through Deep SHAP. As each test scenario contains a dataset with 55 time steps as the input of the two-channel structure, interpretation results for the diagnosis of all datasets included in each test scenario can be obtained. Here, the average interpretation result is calculated, which allows the contribution of each feature (parameter) to the diagnosis result of the model to be determined for that test scenario. Figure 4.52 shows a heatmap visualizing the contribution of each variable when the

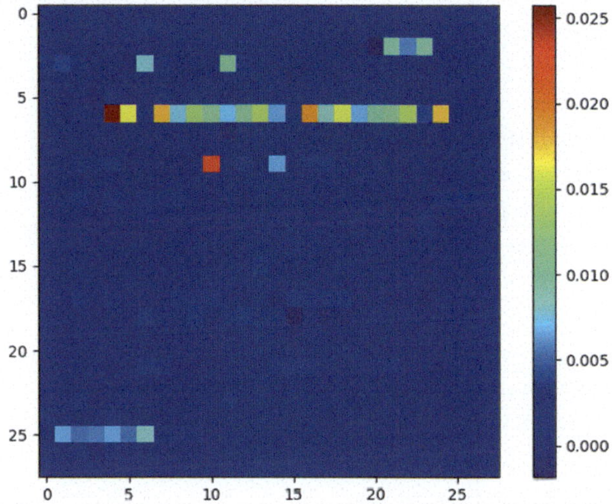

Fig. 4.52 Heatmap from Deep SHAP

model diagnoses test scenario 1 with the CDS label. In the heatmap, red indicates a high contribution.

Table 4.25 lists the three parameters with the highest contribution to the label when the trained model diagnoses each test scenario as the true label. Through such contribution scores to the model diagnosis, the cause of the diagnosis can be better understood for users.

4.6.4.4 Analysis

The interpreted contribution scores can be provided to operators along with the diagnosis results. Such provision of the reasons behind the diagnosis plays a number of roles in this system. First, the operators can check the provided causal parameters and decide whether to accept the diagnosis results of the model. In this example case, it can be seen that the parameter 'LP turbine A vaccum' has the highest contribution score for the model to diagnose test dataset 1 with the CDS label. When the operator is provided with this causal information, they can check the 'LP turbine A vaccum' parameter to confirm that it has a pattern differing from the normal operating state or the expected states of other abnormal events. In addition, the operator can consider whether the parameter is a suitable cause for the diagnosis result of the model, which can support the operator in determining that the diagnosis with the CDS label from the model is valid. In this way, a system that provides both the diagnosis result and the diagnosis cause of the model can secure a more efficient and faster diagnosis process without the need for operators to check the numerous existing monitoring parameters.

Table 4.25 Relevant parameters from Deep SHAP

True label	Relevant parameter	Score
CDS	LP turbine A vaccum	0.026798
	SG feedwater pump turbine B speed control	0.025084
	Intermediate pressure condenser cooling water outlet temperature	0.019749
CHRG	NCP running current	0.040696
	Charging header flow	0.032812
	Charging pump discharge header flow	0.029648
CWS	13.8 kV to site	0.144963
	Condenser make-up valve	0.018840
	18.5 MVA site FDR SPD0209	0.018782
LTDN	Letdown HX outlet flow	0.044578
	Letdown HX outlet temperature	0.042356
	CCW B to SFP CLG HX flow	0.033916
MFW	MFP B SUCT pressure	0.024491
	MFP A SUCT pressure	0.019313
	SG-1 total feedwater flow	0.018084
MSIV	MSL-1B steam flow	0.068966
	MSL-1A steam dump to atmosphere	0.030544
	MSL-1A steam flow	0.030394
POSRV	PZR relief tank temp	0.074451
	PZR relief tank level	0.064210
	RCP-1A seal water flow	0.022176
RCP	RCP-1A seal water flow	0.117724
	RCP-1B seal water flow	0.034537
	SG-1 total feedwater flow	0.014632
SGTL	SG-1 level master controller	0.026942
	SG 1 main feedwater reg valve	0.017940
	RCP-1A seal water flow	0.014504
TCS	HP turbine 1st stage pressure	0.011999
	RCS reference temperature	0.011841
	Load rejection controller	0.010045

NCP normal charging pump; *HX* heat exchanger; *SFP* spent fuel pool; *CLG* cooling; *MFP* main feedwater pump; *SUCT* suction; *PRESS* pressure; MSL main stean line

Figure 4.53 shows the trends of the parameter contributing the most to the diagnosis in the test scenarios with 10 different abnormal events. In the plots, the red line is the parameter trend in the corresponding test scenario, and the black lines are the parameter trends for the test scenarios diagnosed with the remaining 9 abnormal events. It can be seen that the causes of the diagnosis for each abnormal event all have characteristic tendencies. In other words, the model can infer that it has diagnosed the test scenario based on parameters that are understandable to operators. When applying model explanation techniques, the approach that best suits the user's needs should be considered. This suggests that the provision of the diagnosis causes using Deep SHAP may play a role in explaining the diagnosis of the model to the operators.

4.6.4.5 Guided Grad-CAM Results

In the previous section, the example model was analyzed by applying the Deep SHAP technique to interpret the diagnosis. For comparison, the guided Grad-Cam technique is applied in this section to the diagnosis of the example model.

Figure 4.54 shows a heatmap of the guided Grad-CAM interpretation results, visualizing the contribution by each parameter when the model diagnoses test dataset 1 with the CDS label. This figure exhibits a slightly different result from the heatmap in Fig. 4.52, which is the Deep SHAP interpretation results in the diagnosis of the same dataset. The reason why the techniques show different explanation results despite analyzing the model diagnosis of the same dataset is because the approaches of the explanation techniques to interpreting the contribution are different. In other words, explanation techniques can give different reasons for the decisions made about a particular dataset, depending on how they are interpreted.

Through the guided Grad-CAM technique, the parameter measured with the highest contribution to diagnosing the dataset with the CDS label is 'Intermediate pressure condenser cooling water outlet temperature'. This is a different result from that measured by Deep SHAP, which found 'LP turbine A vacuum' to have the highest contribution. However, when checking the trend of the highest contributing parameter as determined by guided Grad-CAM, as Fig. 4.55 shows, it can be seen that the guided Grad-CAM method also presents the diagnosis cause in a way clearly understandable to operators.

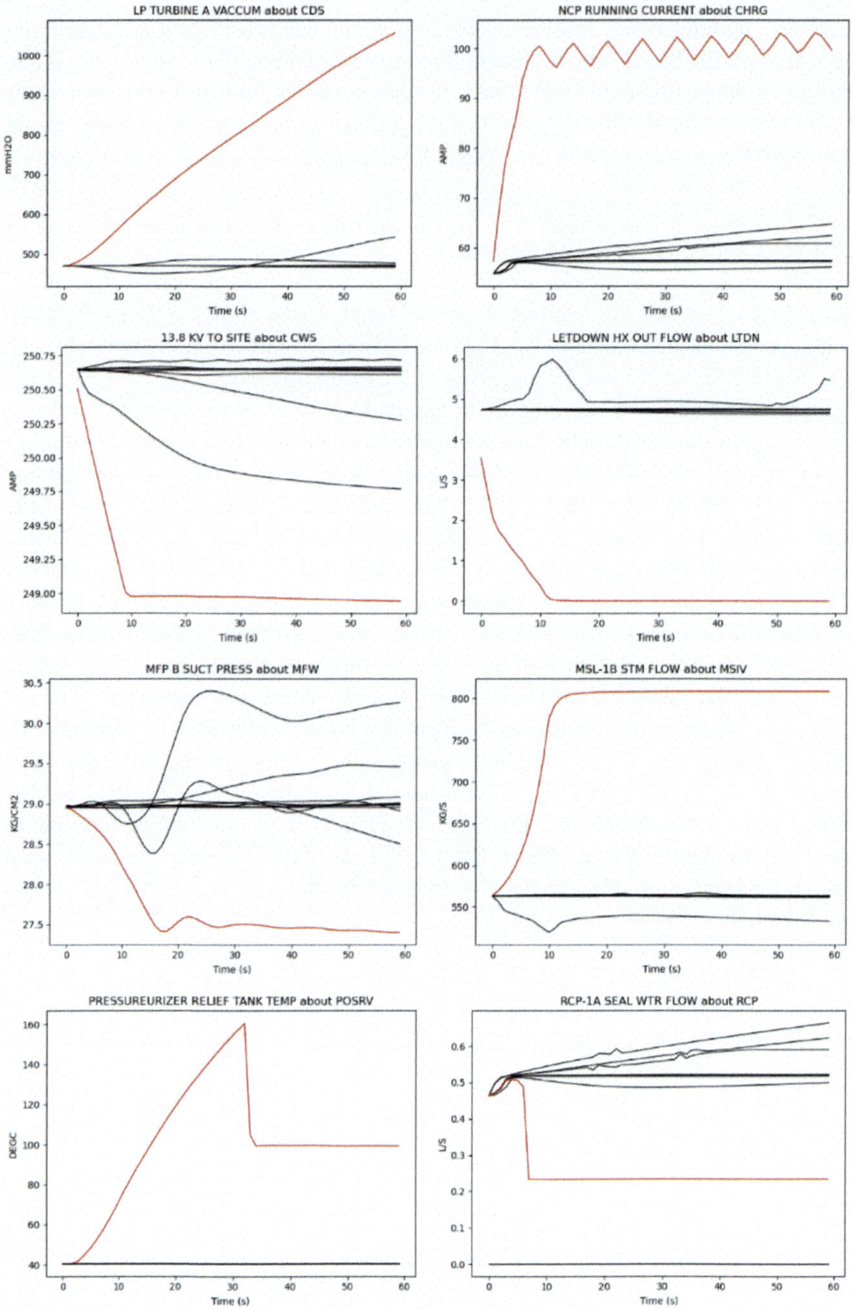

Fig. 4.53 Highest contributing parameter by abnormal event scenario

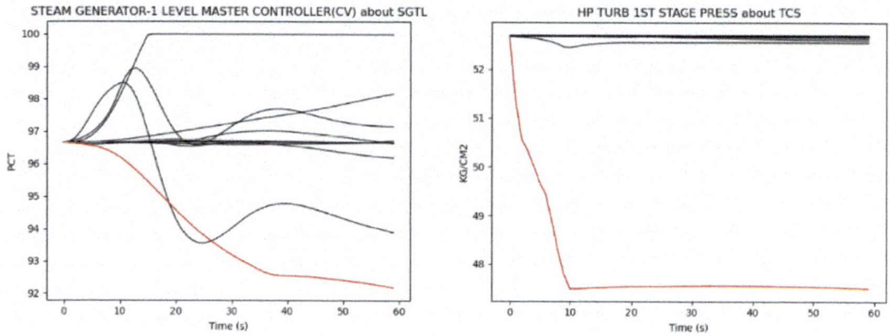

Fig. 4.53 (continued)

Fig. 4.54 Heatmap from guided Grad-CAM

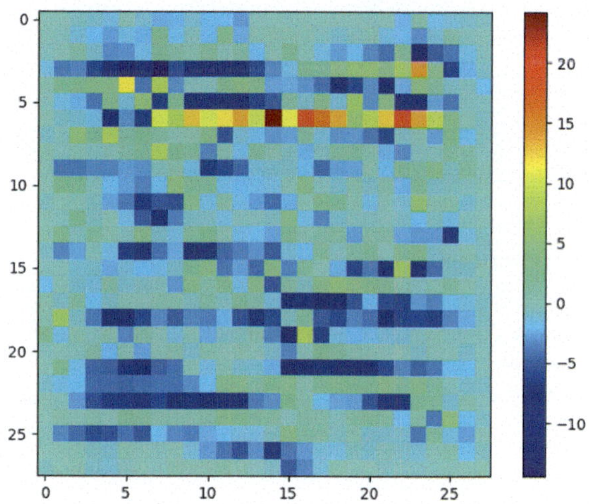

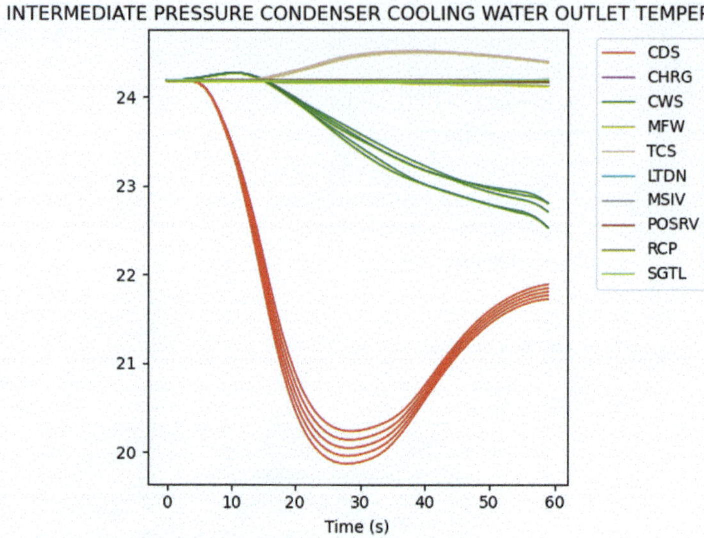

Fig. 4.55 Relevant parameter trend with guided Grad-CAM

References

Alemi AA, Fischer I, Dillon JV, Murphy K (2016) Deep variational information bottleneck. arXiv preprint arXiv:1612.00410

Armstrong JS (2010) Long-Range Forecasting, 2nd. Available at SSRN 666990

Ayodeji A, Liu Y-K, Xia H (2018) Knowledge base operator support system for nuclear power plant fault diagnosis. Prog Nucl Energy 105:42–50

Baraldi P, Di Maio F, Genini D, Zio E (2015) Comparison of data-driven reconstruction methods for fault detection. IEEE Trans Reliab 64(3):852–860

Bengio Y, Simard P, Frasconi P (1994) Learning long-term dependencies with gradient descent is difficult. IEEE Trans Neural Networks 5(2):157–166

Bergstra J, Bengio Y (2012) Random search for hyper-parameter optimization. J Mach Learning Res 13:281–305

Bruna J, Zaremba W, Szlam A, LeCun Y (2013) Spectral networks and locally connected networks on graphs. arXiv preprint arXiv:1312.6203

Chae YH, Lee C, Han SM, Seong PH (2022) Graph neural network based multiple accident diagnosis in nuclear power plants: data optimization to represent the system configuration. Nuclear Engineering and Technology

Chang SJ, Park JB (2018) Wire mismatch detection using a convolutional neural network and fault localization based on time–frequency-domain reflectometry. IEEE Trans Industr Electron 66(3):2102–2110

Che Z, Purushotham S, Cho K, Sontag D, Liu Y (2018) Recurrent neural networks for multivariate time series with missing values. Sci Rep 8(1):1–12

Cho K, Van Merriënboer B, Gulcehre C, Bahdanau D, Bougares F, Schwenk H, Bengio Y (2014) Learning phrase representations using RNN encoder-decoder for statistical machine translation. arXiv preprint arXiv:1406.1078

Choi J, Lee SJ (2020) A sensor fault-tolerant accident diagnosis system. Sensors 20(20):5839

Chung J, Gulcehre C, Cho K, Bengio Y (2014) Empirical evaluation of gated recurrent neural networks on sequence modeling. arXiv preprint arXiv:1412.3555

Ciresan DC, Meier U, Gambardella LM, Schmidhuber J (2011) Convolutional neural network committees for handwritten character classification. In: 2011 International conference on document analysis and recognition, IEEE, pp 1135–1139

Clark C, Storkey A (2015) Training deep convolutional neural networks to play go. In: International conference on machine learning, PMLR, pp 1766–1774

Duval A (2019) Explainable artificial intelligence (XAI). MA4K9 Scholarly Report, Mathematics Institute, The University of Warwick, pp 1–53

Embrechts MJ, Benedek S (2004) Hybrid identification of nuclear power plant transients with artificial neural networks. IEEE Trans Industr Electron 51(3):686–693

Goodfellow I, Bengio Y, Courville A (2016) Deep learning, MIT press

Goodwin P, Lawton R (1999) On the asymmetry of the symmetric MAPE. Int J Forecast 15(4):405–408

Hochreiter S, Schmidhuber J (1997) Long short-term memory. Neural Comput 9(8):1735–1780

Horiguchi M, Fukawa N, Nishimura K (1991) Development of nuclear power plant diagnosis technique using neural networks. In: Proceedings of the first international forum on applications of neural networks to power systems, pp. 279–282, IEEE

IAEA (2006) Development and review of plant specific emergency operating procedures, International Atomic Energy Agency

Ince T, Kiranyaz S, Eren L, Askar M, Gabbouj M (2016) Real-time motor fault detection by 1-D convolutional neural networks. IEEE Trans Industr Electron 63(11):7067–7075

Jolliffe IT (2002) Principal component analysis for special types of data. Springer

Kim H, Arigi AM, Kim J (2021) Development of a diagnostic algorithm for abnormal situations using long short-term memory and variational autoencoder. Ann Nucl Energy 153:108077

Kim K, Kim D-K, Noh J, Kim M (2018) Stable forecasting of environmental time series via long short term memory recurrent neural network. IEEE Access 6:75216–75228

Kim JW, Jung WD, Park JK, Kang DI (2003) A study on the operator's errors of commission (EOC) in accident scenarios of nuclear power plants: methodology development and application

Kim JM, Lee G, Lee C, Lee SJ, (2020) Abnormality diagnosis model for nuclear power plants using two-stage gated recurrent units. Nuclear Eng Technol

Kingma DP, Ba J (2014) Adam: a method for stochastic optimization. arXiv preprint arXiv:1412.6980

Kingma DP, Welling M (2013) Auto-encoding variational bayes. arXiv preprint arXiv:1312.6114

Krizhevsky A, Sutskever I, Hinton GE (2017) Imagenet classification with deep convolutional neural networks. Commun ACM 60(6):84–90

Kuhn HW, Tucker AW (1953) Contributions to the theory of games. Princeton University Press

LeCun Y, Bengio Y, Hinton G (2015) Deep learning. Nature 521(7553):436–444

Lee S-J, Seong P-H (2007) Development of an integrated decision support system to aid cognitive activities of operators. Nucl Eng Technol 39(6):703–716

Lee G, Lee SJ, Lee C (2021) A convolutional neural network model for abnormality diagnosis in a nuclear power plant. Appl Soft Comput 99:106874

Li Q, Cai W, Wang X, Zhou Y, Feng DD, Chen M (2014) Medical image classification with convolutional neural network. In: 2014 13th international conference on control automation robotics and vision (ICARCV), IEEE, pp 844–848

Li H, Lin Z, Shen X, Brandt J, Hua G (2015) A convolutional neural network cascade for face detection. In: Proceedings of the IEEE conference on computer vision and pattern recognition, pp 5325–5334

Liaw A, Wiener M (2002) Classification and regression by randomForest. R News 2(3):18–22

Ma J, Jiang J (2011) Applications of fault detection and diagnosis methods in nuclear power plants: a review. Prog Nucl Energy 53(3):255–266

Mandrekar JN (2010) Receiver operating characteristic curve in diagnostic test assessment. J Thorac Oncol 5(9):1315–1316

Mizokami S, Kumagai Y (2015) Event sequence of the Fukushima Daiichi accident. Reflections on the Fukushima Daiichi nuclear accident. Springer, Cham

Na MG, Park WS, Lim DH (2008) Detection and diagnostics of loss of coolant accidents using support vector machines. IEEE Trans Nucl Sci 55(1):628–636
Nair V, Hinton GE (2010) Rectified linear units improve restricted boltzmann machines. In Icml
Norman DA (1980) Errors in human performance. California Univ San Diego LA JOLLA Center for human information processing
Park SH, Goo JM, Jo CH (2004) Receiver operating characteristic (ROC) curve: practical review for radiologists. Korean J Radiol 5(1):11–18
Park D, Hoshi Y, Kemp CC (2018) A multimodal anomaly detector for robot-assisted feeding using an lstm-based variational autoencoder. IEEE Robot Autom Lett 3(3):1544–1551
Park J, Jeong K, Jung W (2005) Identifying cognitive complexity factors affecting the complexity of procedural steps in emergency operating procedures of a nuclear power plant. Reliab Eng Syst Saf 89(2):121–136
Park J, Jung W (2015) A systematic framework to investigate the coverage of abnormal operating procedures in nuclear power plants. Reliab Eng Syst Saf 138:21–30
Peng B-S, Xia H, Liu Y-K, Yang B, Guo D, Zhu S-M (2018) Research on intelligent fault diagnosis method for nuclear power plant based on correlation analysis and deep belief network. Prog Nucl Energy 108:419–427
Pinheiro VHC, dos Santos MC, do Desterro FSM, Schirru R, Pereira CMDNA (2020) Nuclear power plant accident identification system with "don't know" response capability: novel deep learning-based approaches. Ann Nuclear Energy, 137:107111
Ruder S (2016) An overview of gradient descent optimization algorithms. arXiv preprint arXiv: 1609.04747
Santosh T, Vinod G, Saraf R, Ghosh A, Kushwaha H (2007) Application of artificial neural networks to nuclear power plant transient diagnosis. Reliab Eng Syst Saf 92(10):1468–1472
Scarselli F, Gori M, Tsoi AC, Hagenbuchner M, Monfardini G (2008) The graph neural network model. IEEE Trans Neural Networks 20(1):61–80
Şeker S, Ayaz E, Türkcan E (2003) Elman's recurrent neural network applications to condition monitoring in nuclear power plant and rotating machinery. Eng Appl Artif Intell 16(7–8):647–656
Selvaraju RR, Cogswell M, Das A, Vedantam R, Parikh D, Batra D (2017) Grad-cam: visual explanations from deep networks via gradient-based localization. In: Proceedings of the IEEE international conference on computer vision, pp 618–626
Shrikumar A, Greenside P, Kundaje A (2017) Learning important features through propagating activation differences. In: International conference on machine learning, PMLR, pp 3145–3153
Strumbelj E, Kononenko I (2010) An efficient explanation of individual classifications using game theory. J Mach Learning Res 11:1–18
Tishby N, Zaslavsky N (2015) Deep learning and the information bottleneck principle. In: 2015 ieee information theory workshop (itw), IEEE, pp 1–5
Tolo S, Tian X, Bausch N, Becerra V, Santhosh T, Vinod G, Patelli E (2019) Robust on-line diagnosis tool for the early accident detection in nuclear power plants. Reliab Eng Syst Saf 186:110–119
USNRC (1982) Guidelines for the preparation of emergency operating procedures. Resolution of comments on NUREG-0799. Nuclear Regulatory Commission
Van Niel TG, McVicar TR, Datt B (2005) On the relationship between training sample size and data dimensionality: Monte Carlo analysis of broadband multi-temporal classification. Remote Sens Environ 98(4):468–480
Wang H, Peng M-J, Hines JW, Zheng G-Y, Liu Y-K, Upadhyaya BR (2019) A hybrid fault diagnosis methodology with support vector machine and improved particle swarm optimization for nuclear power plants. ISA Trans 95:358–371
Wen L, Li X, Gao L, Zhang Y (2017) A new convolutional neural network-based data-driven fault diagnosis method. IEEE Trans Industr Electron 65(7):5990–5998
Western Service Corporation (2017) 3KEYMASTER simulator. https://www.ws-corp.com
White IR, Royston P, Wood AM (2011) Multiple imputation using chained equations: issues and guidance for practice. Stat Med 30(4):377–399

Yang J, Kim J (2020) Accident diagnosis algorithm with untrained accident identification during power-increasing operation. Reliab Eng Syst Saf 202:107032

Yao Y, Wang J, Xie M, Hu L, Wang J (2020) A new approach for fault diagnosis with full-scope simulator based on state information imaging in nuclear power plant. Ann Nucl Energy 141:107274

Zhang S (2012) Nearest neighbor selection for iteratively kNN imputation. J Syst Softw 85(11):2541–2552

Zhongming Z, Linong L, Xiaona Y, Wangqiang Z, Wei L (2018) Machine health monitoring using local feature-based gated recurrent unit networks

Zhou J, Cui G, Hu S, Zhang Z, Yang C, Liu Z, Wang L, Li C, Sun M (2020) Graph neural networks: a review of methods and applications. AI Open 1:57–81

Zio E (2007) A support vector machine integrated system for the classification of operation anomalies in nuclear components and systems. Reliab Eng Syst Saf 92(5):593–600

Chapter 5
Prediction

The previous two chapters focused on AI approaches to signal validation and diagnosis in NPPs. In this chapter, prediction is considered, which generally means to predict future NPP parameter trends. But in addition to parameter trends, as mentioned in Chap. 1 this function can also predict (1) the failure of systems and components, (2) occurrence of alarms, (3) occurrence of abnormal situations, and (4) reactor trip. Information provided by the prediction function can be employed for autonomous operation, such as for the following key measures.

- Early request for operator intervention before an adverse situation occurs
- Selection of the optimal control by predicting the plant behavior upon choosing different control means.

Several researchers have proposed models for predicting the future trends of parameters, including thermal–hydraulic design, statistical models, and AI techniques. Among them, thermal–hydraulic design has been applied to predict the response of an NPP system under typical operation or accident conditions. These designs, though, are time-consuming and unsuitable for real-time prediction as they require repeated calculation–evaluation–correction cycles. Statistical methods, such as autoregressive integrated moving average modeling, have also been employed to predict future trends more quickly, but they can only address linear relationships in time trends. Because NPPs consist of nonlinear and multivariate complexes, statistical methods are not appropriate for managing NPP data. More recently, ANNs have been regarded as one of the most relevant approaches for pattern recognition and managing significant amounts of nonlinear data, such as for handwriting recognition, natural language processing, and financial forecasting. In the nuclear industry, Kim et al. was able to predict the next single step of a SG level using a modified backpropagation algorithm (Kim et al. 1993). Moshkbar-Bakhshayesh predicted NPP operating parameters under abnormal and emergency situations with a cascade feedforward neural network (Moshkbar-Bakhshayesh 2019). El-Sefy et al. trained a feedforward backpropagation ANN to simulate the interaction between the reactor core and the primary and secondary coolant systems (El-Sefy et al. 2021). To

J. Kim et al., *Autonomous Nuclear Power Plants with Artificial Intelligence*, Lecture Notes in Energy 94, https://doi.org/10.1007/978-3-031-22386-0_5

predict long-term trends, an LSTM with a multi-input multi-output (MIMO) strategy has been proposed; with such a setup, Nguyen et al. showed the prediction of SG narrow-range water levels for 45 days at an interval of 3 days, or in other words for 15 steps (Nguyen et al. 2020).

This chapter introduces an algorithm to predict NPP parameters following control actions in an emergency situation. It also compares available ANN methods for the model prediction.

5.1 Real-Time Parameter Prediction

Real-time plant parameter prediction refers to the anticipation of future trends from the past values of the parameters, not a future scenario. In other words, the prediction results follow from when there is no further control after the current time instance except automated operation. This section details a real-time prediction model to anticipate the future trends of important or safety-related parameters after a device control by an operator in a given situation (Bae et al. 2021). Figure 5.1 presents a schematic of the prediction model. In the figure, t is the current time step, d is the number of time steps backward in time, also called look-backs, H is the prediction horizon, and N and M are the numbers of input and output parameters, respectively. With the purpose of application to the MCR of an NPP, the prediction model input parameters comprise device states, instrument values from sensors, and other important signals that are available to typical NPP I&C systems. A record of d time steps of the input parameters is provided to the prediction model to reflect that the past values and trends influence the future trends. More specifically, the prediction model receives a matrix of size $N \times d$ as an input and produces a matrix of size $M \times H$ as the output. As this implies a large number of input and output values, the prediction model adopts ANNs along with a multi-step prediction strategy. The prediction strategy works by decomposing the prediction problem and determining the role of the ANNs as well as their inputs and output. During training, a neural network optimizer is used to adjust the inner parameters of the ANNs with the aim to align the training data with a backpropagation algorithm.

The following sections discuss general multi-step prediction strategies, detail the prediction model characteristic when combining each strategy with ANNs, and provide case study results from prediction models trained with emergency operation data including operator actions. A possible application of this real-time prediction model as an operator support system is also explored.

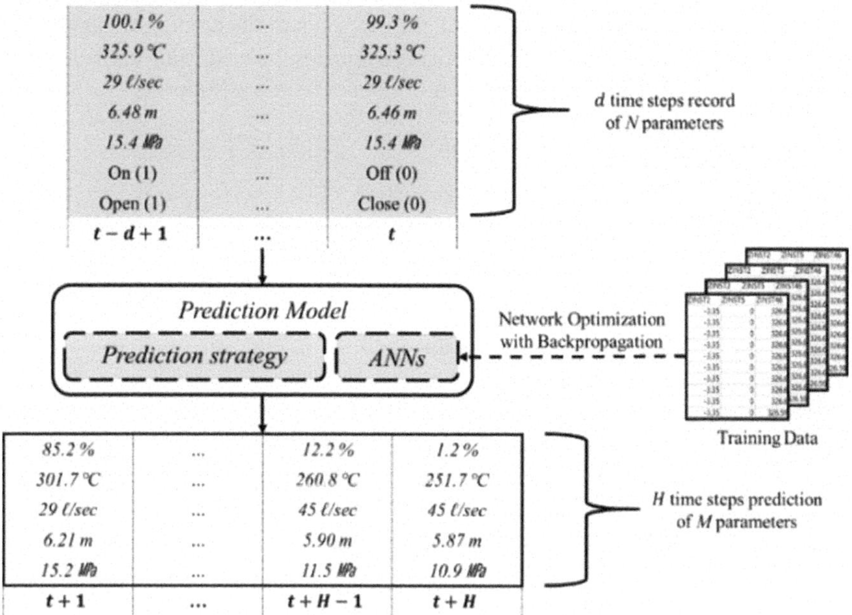

Fig. 5.1 Parameter trend prediction model with prediction strategy and ANNs. Parameters in the ANNs are adjusted by a backpropagation algorithm and training data. Reproduced with permission from Bae et al. (2021)

5.1.1 Multi-step Prediction Strategies

With the historical observation of d time steps, the goal of multi-step ahead prediction is to estimate the observation of the next H time steps in the future (Nguyen et al. 2021; Radaideh et al. 2020; Taieb et al. 2012). For a univariate observation, the multi-step ahead prediction problem can be expressed as follows.

$$\left[\hat{y}_{t+1}, \hat{y}_{t+2}, \ldots, \hat{y}_{t+H-1}, \hat{y}_{t+H}\right] = f\left(y_t, y_{t-1}, \ldots, y_{t-d+2}, y_{t-d+1}\right) \tag{5.1}$$

Here, t is the current time step, y is the historical value, $\hat{y}$ is the predicted value, and f is the prediction model. Clearly, solving the prediction problem greatly increases in difficulty with an increasing prediction horizon H. To handle large prediction horizons, three common prediction strategies are employed: recursive, direct, and MIMO strategies (Taieb et al. 2012), as described below.

The recursive strategy, which is the oldest and most intuitive prediction strategy (Taieb et al. 2012), requires a single prediction model only, f_{REC}. This model is trained to perform a one-step prediction as in the following equation.

$$\hat{y}_{t+1} = f_{Rec}\left(y_t, y_{t-1}, \ldots, y_{t-d+2}, y_{t-d+1}\right) \tag{5.2}$$

After the prediction of one step ahead, the model takes the predicted value of the previous step as a known value and iteratively estimates the next value until H future values are acquired. This future observation can be written as follows.

$$\hat{y}_{t+h} = \begin{cases} f_{Rec}(y_t, \ldots, y_{t-d+1}) & when\, h = 1 \\ f_{Rec}(\hat{y}_{t+h-1}, \ldots, \hat{y}_{t+1}, y_t, \ldots, y_{t-d+h}) & when\, h \in \{2, \ldots, d\} \\ f_{Rec}(\hat{y}_{t+h-1}, \ldots, \hat{y}_{t-d+h}) & when\, h \in \{d+1, \ldots, H\} \end{cases} \tag{5.3}$$

Such an iterative approach reusing the output can ease long prediction horizons. The major drawback, though, is that this prediction strategy is prone to error accumulation because any errors present in the early or intermediate prediction outputs propagate to subsequent predictions.

The second prediction strategy discussed here is the direct strategy, which utilizes multiple prediction models by assigning one per time step (Taieb et al. 2012). Thus, combining with neural networks results in the number of networks equalling the prediction horizon H, in which case H prediction models simultaneously estimate future observations of the next H time steps as $\left[\hat{y}_{t+1}, \hat{y}_{t+2}, \ldots, \hat{y}_{t+H-1}, \hat{y}_{t+H}\right]$.

$$\hat{y}_{t+h} = f_{Dir,h}(y_t, y_{t-1}, \ldots, y_{t-d+2}, y_{t-d+1}) \tag{5.4}$$

In Eq. (5.4), $h \in \{1, \ldots, H\}$. In contrast to the recursive method, this strategy is free from error accumulation because it does not repeat any prediction. However, any interdependencies that are present among the time steps are not accounted for in this case, as all future time steps are estimated by independent prediction models.

The two above strategies can be categorized as single-output strategies. Conversely, as its name implies, the MIMO strategy represents a multiple-output strategy. Here, the prediction model is trained to generate values for the future observations of the next H time steps all at once, as below.

$$\left[\hat{y}_{t+1}, \hat{y}_{t+2}, \ldots, \hat{y}_{t+H-1}, \hat{y}_{t+H}\right] = f_{MIMO}(y_t, y_{t-1}, \ldots, y_{t-d+2}, y_{t-d+1}) \tag{5.5}$$

As such, the MIMO strategy avoids the issues of the recursive and direct strategies. Moreover, a prediction model that adopts the MIMO strategy is able to consider stochastic dependencies among the time-series inputs and outputs. But in this case, the training of the prediction model becomes complicated because the multi-step prediction problem is not decomposed and simplified in this approach. This means that the MIMO strategy does not reduce the problem complexity, which is something a prediction model should achieve.

5.1.2 Plant Parameter Prediction Model with Multi-step Prediction Strategies

As introduced in the previous section, the recursive, direct, and MIMO strategies are selected as candidate multi-step prediction strategies for univariate time series data (Taieb et al. 2012). In the present case, though, the prediction problem is not only multi-step but also multivariate. As a consequence, each multi-step prediction strategy presents additional characteristics following combination with ANNs, as described below.

With the recursive approach, as an iterative prediction strategy, the estimated values are used to make further predictions. Coupled with an ANN, this means that the ANN performs a one-step prediction and recursively uses the predicted values as the next network inputs. Repeating this process over H time steps, plant parameters up to $t + H$ can be estimated. Figure 5.2 depicts the process of this prediction strategy.

In this case, the recursive strategy only needs a single ANN, and thus the neural network training involves a low computational cost compared to other approaches. Moreover, there is no prediction limit from the chosen prediction horizon with this approach. With sufficient accuracy of the trained neural network, prediction over long time scales is possible by simply repeating the one-step prediction continually. Another advantage is that this strategy can reflect interdependencies between adjacent time steps. However, one critical drawback with the recursive strategy is that prediction errors continuously accumulate over repeated predictions (Bae et al. 2019; Taieb et al. 2012; Ryu et al. 2022). Another issue is that, because the one-step prediction output acts as the input of the next step, the inputs and outputs must have an equal number of parameters, and this can drain computational resources. In the current application, the recursive strategy requires the future trend prediction of not only the M target parameters but also the remaining $M - N$ parameters simultaneously.

In contrast to the single network of the recursive strategy, the direct prediction strategy employs multiple neural networks, devoting each one to its own time step like the univariate case. As mentioned above though, the current application requires

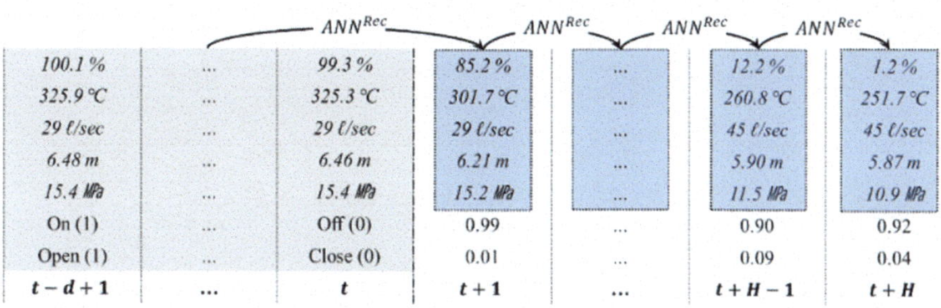

Fig. 5.2 Recursive strategy for multi-step prediction. Arrows indicate one-step predictions, where the outputs are used as the inputs for the next prediction. Reproduced with permission from Bae et al. (2021)

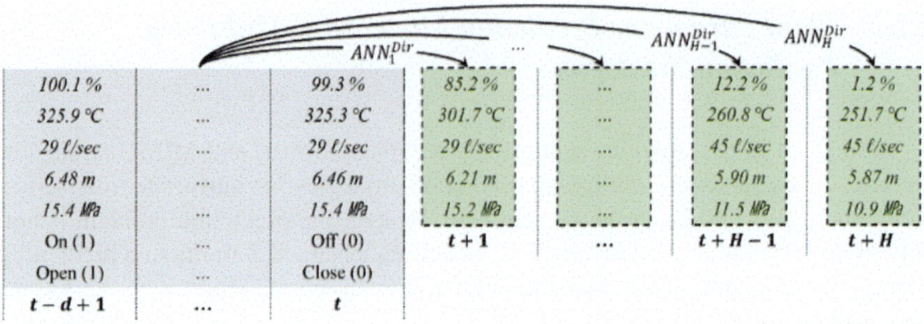

Fig. 5.3 Direct strategy for multi-step prediction. Arrows indicate simultaneous prediction by multiple ANNs, where different ANNs cover each of the time steps separately. Reproduced with permission from Bae et al. (2021)

the networks to predict multiple parameters at dedicated time steps. Figure 5.3 illustrates the simultaneous predictions of future parameter trends by the neural networks following the direct strategy.

The direct strategy is free from the two issues of the recursive strategy. First, it involves no error accumulation from the reuse of the outputs, and second, the number of input and output parameters can differ, meaning that computational resources can concentrate on the target parameter predictions. But one drawback in this case is that the direct strategy demands high computational power for its multiple neural networks to work simultaneously. In addition, neural networks following this strategy can only consider interdependencies among the parameters within the dedicated time steps. And lastly, neural network training in this case is substantially longer than with the recursive strategy, where a set of H ANNs need to be trained to incorporate the direct strategy into the prediction model.

The third candidate prediction strategy is MIMO. This strategy assigns a neural network to each target parameter for prediction; in other words, the number of neural networks equals that of the target parameters, or M in this case. Similar to the direct strategy, multiple neural networks work together to produce the output with MIMO, but in this case by parameter rather than by time step. Figure 5.4 shows the MIMO strategy.

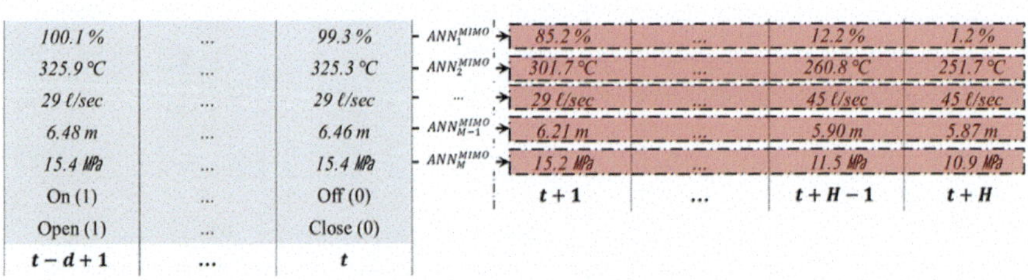

Fig. 5.4 Multi-input multi-output strategy for multi-step prediction. Arrows indicate the prediction of each target parameter by a dedicated ANN. Reproduced with permission from Bae et al. (2021)

This approach shares many characteristics with the direct strategy, but MIMO is distinct regarding the outputs of the neural networks. Each network estimates the future trend of one parameter and considers interdependencies of this target parameter among all time steps. Accordingly, the drawback to this strategy is that interdependencies between different parameters are not covered.

5.1.3 Case Study with Data from an NPP Simulator

To evaluate the feasibility of the multi-step prediction model, a case study is conducted for which a variety of prediction models combining different multi-step prediction strategies and ANNs are constructed. The neural networks used to build the models have different characteristics depending on their particular type of artificial neurons. Three neural networks for future parameter trend prediction are chosen, namely a multilayer perceptron (MLP), vanilla RNN, and LSTM. Coupling these with the three prediction strategies results in nine total prediction models. Keras API (Chollet 2015) is used for network implementation, which is a Python-based high-level application programming interface that runs on the ML framework TensorFlow (Abadi et al. 2016). A CNS is used for data collection considering the confidentiality of real NPP operational data. The target plant is a Westinghouse 990 MWe 3-loop plant (Kwon et al. 1997).

Prior to model construction, the inputs and outputs of the prediction model must first be defined. As the output, 25 target parameters for prediction are selected, all of which must be monitored while operators perform the required EOP during an NPP emergency. The EOPs of the CNS include a CSF tree, which is a type of plant status monitoring procedure. CSFs refer to the particular NPP functions that are essential to maintain for securing plant safety. The CNS has six CSFs: subcriticality of the reactor core, core cooling, heat removal, integrity of the RCS, CTMT integrity, and reactor coolant inventory. As an example, the CSF tree procedure dictates that operators monitor the CTMT integrity through the CTMT pressure, sump water level, and CTMT radiation level. Table 5.1 lists the 25 prediction target parameters that are monitored for the CSFs.

For the model inputs, every parameter that is mentioned at least once in the EOPs for three different accident scenarios is selected; see Table 5.2 for details. This gives a total of 109 input parameters, which include the 25 prediction target parameters (model output). The remaining 84 parameters consist of valve states such as an open or closed main steam isolation valve, component states such as the reactor coolant charge pump status, instrument values from sensors such as the RV water level, and other important signals such as the safety injection (SI) actuation signal.

In the prediction models, the look-backs d is set to 1. Otherwise, setting the prediction horizon H should be done with care considering the trade-off relationship between long-term prediction and model accuracy—lower accuracy accompanies predictions made further into the future. While there is no general rule to setting H (i.e., to determine the optimal prediction horizon), Nguyen et al. (2020) reviewed

Table 5.1 Target plant parameters from the CNS for prediction

	Plant parameter (units)
1	POWER RANGE PERCENT POWER (%)
2	INTERMEDIATE RANGE START-UP RATE (DPM)
3	INTERMEDIATE RANGE NEUTRON LEVEL (A)
4	SOURCE RANGE START-UP RATE (DPM)
5	CORE OUTLET TEMPERATURE (kg/cm^2)
6	LOOP 1 HOT-LEG TEMPERATURE (°C)
7	LOOP 2 HOT-LEG TEMPERATURE (°C)
8	LOOP 3 HOT-LEG TEMPERATURE (°C)
9	PZR PRESSURE (kg/cm^2)
10	SG #1 NARROW LEVEL (%)
11	SG #2 NARROW LEVEL (%)
12	SG #3 NARROW LEVEL (%)
13	FEEDWATER #1 FLOW (m^3/hr)
14	FEEDWATER #2 FLOW (m^3/hr)
15	FEEDWATER #3 FLOW (m^3/hr)
16	SG #1 PRESSURE (kg/cm^2)
17	SG #2 PRESSURE (kg/cm^2)
18	SG #3 PRESSURE (kg/cm^2)
19	LOOP 1 COLD-LEG TEMPERATURE (°C)
20	LOOP 2 COLD-LEG TEMPERATURE (°C)
21	LOOP 3 COLD-LEG TEMPERATURE (°C)
22	CTMT PRESSURE (kg/cm^2)
23	CTMT SUMP WATER LEVEL (m)
24	CTMT RADIATION (mRem/hr)
25	PZR LEVEL (%)

the various selected prediction horizons in literature from the period 2015–2019 in industrial areas (Nguyen et al. 2020). It was found that most studies carried out a single-step prediction only, while a few multi-step prediction studies set the prediction horizon to 3–6 time steps ahead. In the current application, the prediction horizon of the model is set to 20. The time interval between time steps is 30 s, and thus the model predicts the parameter trends up to 10 min in the future. It should be noted that other recent studies in the nuclear industry have applied a prediction horizon of more than 40 time steps for the prediction of NPP parameters (Ryu et al. 2022).

In sum, the prediction problem for the model to solve is the estimation of a 25×20 output matrix (25 prediction target parameters and 20 future time steps) using a 109×2 input matrix (109 input parameters and 2 time steps, namely current and past). The emergency operation data via procedure analysis and the construction of the ANNs for the prediction model are described below.

Table 5.2 Emergency operation scenarios simulated by the CNS

Accident	Accident detail	Operator action	No. of scenarios
LOCA	Leak of reactor coolant due to 10 cm^2, 20 cm^2, 30 cm^2, 40 cm^2, or 50 cm^2 break in the cold leg	Auxiliary feedwater flow control RCP stop PORV shut-off valve open PORV open SI signal reset SI pump stop No action	865
SGTR	Single tube or double tube ruptures in a SG	Auxiliary feedwater flow control PORV shut-off valve open PORV open RCP stop Main steam line isolation Secondary side relief valve manual open Contaminated steam line isolation No action	200
Simple Trip	Unintended reactor trip due to a malfunction of the RPS	Auxiliary feedwater flow control PORV shut-off valve open PORV open RCP stop No action	98

Every scenario has different operator action timings, degrees, and correctly performed prerequisites

The CNS used to generate operational data during an NPP emergency is based on the SMABRE thermal–hydraulic system code, which is a simplified 1D nodalization code based on assumptions and experiments (Miettinen 1985). While the CNS cannot reflect all related phenomena and operator controls, it is able to generate large amounts of emergency operation data in an open and fast manner. The primary side of the simulated plant contains the reactor core, and the secondary side contains the turbines. Heat generated by the reactor core is transferred to the SGs through the reactor coolant, with which the SGs convert the water supplied by the secondary side into steam and pass it to the turbines. A PZR connected to the primary side controls both the volume and the pressure of the reactor coolant, with a PORV installed at the top of the PZR to prevent the over-pressurization of the reactor coolant.

The three different NPP accidents under normal operation considered here are LOCA, SGTR, and a simple reactor trip. The first refers to a leak of reactor coolant due to a variety of possible incidents and can cause the reactor core to overheat. In the simulation, a break in the cold leg, which is the suction line from a SG to the reactor core, is assumed to initiate the LOCA. The second accident, SGTR, is a rupture in a heat exchange tube in a SG that triggers the depressurization of the primary side coolant and the release of radioactive material to the secondary side. In the simulation, a rupture of one or two tubes is assumed. The third accident in this simulation is an unintended reactor trip caused by a malfunction of the RPS. While less severe than a LOCA or SGTR, an unintended reactor trip can cause the plant dynamics to change rapidly.

Considering the huge number of possible operator actions, the CNS EOPs are analyzed to highlight probable operator action scenarios. The EOPs present sequentially organized steps with instructions to check plant parameters, transfer procedures or steps, and control devices. First, the time that each procedural step is reached assuming an appropriate operator response to the accident is identified from our expertise. Second, the particular steps requiring operator actions are distinguished, and third, human errors in these distinguished steps are assumed by the context. For instance, in the E-0 procedure among the CNS EOPs, the 18th step is determined to be reached at 380 s following SGTR. This step requires operators to open a shut-off valve for the PORV, as shown in Fig. 5.5. A possible human error at this step could be that operators open the PORV directly instead of the shut-off valve. In the control interface, the PORV and PORV shut-off valve are located right next to each other. Also, before opening the shut-off valve, the prerequisite action is to first set the PORV to manual mode.

Based on the above analysis, various operator action timings can be considered. In the above example, a number of scenarios indicate that the prerequisites were performed correctly, and the PORV shut-off valve was opened without error at 320, 350, 380, 410, and 450 s following the SGTR. Other scenarios reflect that the PORV was mistakenly opened at 320, 350, 380, 410, and 450 s after the SGTR. As shown in Table 5.2, a total of 1153 emergency operation scenarios are simulated by the CNS. Before using this operational data to train the various prediction models, the parameter values are rescaled via min–max normalization to prevent problems from the different parameter scales.

18.0 Check that the total flow rate of AUX FW is 33l/s or higher. 18.1 PORV : closed · PV-445	18.1 If the pressurizer pressure is less than 164.2 kg/cm2, manually close the PORV. If the valve is not closed, manually close the shut-off valve of the valve.
18.2 PORV shut-off valve : opened · HV-6	18.2 Open the closed valve.

Fig. 5.5 18th step in the E-0 procedure of the CNS EOPs. The left column is the expected response, and the right is what should be done when the correct response is not obtained. Reproduced with permission from Bae et al. (2021)

The MLPs, vanilla RNNs, and LSTMs making up the different prediction models are respectively constructed using the Dense, SimpleRNN, and LSTM classes of Keras API (Chollet 2015). Their overall network structures, referring to the input and output shapes and the numbers of hidden layers as well as cells in each hidden layer, are considered one of the hyperparameters. The neural network structures of the three network types are defined based on the number of trainable parameters and the given multi-step prediction strategy, namely recursive, direct, or MIMO.

As mentioned above, the input matrix size is 109×2, which is too large for the MLP to deal with. The input shape of the MLP is therefore (218,), while that of the other two network types is (2, 109). The output shape is determined by the particular multi-step prediction strategy as follows. The recursive strategy requires the output shape to equal the number of input parameters (109,), the direct strategy requires the output to equal the number of target parameters (25,), and the MIMO strategy requires the output to equal the prediction horizon H (20,).

The trainable parameters are those that are intrinsic to each logical component and can be adjusted during training. Because the trainable parameters are tuned for the given data through backpropagation, the given data can be considered to be compressed into the trainable parameter values. Accordingly, the sizes of the hidden layers of the three networks are carefully chosen such that each network has a similar total number of trainable parameters. Table 5.3 summarizes the different ANN structures of the prediction models.

The Adam optimizer in Keras API (Chollet 2015), which is a stochastic gradient descent method with adaptive estimation of first-order and second-order momentums (Kingma and Ba 2015a), is used to conduct the backpropagation for optimization of the trainable parameters. The Adam optimizer adjusts the trainable parameters of the networks to minimize the MSE between the predicted and real values. The learning rate is set to be constant to determine the step size of the gradient descent per iteration.

Table 5.3 ANN structures of the nine prediction models

Strategy–ANN	ANNs	Input shape	Hidden layers	Cells per hidden layer	Output shape	Trainable parameters
REC-MLP	1	(218,)	8	200	(109,)	347,109
REC-RNN	1	(2, 109)	5	200	(109,)	343.709
REC-LSTM	1	(2, 109)	4	100	(109,)	336,209
DIR-MLP	20	(218,)	8	200	(25,)	330,225
DIR-RNN	20	(2, 109)	5	200	(25,)	335,225
DIR-LSTM	20	(2, 109)	4	100	(25,)	327,725
MIMO-MLP	25	(218,)	8	200	(20,)	329,220
MIMO-RNN	25	(2, 109)	5	200	(20,)	334,720
MIMO-LSTM	25	(2, 109)	4	100	(20,)	327,220

*REC = recursive, DIR = direct

As there are no deterministic rules for setting an optimal learning rate, learning rates of 0.1, 0.01, 0.001, and 0.0001 are explored to train the neural networks.

One major concern in neural network training is the overfitting problem, referring to when a trained neural network is accurate only for the trained data and inaccurate for others. To prevent neural network overfitting, 20% of the data is assigned as validation data, which is not used in the network training. Training sessions are stopped when the prediction accuracy for the validation data does not improve over the previous 200 trials. Furthermore, to prevent the output from relying on minority artificial neurons, some neurons in the network are temporarily detached. Specifically, 30% of the artificial neurons in each hidden layer are randomly selected and isolated from the neural network in every training trial.

To sum up, nine prediction models are trained with the data from 1153 scenarios and then applied to the test data. The test data comprise 35 scenarios and are normalized in the same way as the training data. The test data and training data scenarios differ in terms of the accident initiation, such as the break size of LOCA and the number of ruptured tubes in SGTR, and the operator responses, such as a delayed action and early response. The prediction models estimate 875 trends, or 25 target parameters in the 35 test scenarios, up to 10 min into the future, giving a total of 17,500 points, or 875 trends for 20 time steps. Errors between the real and predicted points and the quantified accuracy of the predicted trends by the different models are discussed below.

Errors between the real and the predicted points are assessed in the form of root mean square error (RMSE), MSE, and MAE metrics, all widely used as regression metrics (Petneházi 2019). The calculated results of the error metrics of the 17,500 points are listed in Table 5.4. Note that the error metrics reflect normalized parameter values between 0 and 1, which as mentioned above is done to remove any parameter scale effects. The prediction model adopting the MIMO strategy and LSTMs trained with a learning rate of 0.01 achieves the lowest error, with RMSE, MSE, and MAE scores of 0.0213, 0.0005, and 0.0087, respectively. The second-best model adopts the direct strategy and LSTMs again with the 0.01 learning rate, showing RMSE, MSE, MAE scores of 0.0246, 0.0006, and 0.0132, respectively. As indicated in Table 5.4, the LSTM-based models outperform the models with MLPs and vanilla RNNs regardless of the prediction strategy. Furthermore, the learning rate is also seen to strongly influence the prediction model performance. For example, the RMSE of the prediction model adopting the direct strategy and MLP ranged widely from 0.0381 to 0.2918 by the learning rate.

Figure 5.6 plots examples of different parameter trend predictions, where the dotted red and solid blue lines indicate the predicted and real trends, respectively. The shaded area shows an error range of 5% based on the maximum and minimum of each parameter. Figure 5.6a shows an example of the predicted trend exactly following the real trend, with the predicted point locating inside the error range at every time step. Since the trends are composed of 20 points, the predicted trends are still mostly accurate with a small number of error points, or points lying outside the 5% error range. For example, Fig. 5.6b–d show predicted trends that follow the real trends despite respectively containing one, two, and three error points. For larger

Table 5.4 Error metrics between the real and predicted points

Learning rate		Recursive strategy			Direct strategy			MIMO strategy		
		MLP	RNN	LSTM	MLP	RNN	LSTM	MLP	RNN	LSTM
	RMSE	**0.0640**	0.0685	**0.0431**	0.2804	0.1695	0.0267	0.2096	0.0581	0.0265
0.0001	MSE	**0.0041**	0.0047	**0.0019**	0.0786	0.0287	0.0007	0.0439	0.0034	0.0007
	MAE	**0.0292**	0.0282	**0.0191**	0.2181	0.0988	0.0146	0.1442	0.0305	0.0119
	RMSE	0.0888	**0.0499**	0.0548	**0.0381**	**0.1110**	0.0260	0.1052	**0.0455**	0.0234
0.001	MSE	0.0079	**0.0025**	0.0030	**0.0015**	**0.0123**	0.0007	0.0111	**0.0021**	0.0005
	MAE	0.0391	**0.0267**	0.0226	**0.0218**	**0.0525**	0.0143	0.0643	**0.0221**	0.0099
	RMSE	0.1718	0.1718	0.0467	0.1455	0.1692	**0.0246**	**0.0829**	0.1694	**0.0213**
0.01	MSE	0.0295	0.0295	0.0022	0.0212	0.0286	**0.0006**	**0.0069**	0.0287	**0.0005**
	MAE	0.1006	0.1006	0.0214	0.0816	0.0991	**0.0132**	**0.0451**	0.0987	**0.0087**
	RMSE	0.1717	0.1716	0.1718	0.2918	0.1693	0.1694	0.3912	0.1693	0.1679
0.1	MSE	0.0295	0.0295	0.0295	0.0852	0.0287	0.0287	0.1530	0.0287	0.0282
	MAE	0.1008	0.1008	0.1005	0.1744	0.0992	0.0991	0.2640	0.0988	0.0986

The lowest error metric values for each prediction model are shown in bold

numbers of error points, though, the predicted trends start to diverge, as shown in Fig. 5.6e, f for four and five error points, respectively. Following this observation, two classification criteria are chosen for the success of parameter trend prediction: an *Accurate* predicted trend is defined as one with all points falling within the 5% error range, and a *Mostly Accurate* predicted trend is defined as one with up to three points falling outside the range.

Based on these criteria, Table 5.5 lists the percentage of successful trend predictions for the 875 trends in the test data. The prediction model with the recursive strategy and LSTMs trained with a learning rate of 0.001 reaches a maximum of 80.7% *Mostly Accurate.* Prediction models with the recursive strategy trained with high learning rates of 0.01 or 0.1 produce straight-line trends only as a consequence of error accumulation, the main drawback of this strategy. Accordingly, when the recursive strategy models are trained with a learning rate of 0.1, the *Accurate* and *Mostly Accurate* percentages are seen to be mostly similar, as in Table 5.5.

The prediction models implementing LSTMs achieve the best results for all prediction strategies. In particular, the models joining the direct or MIMO strategy with LSTMs produce more than 90% *Mostly Accurate* predictions. These two prediction models are also the best in terms of the point error metrics as in Table 5.4. Analyzing the *Accurate* classification, on the other hand, reveals different percentages: 77.7% by the direct strategy model, and 89.1% by the MIMO strategy model.

Based on the results in Tables 5.4 and 5.5, the prediction model with the MIMO strategy and LSTMs represents the most suitable model to estimate the future trends of the target NPP parameters in the current application. This model achieves 89.1% *Accurate* trend predictions for the test data, i.e., 780 of 875 trends, and 95.4% *Mostly*

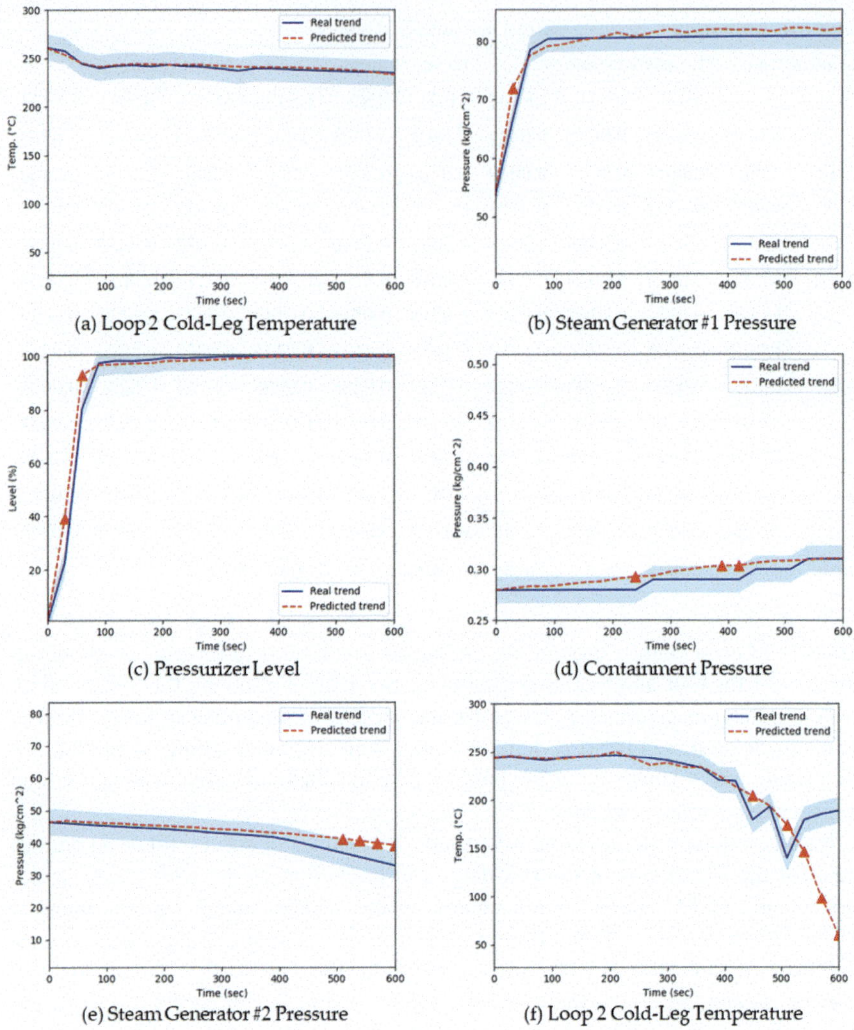

Fig. 5.6 Examples of trend prediction results of selected plant parameters containing different numbers of error points: **a** zero, **b** one, **c** two, **d** three, **e** four, and **f** five, as marked with '▲'. The shaded area denotes the 5% error range. Reproduced with permission from Bae et al. (2021)

Accurate predictions, i.e., 835 of 875 trends. In terms of computation time, it took an average of only 0.06 s to estimate the future trends of the 25 target parameters simultaneously. Figure 5.7 plots a number of trend prediction results from the finest model coupling the MIMO strategy with LSTMs. In the figure, the predicted trends are seen to closely mirror the real trends.

Looking more closely at computation time, Table 5.6 lists the average time for calculation of each pair of prediction strategy and ANN to estimate the future trends of the 25 target NPP parameters. From the simple nature of the MLP cell, the MLP

Table 5.5 Percentages of successful trend prediction by the prediction models

Learning rate		Recursive strategy			Direct strategy			MIMO strategy		
		MLP (%)	RNN (%)	LSTM (%)	MLP (%)	RNN (%)	LSTM (%)	MLP (%)	RNN (%)	LSTM (%)
0.0001	Accurate	**64.5**	**68.1**	71.4	8.8	34.9	76.3	33.6	64.5	86.9
	Mostly accurate	**73.6**	**76.2**	83.7	11.3	41.6	89.6	37.9	73.6	94.5
0.001	Accurate	61.4	61.3	**72.1**	**65.9**	**33.9**	76.0	36.7	**73.1**	89.0
	Mostly accurate	67.2	72.3	**80.7**	**79.9**	**52.3**	89.9	46.9	**82.3**	95.4
0.01	Accurate	36.3	36.6	69.8	22.6	33.7	**77.7**	**50.2**	33.0	**89.1**
	Mostly accurate	43.2	43.2	80.0	40.0	41.6	**92.1**	**62.9**	41.9	**95.4**
0.1	Accurate	36.0	36.1	37.0	12.0	32.9	33.7	24.1	34.4	30.3
	Mostly accurate	43.3	43.4	43.3	12.0	41.5	41.7	26.9	41.7	39.8

The highest percentages for each prediction model are shown in bold

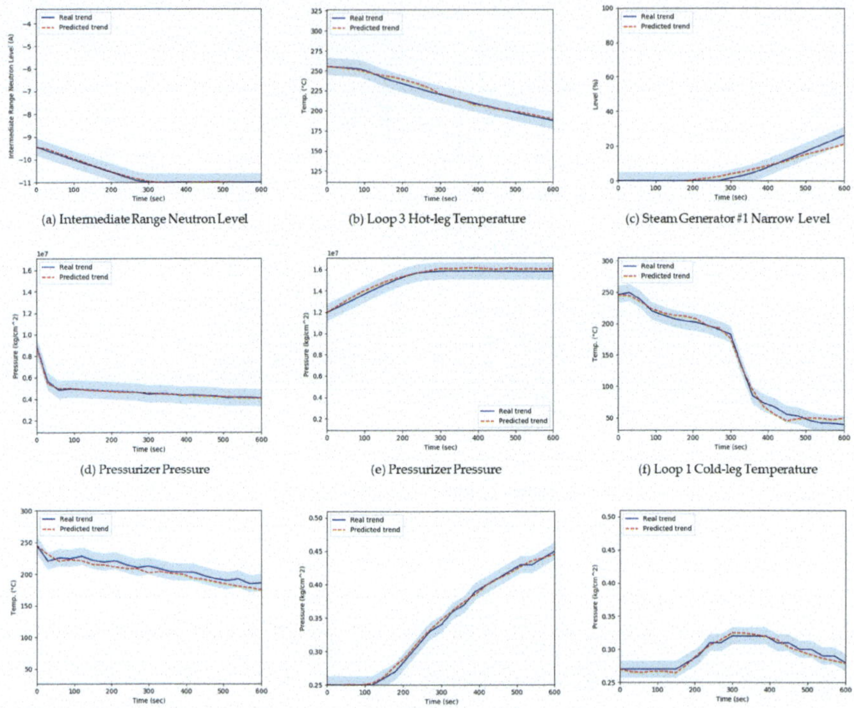

(a) Intermediate Range Neutron Level (b) Loop 3 Hot-leg Temperature (c) Steam Generator #1 Narrow Level (d) Pressurizer Pressure (e) Pressurizer Pressure (f) Loop 1 Cold-leg Temperature (g) Loop 3 Cold-leg Temperature (h) Containment Pressure (i) Containment Pressure

Fig. 5.7 Trend prediction results with the MIMO + LSTM prediction model. **a** Intermediate range neutron level under 25 cm^2 LOCA in the Loop 1 Cold-leg. **b** Loop 3 Hot-leg temperature under 60 cm^2 LOCA in the Loop 1 Cold-leg. **c** SG #1 narrow level under SGTR with double tube rupture. **d** PZR pressure after PORV opening under SGTR with double tube rupture. **e** PZR pressure after PORV shut-off valve opening under the simple reactor trip situation. **f** Loop 1 Cold-leg temperature after resetting the safety signal under 25 cm^2 LOCA. **g** Loop 3 Cold-leg temperature after RCPs stop under 45 cm^2 LOCA. **h** CTMT pressure after PORV opening under SGTR with single tube rupture. **i** CTMT pressure after conducting the prior EOP under 35 cm^2 LOCA. Reproduced with permission from Bae et al. (2021)

models exhibit the lowest computation times. But even the best-performing model combining MIMO with LSTMs is able to record a computation time of merely 61.1 ms, a time considered sufficient for real-time prediction. Model calculations are performed using a laptop computer with an Intel Core i7-8700, 16 GB of RAM, Python version 3.7.6, and TensorFlow version 2.0.0.

5.1.4 Operator Support System with Prediction

Next-generation NPPs have been adopting advanced MCRs with digital features with the aim to improve operator performance and prevent human error. For instance, in

Table 5.6 Average computation times for the 35 test scenarios when the trends of the 25 target parameters are predicted

	Recursive strategy			Direct strategy			MIMO strategy		
	MLP	RNN	LSTM	MLP	RNN	LSTM	MLP	RNN	LSTM
Computation time (ms)	27.9	45.2	46.5	29.6	47.2	48.3	37.2	59.6	61.1

Reproduced with permission from Bae et al. (2021)

the advanced MCR of APR 1400, a large central display and personal computers take the place of the analog indicators, hand switches, and alarm tiles of traditional MCRs. In such digital MCRs, operators monitor the plant status and execute controls from their positions via soft controls like keyboard inputs and mouse devices. The previous paper-based operating procedures are replaced with a computerized procedure system (CPS) that facilitates more efficient task execution. With this system, the task-related parameters can be displayed in a selective manner, and by displaying check-off provisions after the performance of each procedural action, omission errors can be prevented.

One key element of advanced MCRs is the provision of operator support systems, which refer to systems that convey valuable information to operators or automate particular tasks. For such support systems, the important characteristic is that the *final decision-maker* is always a human operator. The general system structure was formalized by Lee et al. based on the cognitive process of MCR operators (Lee and Seong 2014). Operator support systems can then be classified by the cognitive activity they support: monitoring/detection, situation assessment, response planning, and response implementation. For example, operating procedures support activities related to situation assessment and response planning, and alarm systems support operator activities related to monitoring and detection. Along these lines, recent developments include a CPS (Lew et al. 2018), advanced alarm system (Kang and Lee 2022), and abnormal/emergency diagnosis system (Mo et al. 2007; Lee et al., 2021; Shin et al. 2021; Kim et al. 2020). But in terms of supporting activities related to response implementation, such as by developing an operation validation system, relatively limited research has been explored.

The prediction model with neural networks in this chapter has the potential to be utilized as an inference model as part of an operation validation system. Through the future parameter trend estimates of the prediction model, operators can verify an intended control action and detect a potential human error at an early stage. A framework for an operator manipulation validation system has previously been suggested, in which assessments of the CSFs are performed with a prediction model (Bae and Lee 2019). As the primary safety goal in an NPP emergency is to prevent reactor core damage, any action that deteriorates the CSFs, which directly relate to securing the safety goal, could be considered a human error. By forecasting the future trends of the CSF-related parameters, the prediction model enables any possible future degradation of the CSFs to be anticipated. When such a degradation is predicted, the validation system notifies and guides the MCR operators to deliberate the related action, whether planned or already committed. Moreover, the future parameter trends as supportive information can also assist operators in diagnosing accidents and understanding the ensuing plant dynamics.

Three examples of future trend predictions by the best-performing prediction model are shown in Fig. 5.8 for the PZR level once it reaches 0%. Here, the model anticipates that the PZR level will be maintained at 0% upon LOCA occurrence due to the loss of reactor coolant, as shown in Fig. 5.8a. In the case that operators incorrectly open the PORV, the reactor coolant will be depressurized and expended; as shown in Fig. 5.8b, the model correctly predicts for this case that the PZR level will rapidly reach 100%. With no loss of coolant as in a simple reactor trip accident, the reactor coolant volume drops immediately after the trip but then starts to increase again; Fig. 5.8c shows that model reflects this future trend as well. It is noteworthy that the prediction model with MIMO and LSTMs can successfully forecast the future trends of NPP parameters, such as the PZR water level, even under varying operator actions and accident environments. These results demonstrate the model's potential usefulness in helping operators to detect a committed human error, like in Fig. 5.8a, b, as well as to diagnose an accident, like in Fig. 5.8a, c.

Prior to real-world applications, additional work needs to be conducted. The first would be the establishment of a more effective framework for data generation. In early development stages, it is unavoidable to use data from NPP simulators, as real operational data is confidential and in the case of accidents, rare. The prediction model introduced in this section employed a compact NPP simulator and considered a limited number of scenarios, which were analyzed based on the authors' expertise. To minimize uncertainty in the training data, it would be beneficial for the subsequent data generation framework to employ full-scope simulators, which are used in NPP design, construction, management, and operation as well as by regulatory boards. Such a framework should have the capability to generate a wide range of operational scenarios in a short time and prioritize them in terms of probability. The second area for future work to focus on would be to revisit the multi-step prediction strategy in order to expand the prediction horizon to a sufficient length, which would help

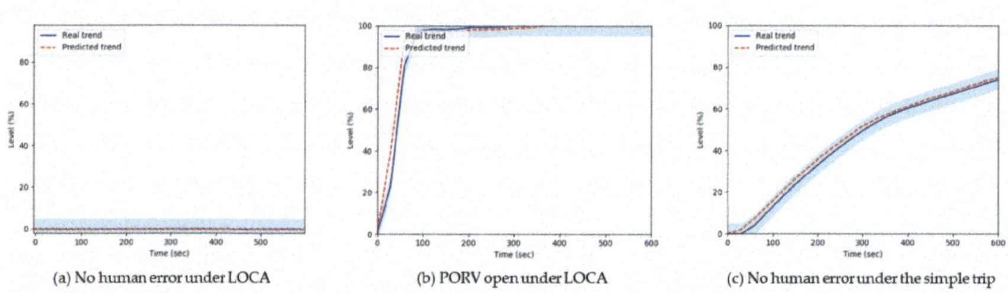

Fig. 5.8 Trend prediction results of PZR water level with the MIMO + LSTM prediction model for different operation actions and accident situations. **a** No human error in a LOCA situation, **b** mistaken opening of the PORV in a LOCA situation, and **c** no human error in the simple trip situation. Each starts with an initial PZR water level of 0%. Reproduced with permission from Bae et al. (2021)

gain more information that supports the trend predictions. A customized multi-step prediction strategy applied to an optimal neural network would grant predictions that are both faster and longer-term. The current prediction algorithm has extended its prediction horizon (Kim et al. 2021) and has implemented an efficient neural network architecture (Ryu et al. 2022). The third requirement for real applications would be to achieve tolerance to noise in the field data. Harsh environments, problems with transmitters, and calibration shifts, among others, all create noise in field data that can markedly decrease neural network prediction accuracy (Shin et al. 2021).

References

Abadi M, Agarwal A, Barham P, Brevdo E, Chen Z, Citro C, Corrado GS, Davis A, Dean J, Devin M (2016) Tensorflow: Large-scale machine learning on heterogeneous distributed systems. arXiv preprint arXiv:1603.04467

Bae J, Kim G, Lee SJ (2021) Real-time prediction of nuclear power plant parameter trends following operator actions. Expert Syst Appl 186:115848

Bae J, Lee SJ (2019) Framework for operator manipulation validation system using plant parameter prediction. In: Korean nuclear society autumn meeting, Korean Nuclear Society

Bae J, Ahn J, Lee SJ (2019) Comparison of multilayer perceptron and long short-term memory for plant parameter trend prediction. Nuclear Technol

Chollet F (2015) keras

El-Sefy M, Yosri A, El-Dakhakhni W, Nagasaki S, Wiebe L (2021) Artificial neural network for predicting nuclear power plant dynamic behaviors. Nucl Eng Technol 53(10):3275–3285

Kang JS, Lee SJ (2022) Concept of an intelligent operator support system for initial emergency responses in nuclear power plants. Nuclear Eng Technol

Kim WJ, Chang SH, Lee BH (1993) Application of neural networks to signal prediction in nuclear power plant. IEEE Trans Nucl Sci 40(5):1337–1341

Kim JM, Lee G, Lee C, Lee SJ (2020) Abnormality diagnosis model for nuclear power plants using two-stage gated recurrent units. Nuclear Eng Technol

Kim H, Jo S, Kim J, Park G, Kim J (2021) Development of long-term prediction algorithm based on component states using BiLSTM and attention mechanism. In: 2021 5th international conference on system reliability and safety (ICSRS), pp 258–264

Kingma DP, Ba J (2015a) Adam: a method for stochastic optimization. CoRR, abs/1412.6980

Kwon KC, Park JC, Jung CH, Lee JS, Kim JY (1997) Compact nuclear simulator and its upgrade plan. In: Training simulators in nuclear power plants: Experience, programme design and assessment methodology Proceedings of a specialists' meeting, p 227

Lee SJ, Seong PH (2014a) Design of an integrated operator support system for advanced NPP MCRs: issues and perspectives. In: Yoshikawa H, Zhang Z (eds) Progress of nuclear safety for symbiosis and sustainability: advanced digital instrumentation, control and information systems for nuclear power plants. Tokyo, Springer Japan

Lee G, Lee SJ, Lee C (2021b) A convolutional neural network model for abnormality diagnosis in a nuclear power plant. Appl Soft Comput 99

Lew R, Boring R, Ulrich T (2018) A computerized procedure system framework for US utilities. Safety and reliability–safe societies in a changing World, pp 427–432

Miettinen J (1985) Development and assessment of the SBLOCA code SMABRE. In: Proceedings in specialists meeting on small break LOCA analyses in LWRs, 2, pp 481–495

Mo K, Lee SJ, Seong P (2007) A dynamic neural network aggregation model for transient diagnosis in nuclear power plants. Prog Nucl Energy 49:262–272

Moshkbar-Bakhshayesh K (2019) Comparative study of application of different supervised learning methods in forecasting future states of NPPs operating parameters. Ann Nucl Energy 132:87–99
Nguyen H-P, Liu J, Zio E (2020) A long-term prediction approach based on long short-term memory neural networks with automatic parameter optimization by Tree-structured Parzen Estimator and applied to time-series data of NPP steam generators. Appl Soft Comput 89:106116
Nguyen H-P, Baraldi P, Zio E (2021) Ensemble empirical mode decomposition and long short-term memory neural network for multi-step predictions of time series signals in nuclear power plants. Appl Energy 283:116346
Petneházi G (2019) Recurrent neural networks for time series forecasting. ArXiv, abs/1901.00069
Radaideh MI, Pigg C, Kozlowski T, Deng Y, Qu A (2020) Neural-based time series forecasting of loss of coolant accidents in nuclear power plants. Expert Syst Appl 160:113699
Ryu S, Kim H, Kim SG, Jin K, Cho J, Park J (2022) Probabilistic deep learning model as a tool for supporting the fast simulation of a thermal-hydraulic code. Expert Syst Appl 200:116966
Shin JH, Kim JM, Lee SJ (2021) Abnormal state diagnosis model tolerant to noise in plant data. Nucl Eng Technol 53(4):1181–1188
Taieb SB, Bontempi G, Atiya AF, Sorjamaa A (2012) A review and comparison of strategies for multi-step ahead time series forecasting based on the NN5 forecasting competition. Expert Syst Appl 39(8):7067–7083

Chapter 6
Control

Control in an autonomous NPP refers to achieving a high-level of automation. To a large extent, NPPs are already automated systems designed to increase electricity availability, reduce accident risk, and decrease operating costs (Wood et al. 2004). Regulatory authorities have required the automation of safety system functions as they must be both exceptionally reliable and rapidly executed to secure public safety. But the level of automation in commercial NPPs at present is not so high, as operator interventions take up a large portion of operation; it can be said that some functions are automated but many are manual. An autonomous NPP, in contrast, aims to minimize the operators' interventions as much as possible. For example, at the automation level of *Operation by Exception* in Table 1.1 of Chap. 1, human operators would only be required for approving critical decisions.

Typical approaches to automatic controllers in current NPPs include PID controllers, programmable logic controllers, and field-programmable gate arrays (Yoo et al. 2004, 2008; She and Jiang 2011; Khatua and Mukherjee 2021). For safety systems, programmable logic controllers are generally used for automatic, fast, and reliable responses to prevent malfunctions from propagating into major accidents. For non-safety systems, PID controllers or controllers that combine two out of the three controller types, e.g., proportional-integral controllers, are the most common among existing NPPs. These controllers generally aim to stabilize a given system within a defined range.

Controllers applying AI techniques have recently been studied in several industrial fields (Wei et al. 2017). Since the 2000s, DL in general has drawn increased attention along with increases in computing power, increases in data sizes, and advances in DL-related research (Du et al. 2020; Zhou et al. 2020). One popular approach is deep RL or DRL, which derives from the combination of RL and DL. In Deep RL, a training mechanism is applied that is very similar to that of humans, and thus a DRL-based controller is able to develop its own experiences through trial and error, similar to humans. This allows DRL-based controllers to perform tasks that classical controllers cannot, such as selecting an operation strategy, operating certain systems, making decisions based on the current conditions, and optimizing operations. For

J. Kim et al., *Autonomous Nuclear Power Plants with Artificial Intelligence*, Lecture Notes in Energy 94, https://doi.org/10.1007/978-3-031-22386-0_6

this reason, DRL-based controllers have been developed in diverse fields including robotics (Ng et al. 2006; Zhang et al. 2020; Kohl and Stone 2004), autonomous vehicles (Bhalla et al. 2020; Yu et al. 2019; Viitala et al. 2020), smart buildings (Wei et al. 2017), power management (Kang et al. 2020; Zhou et al. 2020; Samadi et al. 2020; Kazmi et al. 2018; Rocchetta et al. 2019), the railway industry (Yang et al. 2020), wind turbines (Saenz-Aguirre et al. 2019), traffic signals (Genders and Razavi 2020), and NPPs (Lee and Kim 2021; Dong et al. 2020; Park et al. 2020).

This chapter describes the development of an autonomous operation algorithm applying DRL for two different NPP operational modes.

6.1 Autonomous Control for Normal and Emergency Operations with RL

Automation has been adopted in many industries to reduce human errors and to improve operation reliability (Wood et al. 2004). In the nuclear field, NPPs are operated through manual manipulations by operators in tandem with automated systems (Kim et al. 2020). During full-power operation, most controls are made automatically, but manual controls are still required for start-up and shutdown operations (Lee and Kim 2018). For example, normal operation, which refers to the NPP operational period from start-up to shutdown, includes the following task: "The operator should withdraw the control rods while maintaining the reactor power at 2%". In this task, operators should manually monitor a number of parameters such as the reactor power, concentration of boron, and withdrawal extent of the control rods to maintain the reactor power. Another operational period is emergency operation, which starts at the point of reactor trip due to an abnormal event or accident and ends at the point when the reactor reaches the shutdown cooling entry condition. An example task for operators in emergency operation is as follows: "Cool down the RCS temperature to 55 °C per hour within the pressure–temperature curve", an action taken to prevent damage to the reactor core. In this task, operation becomes burdensome because operators must manually control multiple components such as specific valves and pumps while concurrently considering all operating restrictions. Accordingly, by automating the tasks that require a lot of manipulations, the operators' workload can be lowered, which can reduce human errors (Kim et al. 2016; Kima and Park 2018; Lee et al. 2018).

The tasks that require manipulation and decision-making are indicated in the operating procedures. These tasks can be classified as decision-making, continuous control, and discrete control. Decision-making refers to the task of determining an operating strategy; it does not include any manipulations by the operator. As a representative example, the task requiring operators to determine the rate of power increase during start-up operation is classified as decision-making. The result of the decision, which is the determined rate, affects subsequent manipulations in terms of timing and frequency.

Continuous control focuses on achieving a defined goal but without specifying the related manipulations to achieve the goal, including success criteria and operational limitations. For example, in start-up operation, one operational goal is to increase the power from 2 to 100%. For this, one task requires operators to adjust the boron concentration in the make-up water. The related operating procedure, though, does not indicate by how much the boron concentration should be adjusted because the amount of boron can change depending on the rate of power increase and the reactivity of the nuclear reactor.

Discrete control, on the other hand, specifies explicit conditions including specific states or values for component manipulation. An example of discrete control is the following task: "If the reactor power reaches 40%, operate main feed water pump #2".

To substitute automation for manual manipulation, the characteristics of the tasks need to be understood. For discrete controls, since there is a clear rule, automation can be achieved using a rule-based system with a simple *if–then* logic. However, it is difficult to implement a rule-based system for continuous controls. Likewise, for decision-making tasks, the optimal operating strategies can vary by the operators' decisions, which also makes if–then logic difficult to be applied. To address the cases of manual manipulation for which it is hard to implement the if–then rule for increased levels of automation, DRL presents a promising solution. When given an operating goal, DRL can find an efficient way to achieve that goal through learning (see Sect. 2.2.3 for a brief background on RL). In this section, an autonomous operation algorithm based on DRL is developed and applied to both normal and emergency NPP operations. The proposed algorithm is validated and demonstrated using a CNS via a simulation of a Westinghouse-type NPP.

6.1.1 Case Study 1: Power-Increase Operation

The first target of the autonomous operation algorithm is the power-increase operation, in which electricity is generated by transferring heat from the reactor core to the turbines. As NPPs consist of a primary system and a secondary system, as illustrated in Fig. 6.1, the power-increase operation must be carried out in parallel so that these two systems can be operated simultaneously. In the primary system, operators control the reactor power by adjusting the control rods and boron concentration in accordance with the procedure "Power operation at greater than 2% power". In the secondary system, operators follow the "Secondary system heat-up and start-up" procedure to operate the turbine, which converts the generated power to electricity. The operators take these actions for the primary and secondary systems concurrently to guide the NPP state from the initial conditions of the power-increase operation to the final conditions, as listed in Table 6.1.

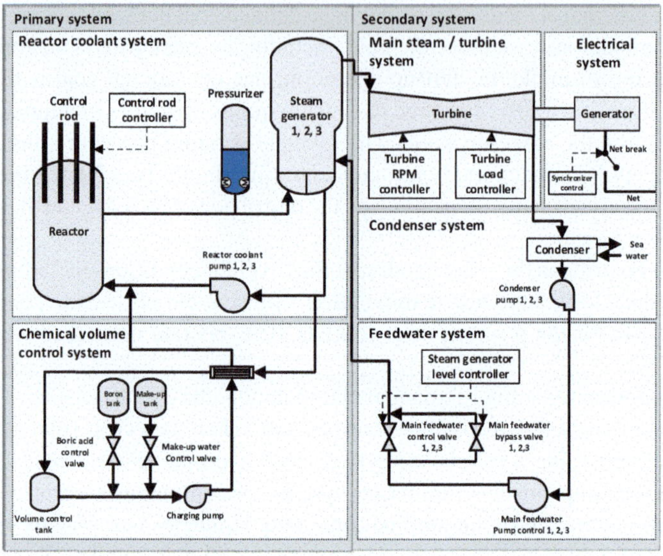

Fig. 6.1 Simplified schematic of the primary and secondary systems with related components. Reproduced with permission from Lee et al. (2020)

Table 6.1 Initial and final conditions of the power-increase operation

Major parameter	Initial condition	Final condition
Reactor power	2%	100%
Electric power	0 MWe	990 MWe
RCS average temperature	294 °C	306 °C
Turbine revolutions per minute (RPM)	0	1800 RPM
Turbine load setpoint	0 MWe	990 MWe
Turbine load rate setpoint	0 MWe/min	2 MWe/min
Boron concentration	637 ppm	457 ppm
Rod position	211 Step (A Bank) 95 Step (B Bank) 0 Step (C Bank) 0 Step (D Bank)	228 Step (A Bank) 228 Step (B Bank) 228 Step (C Bank) 220 Step (D Bank)
Rod controller	Manual	Auto
SG controller	Manual	Auto
Feedwater pump 1	On	On
Feedwater pump 2	Off	On
Feedwater pump 3	Off	On
Condenser pump 1	On	On
Condenser pump 2	Off	On
Condenser pump 3	Off	On
Synchronous connection	Disconnected	Connected

Reproduced with permission from Lee et al. (2020)

6.1.1.1 Timeline and Task Analysis for the Power-Increase Operation

To automate the power-increase operation, it is necessary to clarify which tasks are required for heating up the nuclear reactor and generating electricity. The tasks can be analyzed by task type and time based on the two above-mentioned operating procedures. Table 6.2 sequentially lists the required actions in the order specified in the procedures.

Figure 6.2 shows the timings of the tasks in Table 6.2 as the NPP state changes. This timeline analysis is performed based on interviews with a senior reactor operator with working experience at the reference NPP. While most of the operation steps are made sequentially, some steps can be performed in parallel. For example, as shown in Fig. 6.2e, while an operator is following Step 6 (Adjust the boron concentration to

Table 6.2 Operational tasks for increasing the reactor power

Step	Task type	Action
1	Decision-making	Determine the rate of power increase in %/h
2	Continuous control	Withdraw all control rods to the position of 100% reactor power while maintaining the reactor power at 2% through boration
3	Continuous control	If all the control rods are withdrawn, increase the reactor power from 2 to 6%–10% by reducing the boron concentration
4	Discrete control	If the reactor power is 10%, the turbine RPM setpoint is 1800 RPM
5	Discrete control	If the reactor power exceeds 10%, the acceleration setpoint is 2 MWe/min
6	Continuous control	Adjust the boron concentration to increase the reactor power from 10 to 20%
7	Discrete control	If the reactor power is between 10 and 20%, the load setpoint is 100 MWe
8	Discrete control	If the turbine RPM is 1800 RPM and the reactor power exceeds 15%, push the net-breaker
9	Discrete control	If the reactor power is 20%, start condenser pump #2
10	Continuous control	Adjust the boron concentration to increase the reactor power from 20 to 100%
11	Discrete control	If the reactor power is between 20 and 30%, the load setpoint is 200 MWe
12	Discrete control	If the reactor power is between 30 and 40%, the load setpoint is 300 MWe
13	Discrete control	If the reactor power is 40%, start main feedwater pump #2
14	Discrete control	If the reactor power is between 40 and 50%, the load setpoint is 400 MWe
15	Discrete control	If the reactor power is between 50 and 60%, the load setpoint is 500 MWe
16	Discrete control	If the reactor power is 50%, start condenser pump #3
17	Discrete control	If the reactor power is between 60 and 70%, the load sctpoint is 600 MWe
18	Discrete control	If the reactor power is between 70 and 80%, the load setpoint is 700 MWe
19	Discrete control	If the reactor power is 80%, start main feedwater pump #3
20	Discrete control	If the reactor power is between 80 and 90%, the load setpoint is 800 MWe
21	Discrete control	If the reactor power is between 90 and 100%, the load setpoint is 990 MWe

Reproduced with permission from Lee et al. (2020)

increase the reactor power from 10 to 20%), the operator can concurrently perform Step 8 (If the turbine RPM is 1800 RPM and the reactor power exceeds 15%, push the net-breaker).

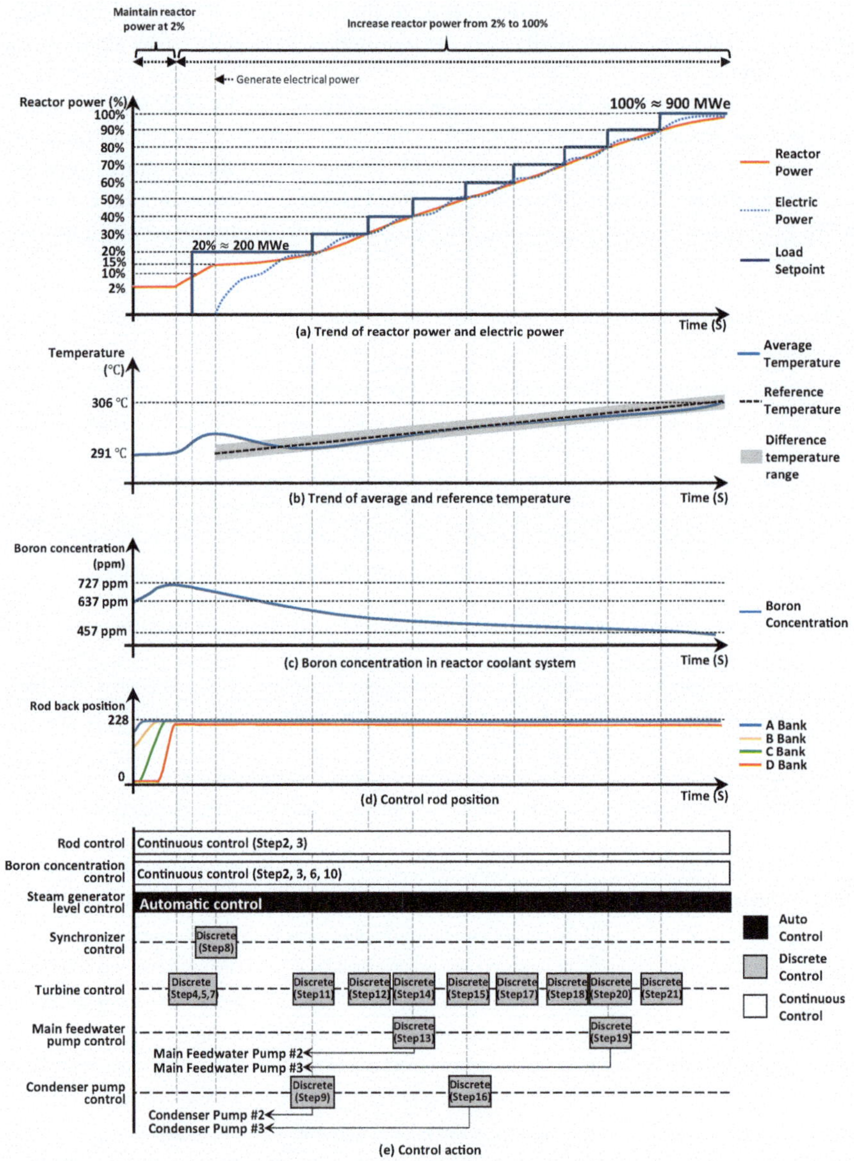

Fig. 6.2 Timeline for increasing the reactor power from 2 to 100%. Reproduced with permission from Lee et al. (2020)

The power-increase operation is divided into two operational ranges: (1) withdrawing control rods and maintaining the reactor power at 2%, and (2) increasing the reactor power from 2 to 100%. In the first operational range, the control rod withdrawal begins as shown in Fig. 6.2d, which may increase the reactor power. In order to maintain the reactor power at 2%, boron is injected to compensate the positive reactivity, as shown in Fig. 6.2c.

The goal of the second operational range is to increase the reactor power up to the full power condition. In this range, operators should raise the reactor power in accordance with the specified power increase rate per hour, which must first be determined by the operators. Here, the difference between the average temperature of the primary system and the reference temperature, which is the temperature of the secondary system, should not exceed ± 1 °C, as illustrated in Fig. 6.2b. To increase the reactor power, operators inject make-up water so that the boron concentration in the RCS gradually decreases. In this case, the amount of the injected make-up water is determined in accordance with the operational constraints of the temperature difference and reactor power.

6.1.1.2 Algorithm Structure for the Power-Increase Operation

Considering the different types of operational tasks discussed above, the developed algorithm for the power-increase operation consists of discrete and continuous control modules, as depicted in Fig. 6.3.

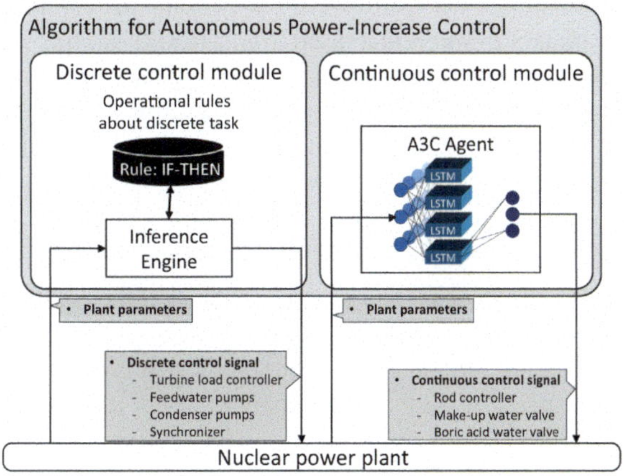

Fig. 6.3 Overview of the algorithm for the power-increase operation. Reproduced with permission from Lee et al. (2020)

Discrete controls can be represented with clear rules, so they can be implemented in the algorithm using a rule-based system. Such a rule-based system can generate an action according to a predefined if–then condition. In the discrete control module, the rule-based system uses converted rules from the operational rules applying if–then logic. The operator tasks of discrete control type are largely four control functions—control of the synchronizer, turbine, main feedwater pump, and condenser pump—which are converted into if–then logic as listed in Table 6.3.

The second module of the algorithm, the continuous control module, aims to manage the reactor power during the power-increase operation based on the specified power-increase rate. Since continuous control in this operation cannot be represented with clear rules, this module applies an asynchronous advantage actor-critic (A3C) agent, which is a kind of DRL. The agent mainly consists of a reward algorithm and an LSTM model, as shown in Fig. 6.4. The reward algorithm calculates the reward for training the agent and evaluates the current plant parameters (Guo 2017). Using the obtained and evaluated plant parameters, the LSTM network selects the appropriate operation strategies among "increase", "decrease", or "stay". Then the A3C agent determines the control actions considering the current goal of the power-increase operation.

The reward algorithm is required to provide the direction of learning to the agent. As the name of the algorithm implies, the direction of learning is provided to the agent in the form of a reward, which is a value determined by the current state of the environment. In particular, the reward in the power-increase operation is divided into two operational criteria: (1) the power increase rate, and (2) the temperature difference between the primary and secondary systems.

The first reward, or power reward, is provided by considering the power increase rate. Since the operational goals differ between the two operational ranges, the reward should be provided separately for each range. Figure 6.5 illustrates the two ranges: the blue area represents the range in which the reactor power is maintained at 2% by $\pm$ 1%, or in other words from 1 to 3%, and the red area represents the range in which the reactor power is increased from 2 to 100%. The transition point, meaning the shift from blue to red, is when the control rods are fully drawn out for the power-increase to 100%. After the transition, the operational goal changes from "power-maintain" to "power-increase". To calculate the first reward, upper and lower boundaries are defined for the reactor power as in Eqs. (6.1) to (6.3). Using these equations, the boundaries for the red area in Fig. 6.5 are outlined linearly with the predefined power increase rate and the current operation time.

$$\text{End of operation time } (t_{100}) = t_2 + \frac{100 - 2}{Pr} \tag{6.1}$$

$$\text{Upper boundary} = \begin{cases} 3 & (t_2 \geq t) \\ \frac{100-3}{t_{100}-t_2}(t - t_2) + 3 & (t_{100} \geq t > t_2) \\ 110 & (t > t_{100}) \end{cases} \tag{6.2}$$

$$\text{Lower boundary} = \begin{cases} 1 & (t_2 \geq t) \\ \frac{100-3}{t_{100}-t_2}(t - t_2) + 1 & (t_{100} \geq t > t_2) \\ 90 & (t > t_{100}) \end{cases} \tag{6.3}$$

Table 6.3 Discrete control module if–then rules for increasing the reactor power from 2 to 100%

Function	Rule number(s)	If–then rule	Input(s)	Output(s)
Synchronizer control	1	If the turbine RPM is 1800 RPM and the reactor power is greater than 15%, push the net-breaker button	Reactor Power, Turbine RPM	Net-breaker Button Control
Turbine control	2	If the reactor power is 10%, the turbine RPM setpoint is 1800 RPM	Reactor Power, Turbine RPM	Turbine RPM Setpoint Control
	3	If the reactor power is greater than 10%, the acceleration setpoint is 2 Mwe/min	Turbine Acceleration	Turbine Acceleration Setpoint Control
	4	If the reactor power is between 10 and 20%, the load setpoint is 100 Mwe	Reactor Power, Load Setpoint	Load Setpoint Control
	5–11	…	…	…
	12	If the reactor power is between 90 and 100%, the load setpoint is 990 Mwe	Reactor Power, Load Setpoint	Load Setpoint Control
Main feedwater pump control	13	If the reactor power is 40% and the state of the main feedwater pump 1 is "activated," start main feedwater pump 2	Reactor Power, Main Feedwater Pumps 1 and 2 States	Main Feedwater Pump 2 Control
	14	If the reactor power is 80% and the state of main feedwater pump 2 is "activated," start main feedwater pump 3	Reactor Power, Main Feedwater Pumps 2 and 3 States	Main Feedwater Pump 3 Control

(continued)

Table 6.3 (continued)

Function	Rule number(s)	If–then rule	Input(s)	Output(s)
Condenser pump control	15	If the reactor power is 20% and the state of condenser pump 1 is "activated," start condenser pump 2	Reactor Power, Condenser Pumps 1 and 2 States	Condenser Pump 2 Control
	16	If the reactor power is 50% and the state of condenser pump 2 is "activated," start condenser pump 3	Reactor Power, Condenser Pumps 2 and 3 States	Condenser Pump 3 Control

Reproduced with permission from Lee et al. (2020)

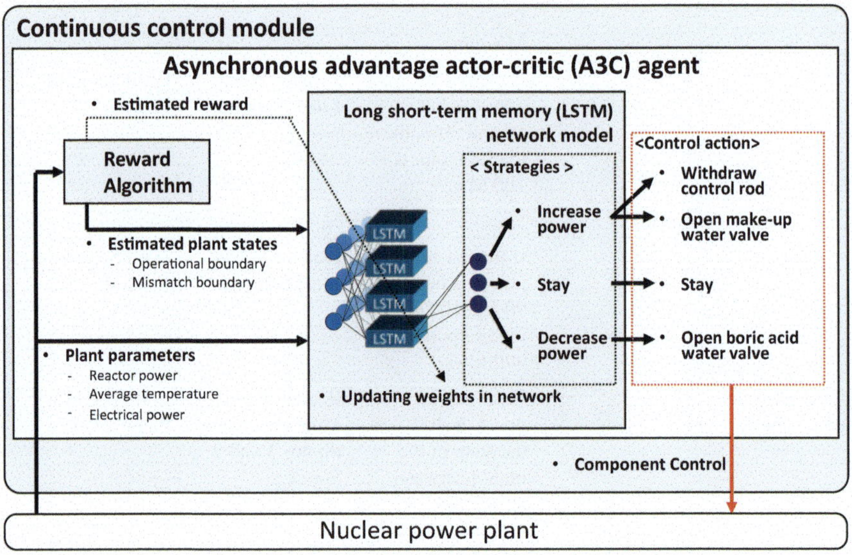

Fig. 6.4 Overview of the continuous control module. Reproduced with permission from Lee et al. (2020)

Pr: Predefined rate of power increase (%/h)
t: Time
t_2: Time at all rods 100% withdrawal
t_{100}: End of operation time

The power reward is then calculated as the distance between the current power at time *t* and the desirable power that is the midpoint between the boundaries.

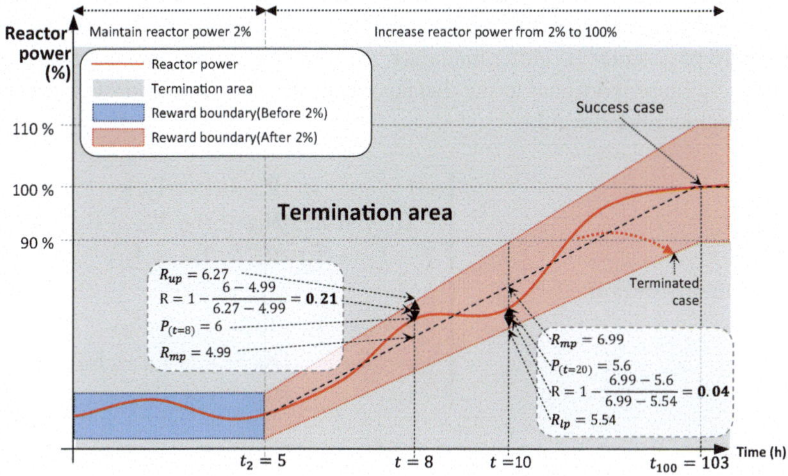

Fig. 6.5 Power reward for the A3C agent. Reproduced with permission from Lee et al. (2020)

$$\text{Power reward } (0 \sim 1) = \begin{cases} 0 & (\mathrm{P} > \mathrm{R_{up}}) \\ 1 - \frac{\mathrm{P}-\mathrm{R_{mp}}}{\mathrm{R_{up}}-\mathrm{R_{mp}}} & (\mathrm{R_{up}} \geq \mathrm{P} > \mathrm{P_{mp}}) \\ 1 & (\mathrm{P} = \mathrm{R_{mp}}) \\ 1 - \frac{\mathrm{R_{mp}}-\mathrm{P}}{\mathrm{R_{mp}}-\mathrm{R_{lp}}} & (\mathrm{R_{mp}} > \mathrm{P} \geq \mathrm{R_{lp}}) \\ 0 & (\mathrm{P} < \mathrm{R_{lp}}) \end{cases} \tag{6.4}$$

P: Current power at time t (%)
$\mathrm{R_{mp}}$: Middle of power reward boundary, i.e., predefined power at time t
$\mathrm{R_{up}}$: Upper power reward boundary
$\mathrm{R_{lp}}$: Lower power reward boundary

A power reward with a negative value indicates that the current power is outside the boundaries. In this case, the operation is considered as a failure and the training is terminated for that episode. The agent then applies different actions in a new episode that is set from the initial operation condition (t = 0). Thus, the power reward plays an important role in determining whether the power-increase operation performed by the A3C agent succeeds or not.

The second reward, or temperature reward, considers the difference between the average temperature and the reference RCS temperature. Regarding the average temperature control, the operation procedure specifies the following recommendation: "Operate the average RCS temperature within ± 1 °C of the reference RCS temperature during the power-increase operation". Considering that the reference RCS temperature is related to the current turbine load (MWe), the temperature reward can be calculated using Eq. (6.5) after electrical power is generated. In case of no electrical power, the temperature reward is zero. Since the temperature reward is derived from the above recommendation operation unlike the power reward, the temperature reward is not directly related to the termination of the episode. If the average RCS temperature passes outside either boundary, the temperature reward is returned with a negative value proportional to the distance from the closest boundary. Figure 6.6 plots the temperature reward.

$$\text{Temperature reward}(-1 \sim 1) = \begin{cases} -1 & (\mathrm{R_{ut}} + 1 < \mathrm{T_{av}}) \\ -\mathrm{T_{av}} + \mathrm{R_{yt}} & (\mathrm{R_{ut}} + 1 \geq \mathrm{T_{av}} > \mathrm{R_{ut}}) \\ 1 - \mathrm{T_{av}} + \mathrm{T_{rf}} & (\mathrm{R_{ut}} \geq \mathrm{T_{av}} > \mathrm{T_{rf}}) \\ 1 & (\mathrm{T_{av}} = \mathrm{T_{rf}}) \\ 1 + \mathrm{T_{av}} - \mathrm{T_{rf}} & (\mathrm{T_{rf}} > \mathrm{T_{av}} \geq \mathrm{R_{lt}}) \\ \mathrm{T_{av}} - \mathrm{R_{lt}} & (\mathrm{R_{lt}} - 1 \leq \mathrm{T_{av}} < \mathrm{R_{lt}}) \\ -1 & (\mathrm{R_{lt}} - 1 > \mathrm{T_{av}}) \end{cases} \tag{6.5}$$

T: Average RCS temperature at time t

$\mathrm{T_{rf}}$: Middle of temperature reward boundary, i.e., reference temperature at time t

$\mathrm{R_{ut}}$: Upper temperature reward boundary ($\mathrm{T_{rf}} + 1$) at time t

$\mathrm{R_{lt}}$: Lower temperature reward boundary ($\mathrm{T_{rf}} - 1$) at time t

With these two rewards, the agent derives the total reward with the arithmetic mean of the power and temperature rewards as Eq. (6.6).

$$\text{Total Reward } (-1 \sim 1) = \frac{\text{Power Reward} + \text{Temperature Reward}}{2} \tag{6.6}$$

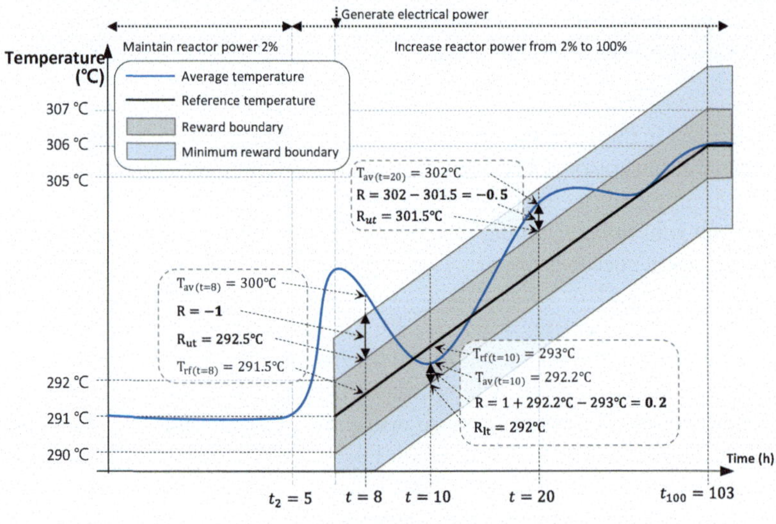

Fig. 6.6 Temperature reward for the A3C agent. Reproduced with permission from Lee et al. (2020)

6.1.1.3 Experiments

For a real-time testbed to train and validate the autonomous operation algorithm for the power-increase operation, a CNS is employed. One main computer and three sub-computers make up the environment for A3C training in parallel, as shown in Fig. 6.7. The main computer embeds 1 main agent and 60 local agents for implementing the algorithm. Each sub-computer can execute 20 CNS simulations at a time, meaning 60 simulations can be run concurrently.

The trend of the rewards is evaluated to check that the A3C agent is well trained from the simulations. Figure 6.8 plots the rewards over 8800 episodes. For a complete power-increase operation (from 0 to 100%) at a rate of 3%/h, the maximum cumulative reward that the agent can optimally achieve through one episode is 4800 (green dashed line in Fig. 6.8). The practical reward value that the agent can achieve is observed to be 3000 (red dashed line in Fig. 6.8). The trend stabilizes as the rewards are returned successful, indicating the completion of the A3C agent training.

Figure 6.9 compares the experimental results with the timeline analysis for the operation strategy shown in Fig. 6.2. The algorithm shows a similar operation pattern as the established strategy. According to the results, the autonomous operation algorithm successfully controls the components to increase the reactor power and generate electricity at the intended rate of power increase. This demonstrates that the A3C agent in the continuous control module can effectively conduct experience-based control after training with the simulator, and also that the discrete control module can manage its related components according to the rules based on the operating procedures. Therefore, the developed algorithm combining a rule-based system and RL is able to successfully conduct the power-increase operation in an autonomous manner.

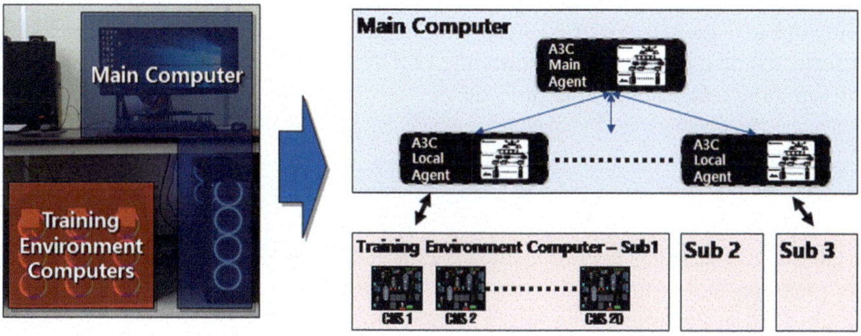

Fig. 6.7 Structure of the training environment for the A3C agent. Reproduced with permission from Lee et al. (2020)

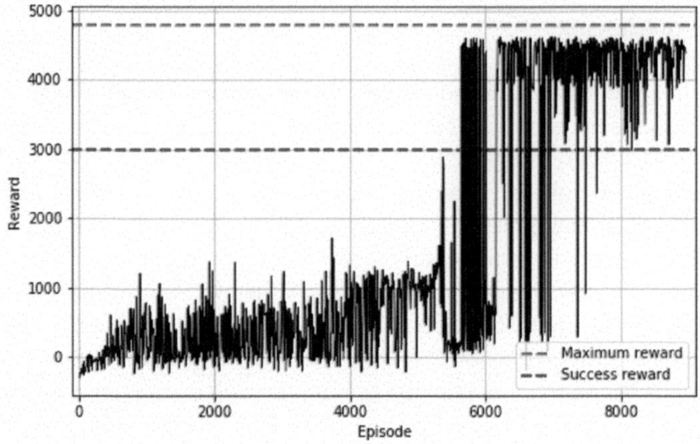

Fig. 6.8 Rewards obtained by the A3C agent. Reproduced with permission from Lee et al. (2020)

6.1.2 Case Study 2: Emergency Operation

Emergency operation refers to the mitigating actions and operations that are conducted following a reactor trip by an initiating event to ensure the integrity of the reactor core and containment building (Yang and Kim 2020). Currently, NPPs adopt highly automated systems that reduce the risk of accidents through immediate management, especially during an emergency situation. Nonetheless, except for the earliest stages of emergency operation, manual manipulations by operators are still required to reduce the pressure and temperature until safety conditions are reached (Lee and Kim 2021).

The second target of the autonomous operation algorithm is emergency operation following a reactor trip caused by a LOCA. In this accident scenario, operators perform accident diagnosis to identify the LOCA and then conduct recovery actions following the related operating procedures.

6.1.2.1 Work Domain Analysis Using Abstraction-Decomposition Space

To automate emergency operation, the systems and components involved in reducing the pressure and temperature must be identified. One good tool for this is the abstraction-decomposition space (ADS) (Rasmussen 1985), a technique that analyzes a given work domain and classifies it into an abstraction level and a decomposition space. Here, an ADS can be used to systematically identify the systems and components that operators are required to manipulate during emergency operation. Moreover, an ADS can also facilitate the identification of the operational goals and constraints of the related systems and components in this operational mode.

In an ADS, the abstraction level represents the work domain as a hierarchical structure consisting of a functional purpose, abstraction function, generalized function,

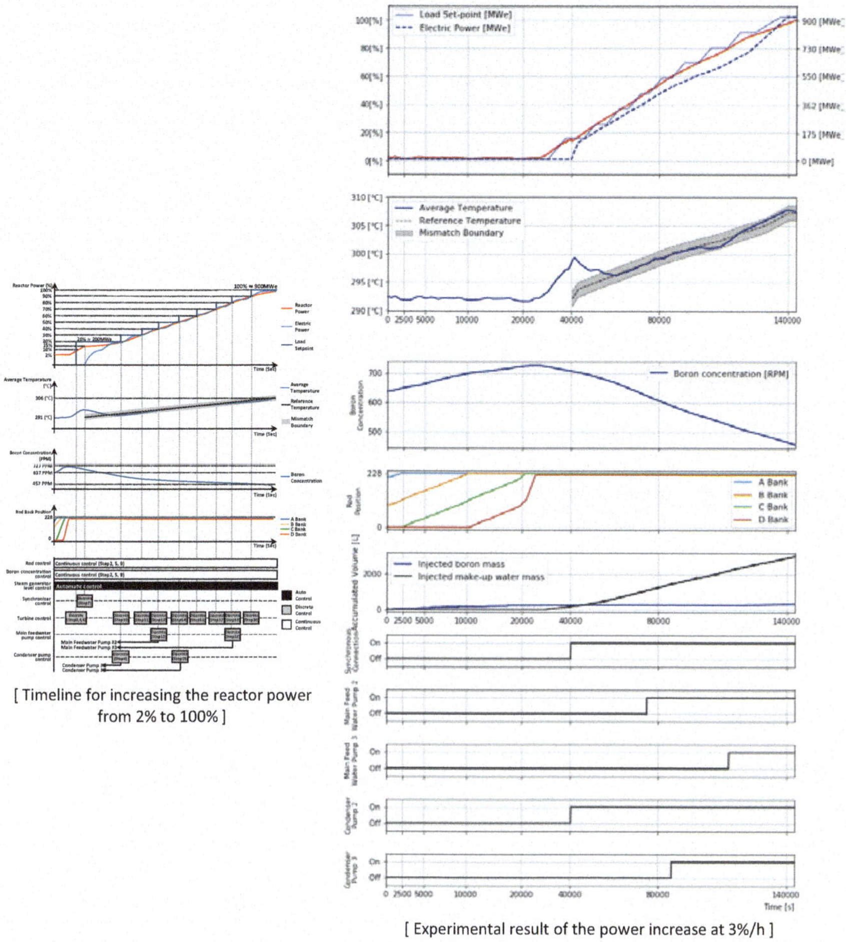

Fig. 6.9 Comparison between the existing operational strategy and the developed algorithm. Reproduced with permission from Lee et al. (2020)

and physical function, in top-down order. The abstraction level is able to discover correlations between the hierarchy as how-what-why.

On the other hand, the decomposition space divides the work domain down to the lowest elements and then maps them to a space composed of the whole system, subsystems, and components. Even if a large system has a complex structure, the decomposition space can review the relationships between the components by stepping down through the spaces of detail to the component space.

Figure 6.10 shows an example of an ADS to control the pressure of the reactor and cooling system. The functional purpose in an ADS is defined as the objective of the given work domain; here, the objective of emergency operation is to prevent core damage, which can be subdivided into decompression and cooling. In the next level, the abstraction function represents the related physical variables such as flow, mass, temperature, and level. As listed in Table 6.4, these variables each have success criteria for achieving the objectives specified at the functional purpose level.

For example, the physical variable of PZR pressure is involved in the success criteria of the RCS pressure control function, namely that the PZR pressure should be below 29.5 kg/cm^2, which is the shutdown operation entry condition, and stay within the P–T curve boundary, as shown in Fig. 6.11.

After the abstraction function level, as shown in Fig. 6.10 the generalized function level includes the operation functions that directly or indirectly affect the basic principle defined in the abstraction function. Specifically, PZR level is affected by operations such as decompressing the PZR and pumping (supplying) coolant. The operation functions that are identified in the generalized function level are related to the operational process required to meet the safety functions. The last level, or physical function level, includes the components that perform each operational systematic process, i.e., the generalized functions such as pumping coolant. For example, in the case of the PZR coolant supply, the physical components that need to be operated to achieve the desired PZR level are SI valve, SI pump, charging valve, letdown valve and orifice valve.

As above, the ADS identifies the components related to reducing pressure and temperature during an emergency situation by analyzing the work domain. Once the components are identified, their control types can be highlighted through the EOPs, as listed in Table 6.5. For instance, manipulation of the steam dump valve is identified as continuous control because when the position of this valve needs to be adjusted,

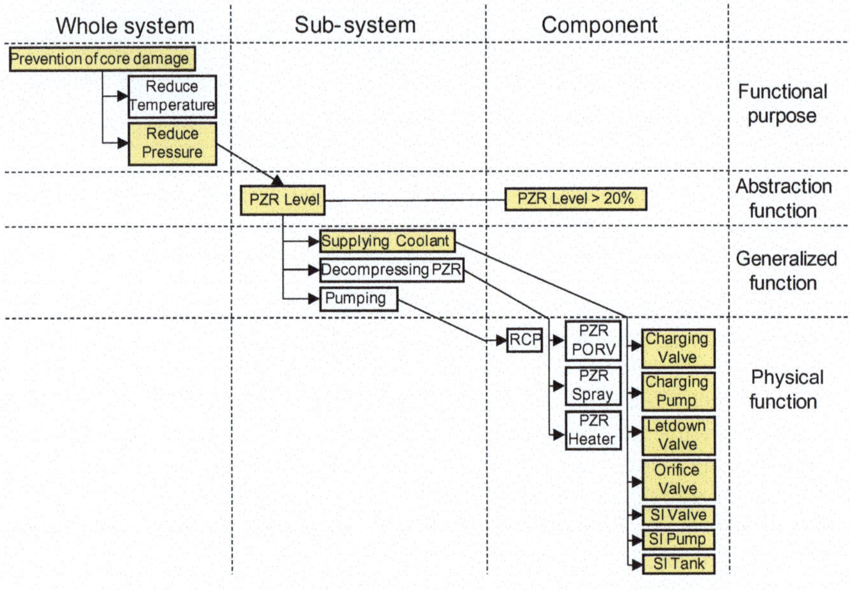

Fig. 6.10 An example ADS to reduce pressure. Reproduced with permission from Lee et al. (2021)

Table 6.4 Required physical parameters and success criteria in the abstraction function level

Physical variable	Success criteria
PZR pressure	Pressure < 29.5 kg/cm^2 Pressure within P–T curve boundary
PZR level	20% < Level < 76%
RCS average temperature	170 °C < average Temperature Temperature within P–T curve boundary 55 °C/h < cooling rate
SG pressure	Pressure <88.2 kg/cm^2
SG level	6% < narrow level < 50%

Reproduced with permission from Lee et al. (2021)

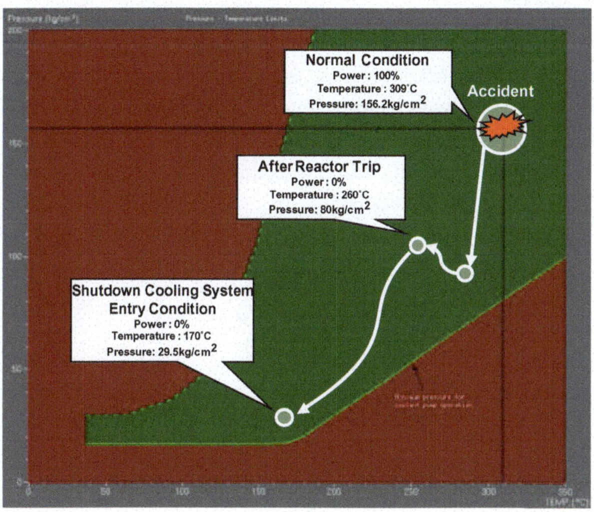

Fig. 6.11 P–T curve boundary and trajectory of the change in pressure and temperature. Reproduced with permission from Lee et al. (2021)

operation procedures provide only the operational goal (e.g., cool the temperature down to the shutdown entry condition within the P–T curve) without clear rules for its manipulation. On the other hand, manipulation of the RCP is given as discrete control. In this case there is a clear rule: if the pressure is below 97 kg/cm^2, switch the RCP off.

6.1.2.2 Algorithm Structure for Emergency Operation

The algorithm for emergency operation, as the schematic shows in Fig. 6.12, is designed considering the control types. The architecture comprises discrete and

Table 6.5 Components required to reduce pressure and temperature

Control type	Component
Continuous control	PZR spray valve, SI pump, SI valve, Aux feedwater valve, Steam dump valve
Discrete control	PZR heater, Charging valve, Letdown valve, Orifice valve, Aux feedwater pump, Main feedwater pump, RCP

continuous control modules, similar to the power-increase algorithm previously introduced (Sect. 6.1.1.2). In the discrete control module, a rule-based system generates discrete control signals through an inference engine that can be logically deducted based on a database consisting of if–then rules. In the continuous control module, control signals are generated by processing physical parameters, such as PZR level or pressure, and component states, such as PZR heater on or off, with a basic DNN.

In more detail, the discrete control module controls the components of discrete control type, namely the PZR heater, charging valve, letdown valve, orifice valve, aux feedwater pump, main feedwater pump, and RCP as listed in Table 6.5. To manage these components, operation rules described in the related EOPs are converted into if–then logic. Table 6.6 shows an example if–then rule specifying the PZR pressure at which the RCPs should be stopped.

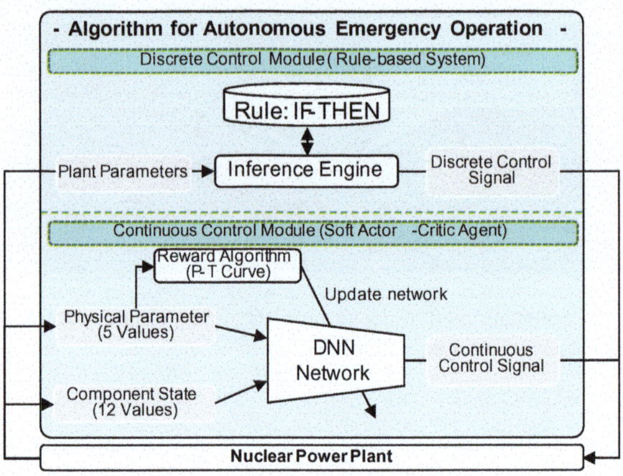

Fig. 6.12 Overview of the algorithm to reduce the primary pressure and temperature during emergency operation. Reproduced with permission from Lee et al. (2021)

Table 6.6 Example if–then logic for the RCP

Function	If–then rule	Input(s)	Output(s)
RCP	If the RCS pressure is below 97 kg/cm^2, stop all RCPs	PZR Pressure	RCP #1, #2, #3

As for the continuous control module, during emergency operation its focus is also to manage components to reduce the pressure and temperature until the shutdown cooling entry condition. This module adjusts components of continuous control type such as the PZR spray valve, aux feedwater valve, and steam dump valve, for which it applies a DNN with a soft actor-critic (SAC) algorithm. Figure 6.13 illustrates a schematic of the continuous control module. The SAC agent works by finding a policy to explore more widely while giving up on avenues that are clearly unpromising. The policy can capture multiple operation paths of near-optimal behavior. Q-values are used to optimize behavior selected from the policy by considering the actual and expected rewards. The DNN operates Q-value and policy networks to capture the particular actions that achieve the operational goals.

A reward algorithm for the SAC agent is designed to reduce the pressure and temperature of the reactor and cooling system down to the shutdown cooling system entry condition. The success criteria shown in Table 6.4 found in the work domain analysis via ADS are used to develop the reward. The reward is calculated as shown in Eqs. (6.7) to (6.10).

$$\text{Calculated cooling temperature } (\mathrm{T_{ct}}) = \mathrm{T_s} - 55\,^{\circ}\mathrm{C} * \frac{(\mathrm{t_c} - \mathrm{t_{tr}})}{3600} \tag{6.7}$$

$$\text{Temperature distance } (\mathrm{T_d}) = |\mathrm{T_c} - \mathrm{T_{ct}}| \tag{6.8}$$

$$\text{Pressure distance } (\mathrm{P_d}) = |\mathrm{P_c} - \mathrm{P_{sc}}| \tag{6.9}$$

$$\text{Total reward} = -(\mathrm{T_d} + P_d) \tag{6.10}$$

$\mathrm{t_c}$: Time [s]
$\mathrm{t_{tr}}$: Time at reactor trip [s]
$\mathrm{T_s}$: Stable temperature after reactor trip (260 °C)
$\mathrm{T_c}$: Temperature at time ($\mathrm{t_c}$)
$\mathrm{P_c}$: Pressure at time ($\mathrm{t_c}$)
$\mathrm{P_{sc}}$: Pressure of shutdown cooling entry condition

Interacting with the simulator every second, the SAC agent receives a total reward as calculated by Eq. (6.10). The range of the expected total reward per second is (–inf ~ 0). A reward close to zero means that the agent satisfies the success criteria. The SAC agent interacts with the simulator until the temperature or pressure moves outside the P–T curve boundary (operation failure) or the agent reaches the shutdown cooling entry condition (operation success). Once the interaction is complete, the simulator returns to the initial operation conditions.

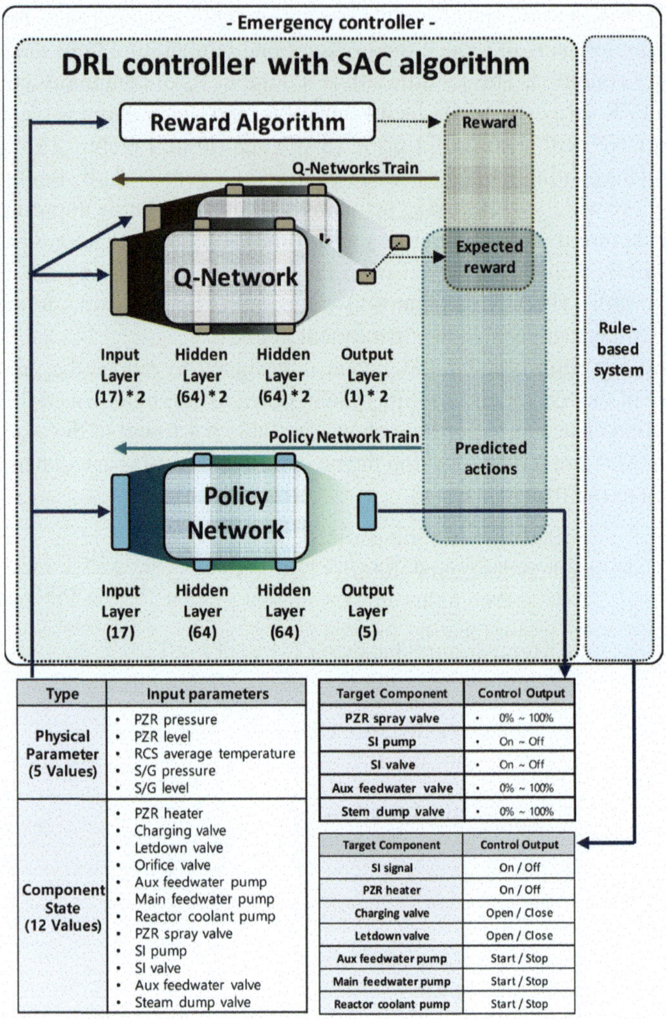

Fig. 6.13 Structure of the SAC agent. Reproduced with permission from Lee et al. (2021)

6.1.2.3 Experiments

The developed autonomous operation algorithm for emergency operation is trained and validated with a CNS in the same manner as that for the power-increase operation (Sect. 6.1.1.3). Training of the SAC agent is performed for more than 800 episodes to complete the emergency operation and is stopped when the average reward saturates; Fig. 6.14 plots the trend of the rewards in the SAC agent training. In a single episode, the theoretical maximum cumulative reward during the entire the emergency operation is 0 (green dashed line in Fig. 6.14). To receive a cumulative reward of zero in

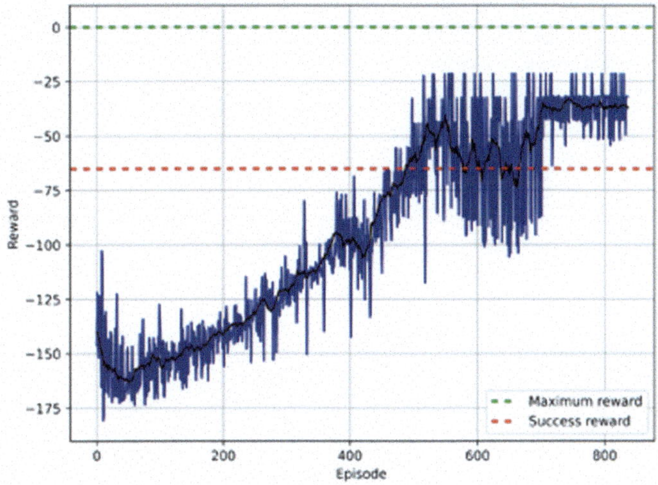

Fig. 6.14 Reward obtained by the SAC agent. Reproduced with permission from Lee et al. (2021)

one episode, the SAC agent should obtain a reward of zero every second. However, since the pressure at the beginning of the operation cannot be the same as the pressure at the shutdown cooling entry condition, the maximum cumulative reward should be selected through experimental observation. Here, a practical maximum reward for emergency operation success is observed to be over −65.

Following successful training, a test is conducted to demonstrate that the developed algorithm for emergency operation can automatically cool down the reactor during a LOCA scenario while satisfying the operational constraints, i.e., staying within the P–T curve boundary at the appropriate cooling rate, 55 °C/hour. As shown in Fig. 6.15, the pressure and temperature are found to be reduced by the algorithm within the operational criteria down to the entry condition of shutdown cooling.

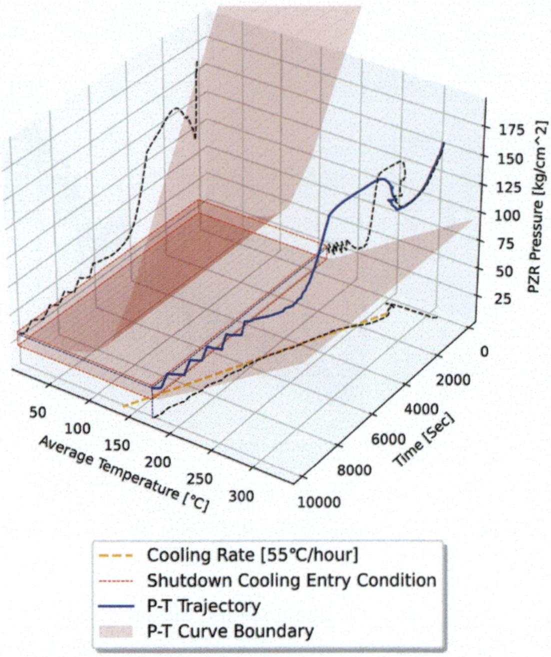

Fig. 6.15 Simulation results of the emergency operation algorithm. Reproduced with permission from Lee et al. (2021)

References

Bhalla S, Ganapathi Subramanian S, Crowley M (2020) Deep multi agent reinforcement learning for autonomous driving. In: Canadian conference on artificial intelligence. Springer, pp 67–78

Dong Z, Huang X, Dong Y, Zhang Z (2020) Multilayer perception based reinforcement learning supervisory control of energy systems with application to a nuclear steam supply system. Appl Energy 259:114193

Du G, Zou Y, Zhang X, Liu T, Wu J, He D (2020) Deep reinforcement learning based energy management for a hybrid electric vehicle. Energy, 117591

Genders W, Razavi S (2020) Policy analysis of adaptive traffic signal control using reinforcement learning. J Comput Civ Eng 34(1):04019046

Guo X (2017) Deep learning and reward design for reinforcement learning (Doctoral dissertation)

Kang C, Huang J, Zhang Z, Liu Q, Xiang W, Zhao Z, Liu X, Chong L (2020) An automatic algorithm of identifying vulnerable spots of internet data center power systems based on reinforcement learning. Int J Electr Power Energy Syst 121:106145

Kazmi H, Mehmood F, Lodeweyckx S, Driesen J (2018) Gigawatt-hour scale savings on a budget of zero: deep reinforcement learning based optimal control of hot water systems. Energy 144:159–168

Khatua S, Mukherjee V (2021) Application of PLC based smart microgrid controller for sequential load restoration during station blackout of nuclear power plants. Ann Nucl Energy 151:107899

Kim AR, Park J, Kim JT, Kim J, Seong PH (2016) Study on the identification of main drivers affecting the performance of human operators during low power and shutdown operation. Ann Nucl Energy 92:447–455

Kim J, Lee D, Yang J, Lee S (2020) Conceptual design of autonomous emergency operation system for nuclear power plants and its prototype. Nucl Eng Technol 52(2):308–322

Kima Y, Park J (2018) Envisioning human-automation interactions for responding emergency situations of NPPs: a viewpoint from human-computer interaction. In: Transactions of the Korean nuclear society autumn meeting

Kohl N, Stone P (2004) Policy gradient reinforcement learning for fast quadrupedal locomotion. In: IEEE international conference on robotics and automation, 2004. Proceedings. ICRA'04. 2004, IEEE. vol 3, pp 2619–2624

Lee D, Arigi AM, Kim J (2020) Algorithm for autonomous power-increase operation using deep reinforcement learning and a rule-based system. IEEE Access 8:196727–196746

Lee D, Kim J (2018) Autonomous algorithm for start-up operation of nuclear power plants by using LSTM. In International conference on applied human factors and ergonomics. Springer, pp 465–475

Lee D, Kim J (2021) Autonomous emergency operation of nuclear power plant using deep reinforcement learning. In: International conference on applied human factors and ergonomics. Springer, pp 522–531

Lee D, Kim H, Choi Y, Kim J (2021) Development of autonomous operation agent for normal and emergency situations in nuclear power plants. In: 2021 5th international conference on system reliability and safety (ICSRS), IEEE, pp 240–247

Lee D, Seong PH, Kim J (2018) Autonomous operation algorithm for safety systems of nuclear power plants by using long-short term memory and function-based hierarchical framework. Ann Nucl Energy 119:287–299

Ng AY, Coates A, Diel M, Ganapathi V, Schulte J, Tse B, Berger E, Liang E (2006) Autonomous inverted helicopter flight via reinforcement learning. Experimental robotics IX. Springer

Park J, Kim T, Seong S (2020) Providing support to operators for monitoring safety functions using reinforcement learning. Prog Nucl Energy 118:103123

Rasmussen J (1985) The role of hierarchical knowledge representation in decisionmaking and system management. IEEE Trans Syst Man Cybern 2:234–243

Rocchetta R, Bellani L, Compare M, Zio E, Patelli E (2019) A reinforcement learning framework for optimal operation and maintenance of power grids. Appl Energy 241:291–301

Saenz-Aguirre A, Zulueta E, Fernandez-Gamiz U, Lozano J, Lopez-Guede JM (2019) Artificial neural network based reinforcement learning for wind turbine yaw control. Energies 12(3):436

Samadi E, Badri A, Ebrahimpour R (2020) Decentralized multi-agent based energy management of microgrid using reinforcement learning. Int J Electr Power Energy Syst 122:106211

She J, Jiang J (2011) On the speed of response of an FPGA-based shutdown system in CANDU nuclear power plants. Nucl Eng Des 241(6):2280–2287

Viitala A, Boney R, Kannala J (2020) Learning to drive small scale cars from scratch. arXiv preprint arXiv:2008.00715

Wei T, Wang Y, Zhu Q (2017) Deep reinforcement learning for building HVAC control. In: Proceedings of the 54th Annual Design Automation Conference, pp 1–6

Wood RT, Neal JS, Brittain CR, Mullens JA (2004) Autonomous control capabilities for space reactor power systems. In: AIP conference proceedings. American Institute of Physics, vol 699(1), pp 631–638

Yang Z, Zhu F, Lin F (2020) Deep-reinforcement-learning-based energy management strategy for supercapacitor energy storage systems in urban rail transit. IEEE Trans Intell Transp Syst

Yang J, Kim J (2020) Accident diagnosis algorithm with untrained accident identification during power-increasing operation. Reliab Eng Syst Saf 202:107032

Yoo J, Cha S, Son HS, Kim CH, Lee JS (2004) PLC-Based safety critical software development for nuclear power plants. In: International conference on computer safety, reliability, and security. Springer, pp 155–165

Yoo J, Cha S, Jee E (2008) A verification framework for FBD based software in nuclear power plants. In: 2008 15th Asia--Pacific software engineering conference. IEEE, pp 385–392

Yu C, Wang X, Xu X, Zhang M, Ge H, Ren J, Sun L, Chen B, Tan G (2019) Distributed multiagent coordinated learning for autonomous driving in highways based on dynamic coordination graphs. IEEE Trans Intell Transp Syst 21(2):735–748

Zhang W, Zhang Y, Liu N (2020) Map-less navigation: a single drl-based controller for robots with varied dimensions. arXiv preprint arXiv:2002.06320

Zhou S, Hu Z, Gu W, Jiang M, Chen M, Hong Q, Booth C (2020) Combined heat and power system intelligent economic dispatch: A deep reinforcement learning approach. Int J Electr Power Energy Syst 120:106016

Chapter 7
Monitoring

Among the various functions undertaken by the autonomous NPP, the monitoring function has a twofold purpose: (1) monitoring whether the plant status meets its operational limits or constraints, and (2) monitoring whether the autonomous system itself can manage the NPP properly.

First, the monitoring function can focus on the several conditions that the various NPP operations should satisfy. One example is LCO required by regulation. The LCO is the lowest functional capability or performance level of equipment required for the safe operation of the facility (NRC 2012). If an LCO is violated in any operation mode, operators should conduct immediate actions provided by the technical specifications of the equipment. Another example is the status of the safety functions in emergency operation. In this operation mode, the plant status can be evaluated via the health status of the safety functions, such as reactivity control, RCS inventory control, RCS pressure control, etc. These health statuses can be categorized into normal, abnormal, and severe following criteria defined in the relevant procedure. In the case that all the CSFs not damaged by the initiating event are evaluated as normal, the indication is that the plant is being managed properly in the emergency situation.

Second, the monitoring function can also check the health of the autonomous system itself, i.e., perform self-diagnosis. This allows any failure of the autonomous system functions or any error committed by the system to be detected. Once a problem is detected by the monitoring function, the information is transferred to the decision-making function of the system and used to select an alternative strategy.

Although many conditions can be monitored in the autonomous NPP, this chapter introduces two approaches. The first is a system to detect errors committed by autonomous systems as well as operators, and the second is a method to monitor for violations of LCOs.

J. Kim et al., *Autonomous Nuclear Power Plants with Artificial Intelligence*, Lecture Notes in Energy 94, https://doi.org/10.1007/978-3-031-22386-0_7

7.1 Operation Validation System Through Prediction

Human errors and organizational factors account for a large part of the safety of NPPs. In fact, it has been revealed that human error can take up more than half of the core damage frequency of plants (Gertman et al. 2002), and moreover, it is well known that human factors were the direct or indirect cause of the three major nuclear accidents in history (Stanton 1996; NRC 1979; Infield 1987).

The TMI accident in particular led to improvements in the various human–system interfaces (HSIs) in NPPs with the goal to prevent human error (NRC 1979; NRC 1980). Most of the notable improvements to the HSIs have focused on MCR interface design and operator support system development, as plants are largely operated and maintained by operators in the MCR. As mentioned previously, the purpose of operator support systems in NPPs is to reduce human errors and improve human performance by substituting or partially substituting operator tasks or by assisting the operators in performing their tasks. From a technical standpoint, introducing such computerized support systems to modern NPPs is relatively easy as the I&C systems for various plant functions are being rapidly digitalized (Lee and Seong 2014). As introduced in other chapters of this book, various technologies based on AI are being studied, and they are expected to contribute to increasing the safety of NPPs.

Operator support systems can be classified as a kind of automation system. An appropriate level of automation technology is believed to reduce operator error or compensate for the effects of operator error. But before realizing the advantages of such level of automation in actual plants, several associated issues need to be addressed. One major example is that over-relying on an automated system can lower not only the situational awareness but also the skill proficiency of operators (Endsley 1996; Endsley and Kaber 1999; Lee and See 2004; OHara et al. 2010; Parasuraman and Riley 1997; Kaber and Endsley 2004). In this section, an intelligent support system that is concealed from operators is introduced to compensate for this automation-related issue.

In terms of human error, emergency operation of an NPP represents an environment in which the operators are prone to error due to the dynamically changing plant conditions and great stress. But in emergency situations in particular, the potential for errors should be reduced and their effects should be minimized, as in this case a wrong action by an operator can cause critically serious consequences. Moreover, if a human error is made, recovery actions must be taken in a timely manner or the impact of the error may aggravate the situation over time. This means that detecting an occurred error and conducting coping actions as soon as possible are essential. In NPP emergency operation, a work environment with high task loads and utmost stress, a support system for reducing operator workload without adding confusion or burden could prove beneficial.

The developed operator support system in this section is called the CIA, or concealed intelligent assistant. It works by detecting human error in an NPP emergency situation and providing the relevant information to the operators. The *concealed* aspect of the CIA is that the system is not visible to operators during normal duties, only appearing when an error occurs. The CIA system employs AI technologies such as DNNs, and thus its development provides a further opportunity to examine the potential of AI technology to partially replace or assist the role of NPP operators.

7.1.1 CIA System Framework

If the goal is to provide information about a human error to the operators in an emergency situation, then the first step is the determination of whether an operator behavior is an error or not, for which appropriate judgment criteria are necessary. For this determination, the developed system implements two-stage filtering following the basic principle of human error determination shown in Fig. 7.1. A general schematic of the CIA with its two filtering modules is illustrated in Fig. 7.2. The first criterion in the determination of a human error is whether an operator behavior complies with the relevant operation procedure, based on the fact that operators strictly follow EOPs to cope with an emergency. The first filter, the procedure compliance check (PCC) module, makes this determination. But since emergency operation situations are likely highly dynamic, considering static procedures as the complete standard is insufficient. In the case that an action does not adversely affect plant integrity, even when it is technically a procedural non-compliance, notifying the operator in the midst of the emergency situation would be inappropriate from a safety as well as operator burden point of view. In principle, no additional workload should be assigned to the operators as related to any activity that does not pose a threat to safety in the critical course of the emergency. Additional determination is therefore necessary, namely how a non-compliant action may affect the integrity of the

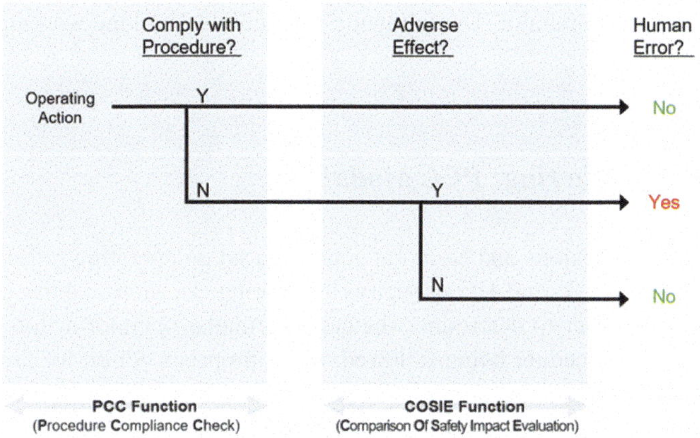

Fig. 7.1 Decision tree for human error. Reproduced with permission from Ahn et al. (2022)

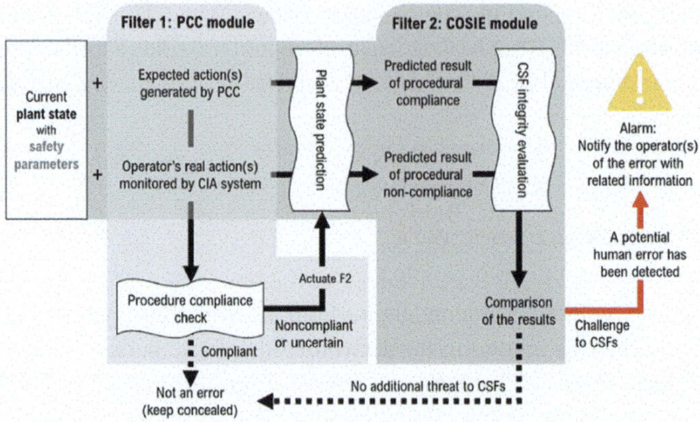

Fig. 7.2 Overall framework of the CIA system. Reproduced with permission from Ahn et al. (2022)

NPP. This is implemented as CIA's second filter, the comparison of safety impact evaluation (COSIE) module.

As shown in Fig. 7.2, the PCC module as the first filter obtains the expected action (EA) according to the procedure and compares it with the real action (RA) by the operators. If the RA differs from the EA, the PCC regards the RA as a procedural non-compliance and initiates the COSIE module for a second layer of filtering. The COSIE module analyzes the safety impacts that the EA and RA received from the PCC have on the NPP. For this, the COSIE module utilizes a DL algorithm as a tool for the diagnosis of potential risk based on the prediction of plant safety variables. In other words, the second filter works by predicting the future state of the NPP according to the given EA and RA and then comparing the two predicted results in terms of safety. An RA that is evaluated to threaten plant safety more than the EA is flagged as a potential threat. In this way, the CIA system can determine a human error and notify the operator. The following subsections detail the workings of each module.

7.1.2 *Step 1 Filtering: PCC Module*

Operators make decisions and carry out actions based on operating procedures that guide all cognitive tasks and actions such as monitoring, decision-making, and equipment control. In order to determine whether a particular operator action (i.e., RA) conforms to the procedure being followed, the appropriate action for the situation (i.e., EA) should first be determined, after which the two can be compared.

The overall framework of the PCC module to determine whether operator actions comply with the procedure is depicted in Fig. 7.3. To carry out its function, this module monitors the plant state, receives information on the current stage of the procedure from the computerized procedure system, and derives the appropriate task

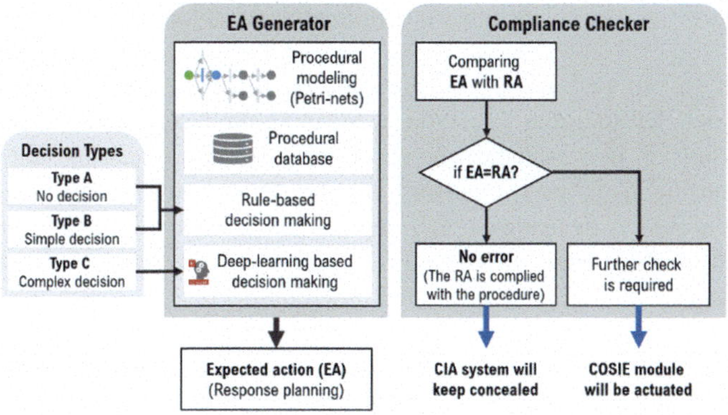

Fig. 7.3 Procedure compliance check (PCC) module framework. Reproduced with permission from Ahn et al. (2022)

to be performed at the current stage, or in other words the EA. At this time, the method of judgment differs according to the decision type, classified as Type A, B, and C in Fig. 7.3, to determine the necessity of execution. In this judgment process following the procedure directives, in the case of Type B (simple decision), rule-based judgment can be applied.

However, in the case of Type C including complex decisions (CDs), a simple rule-based method is inapplicable. For these types of decisions, a CD model adopting a DL algorithm is implemented to predict the judgment of an operator when making a CD. To do so, the model judges whether a particular trend is increasing, maintaining, or decreasing with respect to time-series data and uses this as its basis for predicting the operator's decision in the given situation. If it is judged through the CD algorithm that the task predicted to be performed by the operator does not comply with the procedure or if the PCC module requires additional information, the COSIE module is activated. The below subsections discuss the overall framework of the PCC module, related procedure and task analyses, application of a procedure-based task/action decision logic, and CD methods using DL algorithms.

7.1.2.1 Analysis of EOPs

Emergency operating procedures are generally written in a simple logic form such as if–then-else, as the example in Table 7.1 shows, to reduce the mental burden of following the procedure and to support operator decision-making. As exhibited in Fig. 7.4, EOPs map out the appropriate process of operation in each particular situation, where the flow of operation is presented as a sequence of procedural steps. As implied in their name, EOPs are followed in an emergency situation following a reactor trip, during which all operator actions are dictated by the appropriate EOP. In short, operators examine the integrity of the NPP by comparing the setpoints of the automatic systems to the current plant state as instructed by the EOP, and then maintain or restore the safety functions following specific EOP steps based on the symptoms.

As briefly mentioned above, decisions can be classified into three types, which are further detailed in Table 7.2. Type A, meaning no decision, refers to when the entry into the step itself satisfies the executing condition, meaning that operators proceed with the given step or perform a given action without any judgment. An example of a Type A decision is "E-0 step 2.0 RNO, Manually trip turbines" as in Table 7.2. At this step, the current state is one in which the turbine is not tripped in the previous step. E-0 step 2.0 is entered based on the determination that a turbine trip is needed, where this determination is made in the previous step. Therefore, since the condition for performing the task in this step is satisfied just by entering the step, operators may simply perform the instructed task without any additional decision-making after entering the step. The second decision type, or Type B simple decision, involves a basic comparison between a reference value of a parameter given in the procedure and the indicated value of the current plant state. Other Type B decisions include the simple determination of a device state, such as on or off, or open or closed. For example, response planning for the control of a SG level requires operators to first check the EOP instruction, "Is the steam generator level 6% or higher?", which corresponds to a simple decision. The third decision type is Type C indicating a CD that involves continuous observations of any changes in parameter trends. For example, if an EOP step requires the operator to make a determination about the statement "the RCS pressure is increasing", the related parameters should be observed for a certain period of time with no set criteria.

Type A and Type B decisions can be handled by the support system with relatively simple rule-based approaches. Type C, on the other hand, cannot follow a simple rule as it requires judgments of time-series parameter trends that depend on the window size and numerous situational variables. Figure 7.5 plots the average temperature history of the RCS in an arbitrary condition as an illustrative example. As the figure shows, monitoring the temperature value at different points in time or for different amounts of time in a separate manner may result in different RCS trend judgments. Well-trained and experienced plant operators are likely to make the proper judgement though, and thus, opposed to designing a support system that attempts to judge complex trends following fixed rules, a system that can learn how actual experts arrive at their decisions in specific circumstances may be more appropriate. For this, applying a DL algorithm to a neural network offers a valuable solution as part of the PCC module with the goal to determine procedure compliance in NPP emergency situations by predicting the complex judgments made by operators.

Table 7.1 Example of the logical construct of an EOP step

[E-1 Loss of coolant accident, step 14] If RCS hot-leg temperature is less than 185 °C Then, Reset SI actuation signal Else, go to step 15

RCS reactor coolant system; *SI* safety injection
Reproduced with permission from Ahn and Lee (2020)

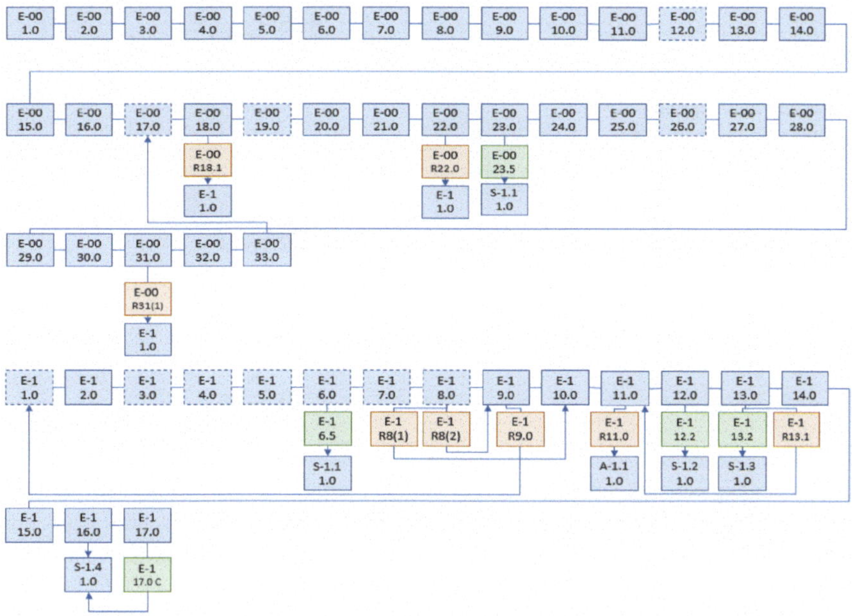

Fig. 7.4 Schematic of the flow of EOP steps. Reproduced with permission from Ahn and Lee (2020)

Table 7.2 Decision types within the response planning task in following an EOP

Decision types	Description	Example
Type A (No decision)	Action without decision-making	[E-0 step 2, RNO] "Manually trip turbines"
Type B (Simple decision)	Action requiring checking a parameter/component state	[E-0 step 18, RNO] "If pressurizer pressure is less than 164.2 kg/cm^2, then close PORVs manually"
Type C (Complex decision)	Action requiring monitoring a parameter/component state	[E-0 step 31] "If RCS pressure is stable or increasing, then stop RHR pumps and keep standby"

PORV power-operated relief valve; *RHR* residual heat removal
Reproduced with permission from Ahn and Lee (2020)

7.1.2.2 Procedure Modeling Using Colored Petri Nets

A Petri net model is a common choice to model a multi-structural procedure, such as the emergency operation of an NPP considered here. The basic concept is that the entities constituting a system work independently of a shared administrator or time, or in other words, each element acts by its own judgment. This makes the model suitable

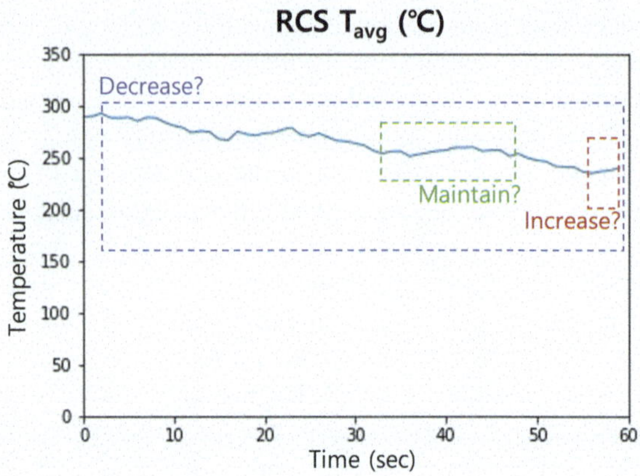

Fig. 7.5 Example time-series variable showing different possible trend interpretations. Reproduced with permission from Ahn and Lee (2020)

to represent systems in which events occur either simultaneously or asynchronously and which have either distributed or decentralized elements. The events each occur once a predetermined condition is met, and accordingly, a set of these event states can represent the whole system. Both concurrent and asynchronous characteristics can be expressed by the model because the events that do not share conditions for occurrence are independent from one another. The base of a Petri net model is the place/transition nets (PN) that only express the relationship of event occurrence. The three major model components are place, transition, and arc that respectively describe an initial condition, an event, and the relationship between the two. Other models can be added to the base model for various performance evaluations, and an extension model suitable to the target system to be analyzed can be built. Here, colored Petri net (CPN) modeling is applied, where color information is included to express more complex conditions and logic.

The CPN reflects various types of states with color information, making it well matched for handling complex processes such as NPP EOPs. The first step to convert a procedure document into a CPN model is to prepare a procedure database, for which a procedure analysis is conducted to classify the entry conditions and required tasks at each procedural stage. Then to create the CPN model, the procedural steps are configured as the model "places" and the required tasks are configured as the model "transitions". For a transition to occur, several conditions of the system should be satisfied, information about which is stored in each place. Once the conditions are satisfied, a colored token is indicated. Each place can be classified as one of six types, where the token color represents the plant state, as shown below.

Token color set—Place states

1. Green: Current – appropriate
2. Red: Current –wrong
3. Yellow: Following
4. Blue: Continuous
5. Purple: Concurrent

Here, the green token reflects that the current step is the appropriate one based on a proper following of the procedure. Red reflects that the current step is not appropriate based on a prior procedural mistake or mistakes, including a wrong step transfer. The yellow token represents the expected next step that the operator should transfer to, and the blue token represents a continuous step that can be performed once its operating conditions are satisfied. Purple represents a concurrent step, which means that the current procedural step is still ongoing even after the operator transfers to the subsequent step. The sixth type is no token, which is used to reflect a standby step that does not correspond to the states of other steps.

An example of the CPN modeling approach is shown in Fig. 7.6. The circles represent the steps of the EOP (place) and the rectangles represent the actions by operators (transition). A procedural non-compliance can be detected with a CPN model by checking whether the actual operator performance matches the correct sequence. In the top row of the figure, the green circle represents that step 15 of the EOP has been appropriately conducted, and step 16 (yellow) is the expected next step. In the middle row, the operator transfers to step 15R, not step 16, and this is highlighted with a red token representing an error. In the bottom row, the operator correctly transferred to step 16, now appearing as green, which makes step 15, as shown in purple, a concurrent procedural step indicating that step 15 is not yet completed. The EAs are derived from the places in the model having a green, blue, or purple token; in other words, the current-appropriate, continuous, and concurrent states are treated as those to be conducted.

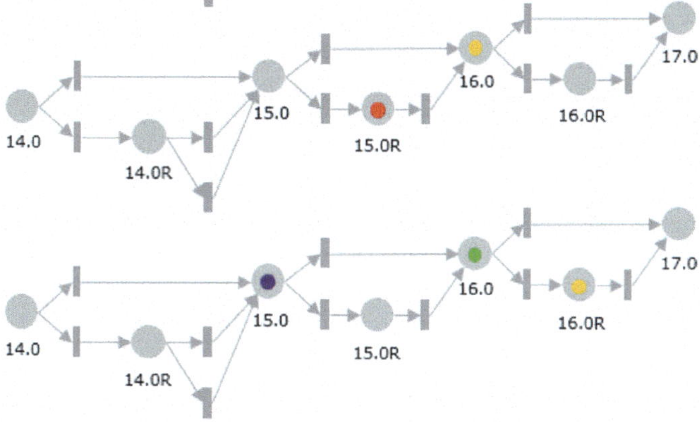

Fig. 7.6 Example of modeling a procedure using CPNs. Reproduced with permission from Ahn and Lee (2020)

7.1.2.3 Applying a DL Algorithm to Complex Decisions

Potential operator mistakes are determined by comparing the response implementation task performed by the operator with the response planning predicted by the system. As mentioned above, CDs corresponding to Type C are not that simple. Time-series information collected over a period of time is needed to determine parameter trends, but since procedures in certain cases do not provide specific criteria for determining the parameter trends, the various tasks that correspond to monitoring the trends are influenced by the state recognition abilities of the operators. Therefore, the required time to discern parameter trends varies from operator to operator. Along these lines, when the tendencies of the parameters are unclear or ambiguous, operator judgment is based not only on the current condition information but also on the personal experience and knowledge of the operators. Consequently, applying a simple mathematical analysis method to the system to make a judgment in such cases is inappropriate.

Operators conduct NPP operations after extensive education over years of training and operating experience, leading to rich empirical knowledge. This fact itself can provide a way to apply DL models to the problem of operator judgment prediction in complicated situations when implementing rule-based logic is difficult, such as the case of forecasting future parameter trends. Judgments by expert operators are likely to be credible, even in complicated situations requiring CDs. Therefore, by developing a prediction model for operator response based on expert operator judgments, a specific CD such as one in a monitoring situation can be considered accurate or in error by comparison with the prediction model. To realize such a model for operator judgment prediction, the model training requires a large amount of appropriate data on how real operators make decisions in particular situations. In this case, an appropriate labeling of the situations is necessary because the decisions stem from human judgment. A DL algorithm sufficiently trained with wide data for various situations can make it possible to predict the decisions made by operators in certain situations.

The overall development process of the model is shown in Fig. 7.7. First, training datasets are constructed from randomly generated simulator data showing various parameter trends, such as RCS temperature as shown in Fig. 7.7, and the data is labeled as increasing, maintaining, or decreasing based on operator decisions about the trends. For this, a process called virtual operator labeling is employed, and the training data represent the sum of the results of the Monte Carlo method and the responses to student surveys. The labeled data with varying ratios of answers are then used for DL model training and testing. Following successful training, the so-called CD model is able to predict how operators will judge parameter trends. This prediction is then used in the PCC module.

7.1.2.4 Complete CD Model

Figure 7.8 depicts the overall architecture of the CD model having a total of eight hidden layers. The first three layers are densely-connected neural network layers

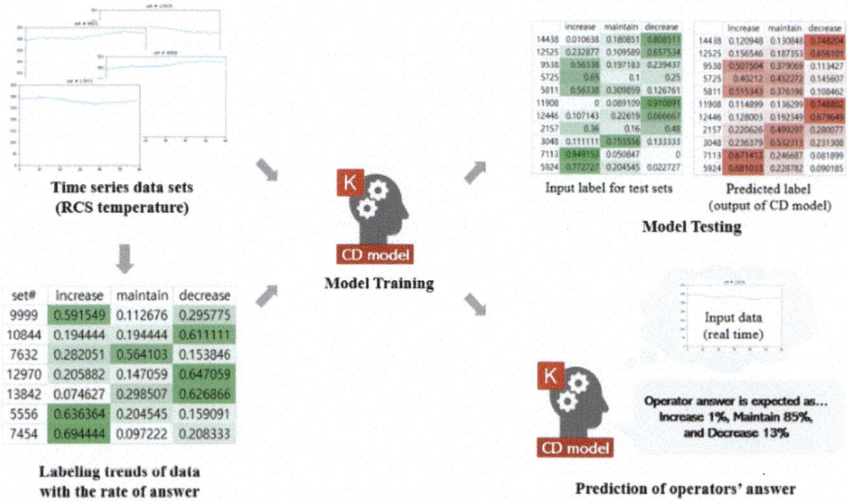

Fig. 7.7 Complex decision (CD) model training. Reproduced with permission from Ahn and Lee (2020)

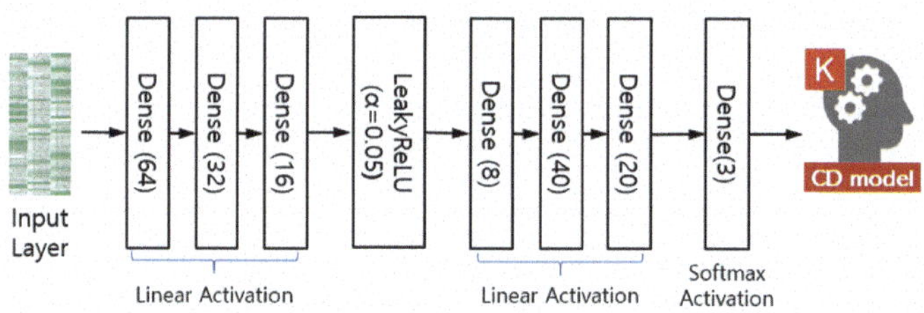

Fig. 7.8 Architecture of the CD model network. Reproduced with permission from Ahn and Lee (2020)

with 64, 32, and 16 neurons, respectively. These are followed by an activation layer with the LeakyReLU ($\alpha = 0.05$) function for data nonlinearity and complexity. The next three layers also have a dense structure with 8, 40, and 20 neurons, respectively, and the last layer applies softmax activation to produce output in a probabilistic form. A total of 15,000 sets of time-series data are used in the input layer, which are split into 80% for training and 20% for testing (i.e., 12,000 and 3000 datasets, respectively). Each dataset contains an RCS temperature history for 60 s with trends labeled via virtual operator labeling for a random monitoring time. The results of the model testing show an 87% accuracy of the model in predicting the correct operator decisions about the trends.

The DL model is built using Python (version: 3.6) with Tensorflow and keras utilities and then compiled with the RMSprop optimizer with a categorical cross-entropy loss function with metrics of accuracy. Table 7.3 shows the test results in heat map form. The left columns represent the correct answer, or operator judgment, and

Table 7.3 Sample heat map for the test results

Data set	(L) increase	(L) maintain	(L) decrease	(P) increase	(P) maintain	(P) decrease
12446	0.107143	0.22619	0.666667	0.128003	0.192349	0.679649
2157	0.36	0.16	0.48	0.220626	0.499297	0.280077
3048	0.111111	0.755556	0.133333	0.236379	0.532313	0.231308
5924	0.772727	0.204545	0.022727	0.681033	0.228782	0.090185
11627	0.035714	0.142857	0.821429	0.114551	0.119245	0.766203
14150	0.245614	0.192982	0.561404	0.153697	0.50229	0.344013
1677	0.157895	0.684211	0.157895	0.219022	0.513831	0.267147
9281	0.585366	0.219512	0.195122	0.442196	0.427004	0.1308
6541	0.65	0.125	0.225	0.60612	0.273949	0.119931
12123	0.043956	0.120879	0.835165	0.11594	0.144021	0.74004
4215	0.1	0.833333	0.066667	0.230494	0.528783	0.240723
14729	0.157143	0.042857	0.8	0.085654	0.125699	0.788647
9295	0.318182	0.136364	0.545455	0.241787	0.490792	0.267421
12855	0.155556	0.177778	0.666667	0.129679	0.151835	0.718486
9483	0.734694	0.081633	0.183673	0.528889	0.370692	0.10042
11056	0.3125	0.5	0.1875	0.244844	0.475878	0.279278
. 14333	0.268657	0.014925	0.716418	0.108961	0.097167	0.793872

The green columns are the labeled results, (L) operator answer, and the red columns are the prediction results, (P) model output. Reproduced with permission from (Ahn and Lee 2020)

the right columns represent the CD model prediction. The table reflects a similar heat map pattern on both sides, implying that the CD model predicts operator judgments with good accuracy. As complex operator decisions involve considerations of the various relationships among related parameters, the CD model is better suited to predicting complicated situations than rule-based models.

7.1.3 *Step 2 Filtering: COSIE Module*

In the case that an operator conducts an action that does not comply with the procedure, conveying warnings or alerts that the operator must respond to would take up cognitive resources. Considering the potential burden of additional tasks during emergency operation, it is clear that any additional task such as responding to an alarm or alert must be related to eliminating potential safety threats. Accordingly, alerts should only be provided to operators when their action results in a potential threat to safety. As the second filter of the CIA system, the COSIE (comparison of safety impact evaluation) module works by evaluating the actions that might adversely affect the plant integrity. Figure 7.9 shows a general schematic of this module. Since the integrity of an NPP is maintained through its CSFs, the parameters related to the

CSFs are monitored and predicted by a predictor function in the COSIE module to determine their impact on safety. Table 7.4 lists some example parameters related to the CSFs of the reference plant, a Westinghouse 3-loop 990 MWe PWR from Westinghouse Owners Group (WOG).

The COSIE module includes a prediction function for future safety parameter changes according to action/no action in the current state of the plant, based on the work presented in a preceding chapter of this book. The prediction function is implemented with an AI algorithm, where the plant state prediction model is built using layered LSTMs. Details and comparative results with other prediction strategies and inference models can be found in Chap. 5. As part of the CIA system discussed here, the result of the prediction function is scored through the safety impact evaluator.

As depicted in the schematic in Fig. 7.9, the safety impact evaluator in the COSIE module produces CSF integrity judgment scores based on reference values from CSF status trees. An example of this scoring process is shown in Fig. 7.10. The COSIE module is actuated by the PCC module (see Fig. 7.2) and provided with the EA and RA information. The DL algorithm-based predictor predicts the plant state for 10 min in the future from inputs of the current plant state and the combination of safety variables with an action set. The action set contains the single or multiple actions dictated by the procedure, EAs, or those actually performed by the operators, RAs. The result of the prediction is given as a number that reflects the value of each safety variable, for example the value of RCS pressure after 10 min. The prediction process is separately performed for the EAs and RAs, i.e., for action sets assuming full procedural compliance and for the action sets actually performed that may contain procedural non-compliance. The CSF integrity scores are then calculated for each result and compared to check whether the non-compliance is an error. Equation (7.1) shows the CSF integrity score calculation, which adds up the scores of the individual CSFs. According to the criteria provided by the CSF status trees, CSF scores of 0,

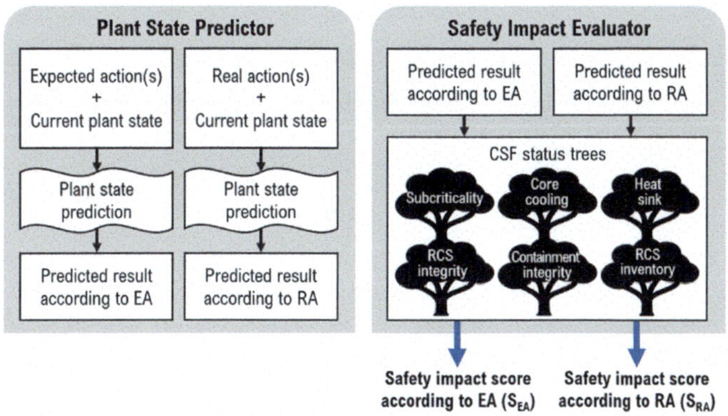

Fig. 7.9 Comparison of safety impact evaluation (COSIE) module system framework. Reproduced with permission from Ahn et al. (2022)

Table 7.4 Critical safety functions and related parameters of the reference system

#	Name	Related parameters	Units
1	Subcriticality	Power range percent power Intermediate range detector start-up rate Intermediate range neutron level Source range detector start-up rate	% DPM A DPM
2	Core cooling	Core outlet temp RCS hot-leg temp. (#1–3) RCS pressure	°C °C kg/cm^2
3	Heat sink	S/G narrow range level (#1–3) Feedwater flow (#1–3) S/G pressure (#1–3)	% m^3/hr kg/cm^2
4	RCS integrity	RCS cold-leg temp. (#1–3) RCS pressure	°C kg/cm^2
5	Containment integrity	Containment pressure Containment sump water level Containment radiation level	kg/cm^2 M mRem/hr
6	RCS inventory	Pressurizer (PZR) level	%

Reproduced with permission from Ahn et al. (2022)

1, 2, and 3 respectively represent the normal state, departure, violation, and heavy violation (Fig. 7.10). The lower the S_{total} score, then, the higher the integrity. If $S_{total,RA} > S_{total,EA}$, or in other words if the total score of the action set actually performed by the operator is high, then the consequent challenge to the CSFs is predicted via comparison to the score from complying with the procedure. In such a case, the CIA system flags the action set as a human error and provides a warning to the operator.

$$S_{total} = \sum_i S_i \tag{7.1}$$

$$S_i = \begin{cases} 0 & normal status & \cdots & green \\ 1 & departure & \cdots & yellow \\ \\ 2 & violation & \cdots & orange \\ 3 & heavy violation & \cdots & red \end{cases} \tag{7.2}$$

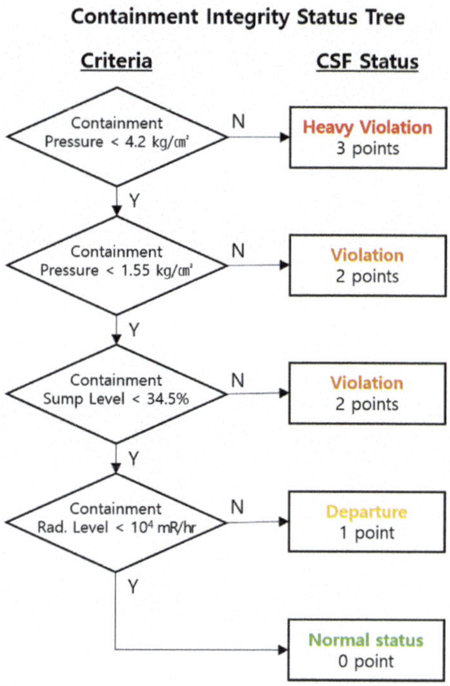

Fig. 7.10 Example of a CSF status tree for containment integrity. Reproduced with permission from Ahn et al. (2022)

7.1.4 CIA System Prototype

7.1.4.1 CIA System Overall Description

A prototype of the CIA system is developed with the use of a CNS along with the development of a computerized procedure for the CNS to collect information on the progress of EOPs. Simulations are conducted through the CNS interface and the computerized procedure system interface, and the progress of the CPS along with the plant parameters are sent to database storage in shared memory. The CNS operates on Linux CentOS 7 and transmits information between systems through the user datagram protocol.

Figure 7.11 shows a flowchart of the CIA system process. The PCC and COSIE modules, making up the core of the system, act by monitoring the plant parameters and the progress of the given procedure via the CNS database. The PCC module actuates each time an operator performs an operation or a procedure step changes. If the PCC module judges any action to be non-compliant or unclear, the COSIE module first predicts and evaluates the future CSF trends and then determines whether the action is a potential threat or not. If judged as a potential threat, the action is treated as a human error and a warning with related information is given to the operator. If the action is judged as proper or as one that poses no potential threat, no notification is sent to the operator.

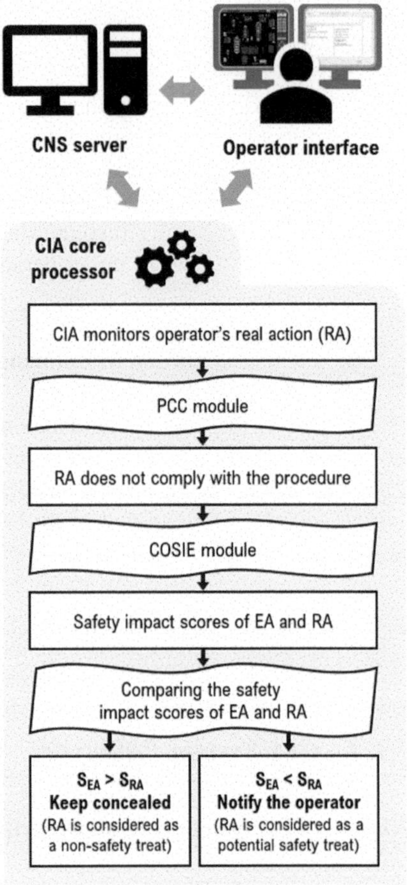

Fig. 7.11 Overall process of operating action validation with the CIA system. Reproduced with permission from Ahn et al. (2022)

7.1.4.2 CIA System User Interface

As mentioned above, the CIA system remains hidden unless a human error is detected. Therefore, the CIA user interface has only one purpose, which is to provide warnings and relevant information to the operator upon the occurrence of a human error. The layout of the interface is shown in Fig. 7.12. The provided information highlights the particular action that did not comply with the procedure as well as the specific contents of the procedure being performed at the time of the human error. The interface also conveys visual information on the particular safety function that was evaluated to be impaired. Clicking the CSF button reveals the prediction result of the related safety variable. Based on this provided information, the operator can recognize the occurrence of the error and decide whether to perform appropriate recovery measures according to the predicted results. It should be noted here that the CIA system does not override the operator; rather, it merely provides information, and thus for all operational actions, operators are maintained as the final decision-makers.

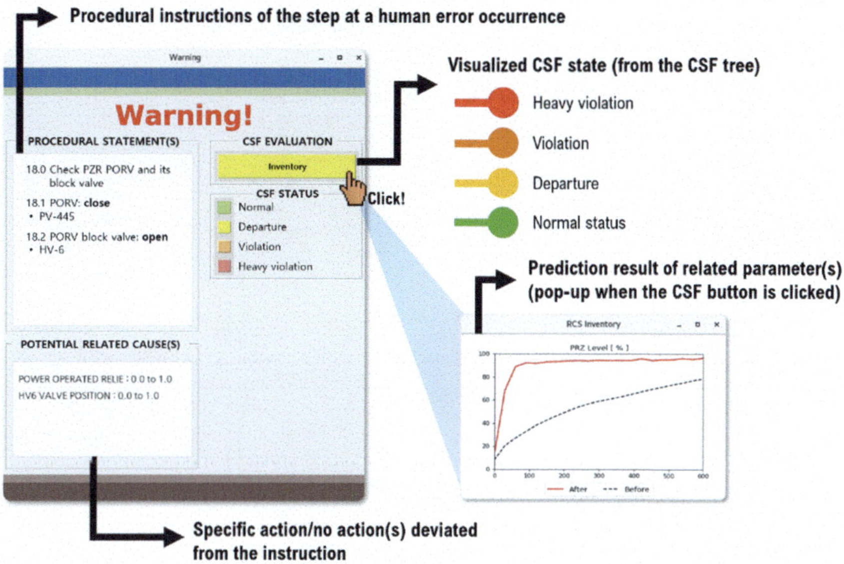

Fig. 7.12 Human error warning interface of the CIA system. Reproduced with permission from Ahn et al. (2022)

7.1.5 Case Study

7.1.5.1 Test Scenario Descriptions

In a case study to verify the CIA system operation, three scenarios are prepared for evaluation. The first reflects an operator action that complies with the procedure; the second reflects an operator action that does not comply with the procedure but the integrity of the CSFs remains intact; and the third reflects an operator action that does not comply with the procedure and is expected to harm the CSFs. The operating paths of the CIA system for these scenarios are illustrated in Fig. 7.13.

7.1.5.2 Test Result for Scenario #1: Procedural Compliance

In the first test scenario, the operational action complies with the procedure and thus the PCC module should not judge it as a human error. The scenario simulates Step 19.0 of the E-0 procedure of the reference plant. Table 7.5 lists the instructions for this step.

This step instructs the operator to check the conditions for stopping the RCPs and to stop all RCPs if necessary. In a LOCA with a fracture size of 50 cm^2, the corresponding stage is entered within 465 s following the accident, at which point at least one charge pump is operating and the RCS pressure falls below the condition

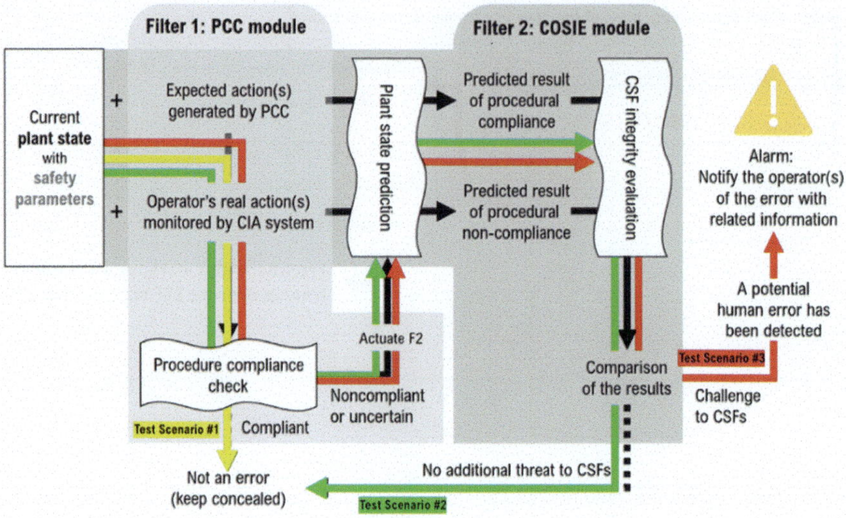

Fig. 7.13 CIA operating paths for the three test scenarios. Reproduced with permission from Ahn et al. (2022)

Table 7.5 Instructions in E-0 procedure step 19

Response	Response not obtained
19.0 Check if RCP should be stopped	Go to step 20.0
19.1 Charging pump status: at least 1 pump is operating	Go to step 20.0
19.2 RCS pressure < 97 kg/cm^2 (RCS Pressure < 104 kg/cm^2, if containment is in abnormal state)	
19.3 Stop all RCPs	

Reproduced with permission from Ahn et al. (2022)

to stop the RCPs. Following step 19.3, the operator must stop all the RCPs and then press the 'complete' button on the computerized procedure system. In this case, the PCC module should judge the measure to be proper as it followed the procedure. The test result of the first scenario is shown in Fig. 7.14. As seen in the result, the PCC module confirms that the operator's action complied with the procedure. Then based on the system design, the COSIE module is not actuated, and the whole CIA system remains concealed and gives no notification to the operator.

7.1.5.3 Test Result for Scenario #2: Procedural Non-Compliance, CSFs Not Challenged

In the second test scenario, an operator non-compliance that does not affect the integrity of the CSFs is simulated. Here, Step 15 of the E-0 procedure is considered; Table 7.6 lists the corresponding contents. After the occurrence of a single-tube

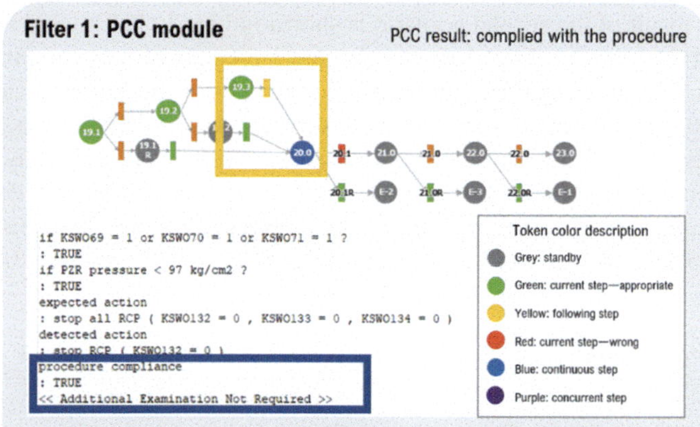

Fig. 7.14 Test result of the PCC module for scenario #1. Reproduced with permission from Ahn et al. (2022)

SGTR, E-0 step 15 is reached within 210 s in the simulation, at which point the operator arbitrarily isolated the SG based on the assumption that it was an SGTR accident. Normally, SG containment measures are performed only after entering Category 3 procedures, but in this case the operator performed the measure before reaching the proper stage.

Step 15 of the E-0 procedure is to check the arrangement of the valves related to the auxiliary water supply and take action as necessary. This step is independent of the one to isolate the faulty SG. But following the occurrence of an SGTR accident, the secondary radiation level increases and the PZR pressure changes after some amount of time, and through these tendencies the occurrence of an SGTR accident can be inferred before the accident diagnosis stage. An operator prematurely taking the action to arbitrarily close the main steam isolation valve in this situation is a violation of the procedure; however, as concerns the CIA system, the action has no negative effect on the integrity of the CSFs, and thus the COSIE module should not flag the action as an error and not warn the operator in this test scenario.

Table 7.6 Instructions in E-0 procedure step 15

Response	Response not obtained
15.0 Check the aux. feedwater valves alignments: 15.1 Aux. feedwater flow control valves: open – HV-313, HV-314, HV-315 15.2 CST–Aux. feedwater pump supply valves: open – HV-302	Manually align the valves as needed

Reproduced with permission from Ahn et al. (2022)

The test result of the second scenario is shown in Fig. 7.15. The first filter finds that the action performed by the operator was a procedural non-compliance, but the second filter finds that the action did not negatively affect the related safety function. By prematurely closing the main steam isolation valve of the abnormal SG, the pressure of SG #2 is predicted to decrease slightly, but the safety function integrity score from the CSF tree standard does not change because the decrease is small. Therefore, even though the COSIE module is actuated in this case, the operator continues to perform their tasks as normal without interruption from the CIA system.

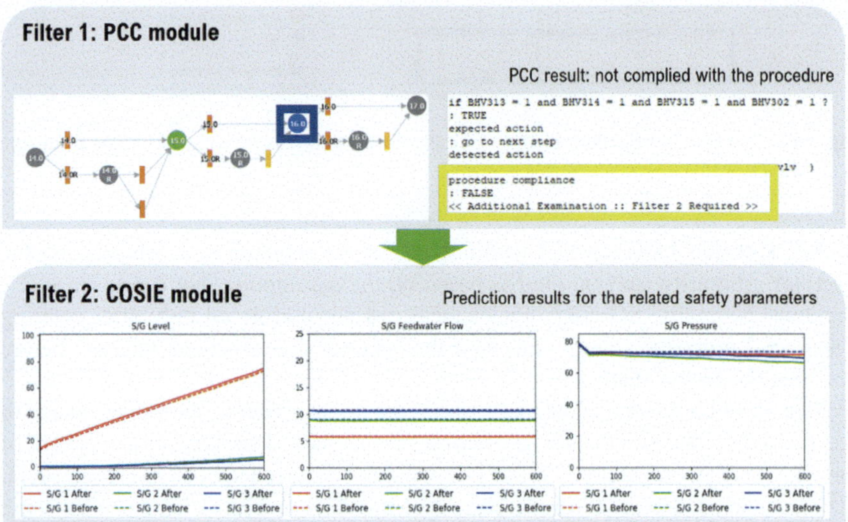

Fig. 7.15 Test results of the PCC and COSIE functions for scenario #2. Reproduced with permission from Ahn et al. (2022)

7.1.5.4 Test Result for Scenario #3: Procedural Non-compliance, CSFs Challenged

If an operator action does not comply with the procedure and degrades the CSFs, the CIA system should flag the action as an error and provide a warning along with related information to the operator. The third test scenario simulates this situation with step 18 of the E-0 procedure entered after an accident has occurred. Table 7.7 lists the instructions in this procedure.

Step 18 of the E-0 procedure is to check the valve arrangement of the PZR and to shut off the PORV. In the simulated test scenario, the PORV block valve should be opened and the PORV should be closed, but the operator mistakenly opened the PORV instead of the PORV block valve. This is clearly both a clear procedural non-compliance as well as human error because opening the PORV in a condition where

Table 7.7 Instructions in E-0 procedure step 18

Response	Response not obtained
18.0 Check PZR valves alignment: 18.1 PORV: close – PV-445 18.2 PORV block valve: open – HV-6	18.1 If PZR P < 164.2 kg/cm^2, close PORV manually. If the PORV cannot be closed, close the PORV block valve instead 18.2 Open the closed valve

Reproduced with permission from Ahn et al. (2022)

the PZR pressure is not high will negatively affect the CSFs. The simulation results in Fig. 7.16 for the third test scenario show that the CIA system flags the action as a human error and displays a warning in a window with relevant information, as shown in Fig. 7.17. As part of the warning, the CIA interface provides the instructions in the violated step of the procedure, the action that caused the violation, and the CSF impact assessment results. In the simulated scenario, the system predicts that the integrity of the RCS will degrade and provides the related information as visualized in yellow. Moreover, in the interface, the corresponding CSF is implemented as a clickable button that displays the prediction results in a pop-up window. This way, the operator can see that accidentally opening the PORV will lead to an inappropriate rise in the water level of the PZR, which will challenge the RCS integrity.

The test scenarios demonstrate that the CIA system only interacts with operators when an action is expected to negatively affect the safety functions of the plant. When met with a CIA warning, operators have the chance to reconsider the appropriateness of the action they performed and to recover it if necessary. In dynamic situations such as NPP emergency operation, potential threats from a human error will increase over time if not addressed quickly, meaning that the burden to recover the error will

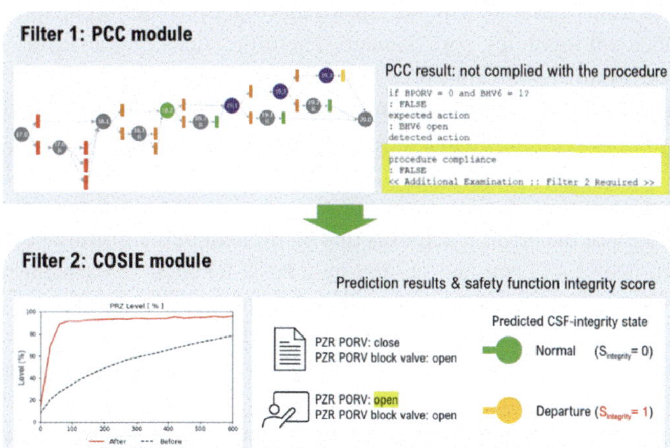

Fig. 7.16 Test results of the PCC and COSIE modules for scenario #3. Reproduced with permission from Ahn et al. (2022)

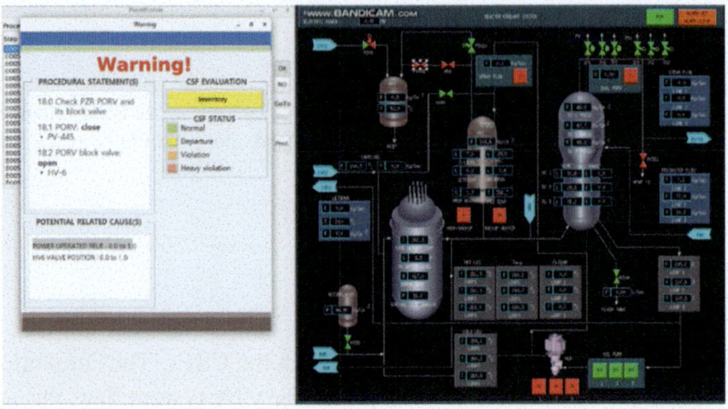

Fig. 7.17 CIA interface warning screen with operator error alarm in scenario #3. Reproduced with permission from Ahn et al. (2022)

also increase with time. In this light, the CIA system can contribute to reducing the potential workload required for human error recovery in an emergency.

7.1.6 Summary and Scalability of the Operation Validation System

In an NPP emergency situation, operators carry out their duties by strictly following EOPs that instruct the measures to take to secure the safety of the plant. Situations like this with great psychological burden are accompanied by a relatively high potential for human error. In particular, NPP emergency situations require both rapid and accurate responses from operators with the knowledge that even a minor mistake can lead to a major accident involving core damage. Therefore, detecting an error as quickly as possible and taking appropriate recovery measures is essential, but considering the unfamiliar environment of an actual emergency situation, the possibility exists that an error may not be recognized in a timely manner. In such a situation, the developed CIA system, as a type of operator support system, can greatly reduce the cognitive burden of operators in emergency situations and provide an opportunity for the optimal recovery of the error. It achieves this by immediately informing the operator of the occurrence of an error and presenting a basis for their subsequent judgment. Another advantage is that the CIA system remains concealed in the absence of human error and is thus free from many automation-related issues. In other words, the CIA system is designed to not degrade the skill, situational awareness, or authority of operators.

Providing various forms of additional information to NPP operators in critical situations can create compounding cognitive burden and thus detrimental effects. Accordingly, prior to introducing a new system, an appropriate balance should be achieved in terms of its overall impact on safety. The CIA system is designed to minimize increases in the cognitive burden of operators by providing additional information only when an error is expected to negatively affect plant safety; but the extent to which the developed system can actually reduce operator cognitive load remains to be evaluated. Another point of consideration is the use of the CIA system in the course of regular or irregular operator training sessions. The opportunity to gather and analyze information from the operation logs on the particular actions that involve frequent operator errors would be valuable in terms of discerning whether the cause of frequent human errors is rooted in safety culture, such as the operators' mindset, or from other human factors. Such insight could contribute to strengthening the safety culture of NPPs and related organizations, thereby helping to prevent serious accidents and achieve higher levels of nuclear safety.

7.2 Technical Specification Monitoring System

The monitoring of NPP operations in compliance with the plant technical specification (Tech Spec) represents an essential yet demanding task for operators. Challenges stem from the large volume of the document, complexity of the contents, high workload in handling the document, diversity of possible interpretations, and time dependence of the related activities.

To address these difficulties, numerous operator support systems have been designed to assist operators in the performance of their Tech-Spec-related activities. But despite the good capabilities of the developed systems, a significant potential for improvement still exists. First, many previous systems (Ragheb et al. 1988; Lidsky et al. 1988; Paiva and Schirru 2015) mainly utilized rule-based techniques to implement LCO monitoring in a Tech Spec. However, rule-based approaches are not typically an efficient means for monitoring all LCOs. Second, several previous studies (Lidsky et al. 1988; Ragheb et al. 1988; Paiva and Schirru 2015) did not include all the functions necessary for operators to successfully apply a Tech Spec. More specifically, while these approaches were able to focus on monitoring different types of parameters and determining the LCOs, the application of a Tech Spec requires numerous additional activities such as examining the operability of instruments, conducting follow-up actions, and confirming that the follow-up actions are completed, in addition to the parameter monitoring and LCO determination activities.

As complements or alternatives to rule-based systems, recent AI techniques open the door to strengthening the performance of systems by way of more powerful diagnosis associated with human intelligence-related activities. In comparison to the low-level ML techniques and expert systems produced in the early stages, current AI techniques exhibit substantial improvements in solving problems in various areas. In the nuclear industry, as discussed throughout this book AI techniques have great potential for operator support, especially for error-prone tasks and those that follow complex procedures or require skilled expert knowledge.

This section presents the concept of a Tech Spec monitoring system (TSMS) using modern AI techniques. The design process starts by analyzing the Tech Spec of an NPP and identifying the high-level design requirements for the system. Then based on these design requirements, the specific functions of the TSMS are defined. Following a review of available AI, the most suitable AI techniques are selected for each system function, after which the TSMS functions are implemented applying the selected techniques. As a demonstration, a TSMS prototype is developed using a CNS of a Westinghouse 990 MWe three-loop PWR and tested.

7.2.1 *Identification of Functional Requirements*

The first step in the development of the TSMS is a task analysis to grasp the operator tasks that are detailed in a Tech Spec in order to determine the functional requirements of the system. For this, the standard Tech Spec for Westinghouse-type plants provided in (NRC 2012) is examined. Based on the task analysis, the high-level functional requirements of the TSMS can be recognized.

7.2.1.1 Structure of Tech Spec

Figure 7.18 depicts an example of the organization of the Tech Spec for Westinghouse-type plants (NRC 2012). The document largely covers the following four content areas: LCOs, applicability, actions, and surveillance requirements (SRs). Brief explanations of these parts are given below.

LCOs represent the minimum requirements for ensuring safe plant operation. Applicability represents the specified conditions or modes under which the LCOs should be assured. The different operation modes of the reference plant are listed in Table 7.8 along with some related parameters. Actions refer to the tasks that must be performed under certain conditions when an LCO is not assured. Lastly, SRs are

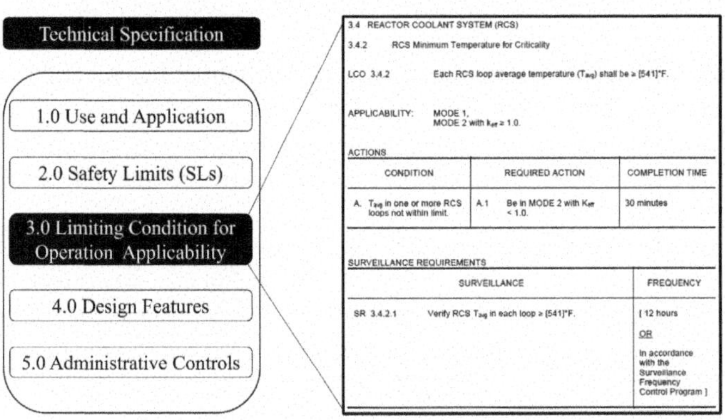

Fig. 7.18 Example of standard Tech Spec for Westinghouse-type plants. Reproduced with permission from Lee and Kim (2022)

Table 7.8 Plant modes of a Westinghouse-type plant

Mode	Title	Reactivity condition (k_{eff})	% Rated thermal power	Average reactor coolant temperature (°C)
1	Power operation	≥ 0.99	> 5	–
2	Startup	≥ 0.99	≤ 5	–
3	Hot standby	<0.99	–	$\geq 177°$ C
4	Hot shutdown	<0.99	–	$94 \leq T_{avg} < 177°$ C
5	Cold shutdown	<0.99	–	$\leq 94°$ C
6	Refueling	–	–	–

Reproduced with permission from Lee and Kim (2022)

the requisite activities related to testing, calibration, and inspection to satisfy that the quality and reliability of the structures, systems, and components of the plant are preserved.

In the example in Fig. 7.18, the LCO is the average temperature T_{avg} of all the RCS loops, which needs to be maintained above 283 °C (541 °F). This LCO applies to Modes 1 and 2 with $k_{eff} \geq 1.0$. If the LCO is not met, the operators should conduct the required actions to enter Mode 2 with $k_{eff} < 1.0$ within a period of 30 min.

7.2.1.2 Types of Monitored Variables

The different types of monitored variables need to be distinguished as the most suitable AI method to be used may differ depending on the monitoring type. From the task analysis of the Tech Spec, six types of variables are defined: parameters, calculated parameters, graph-related parameters, system and component statuses, condition of instrumentation, and operator inputs. Table 7.9 lists the different types of variables with examples.

7.2.1.3 Identification of High-Level Functional Requirements for the TSMS

As mentioned above, the task analysis involves reviewing the tasks related to applying a Tech Spec and using the results to define the high-level functional requirements for the TSMS. As implied by Fig. 7.19, it is essential for the TSMS to first define the current operation mode because it determines the particular LCO application. In normal operation, operators continuously monitor the plant variables based on the SRs. But in the event that the plant state does not fulfill any LCO, follow-up actions should be identified, executed, and completed within the designated time. If the follow-up actions cannot be finished within the available time, then the operators need to conduct other follow-up actions or apply other LCOs.

Table 7.9 Types of monitored variables and examples

Monitored Type	Examples in Tech Spec
Parameters	Verify PZR pressure is greater than or equal to the limit
Calculated parameters	Verify the shutdown margin to be within limits
Graph-related parameters	Verify that RCS pressure, RCS temperature, and RCS heat up and cool down rates are within the P–T limit curve
System and component statuses	Verify that, for each emergency core cooling system throttle valve listed below, each position stops at the correct position
Condition of instrumentation	Perform a channel check for instrumentation
Operator inputs	Perform visual inspection of exposed interior and exterior surfaces of shield building

Reproduced with permission from Lee and Kim (2022)

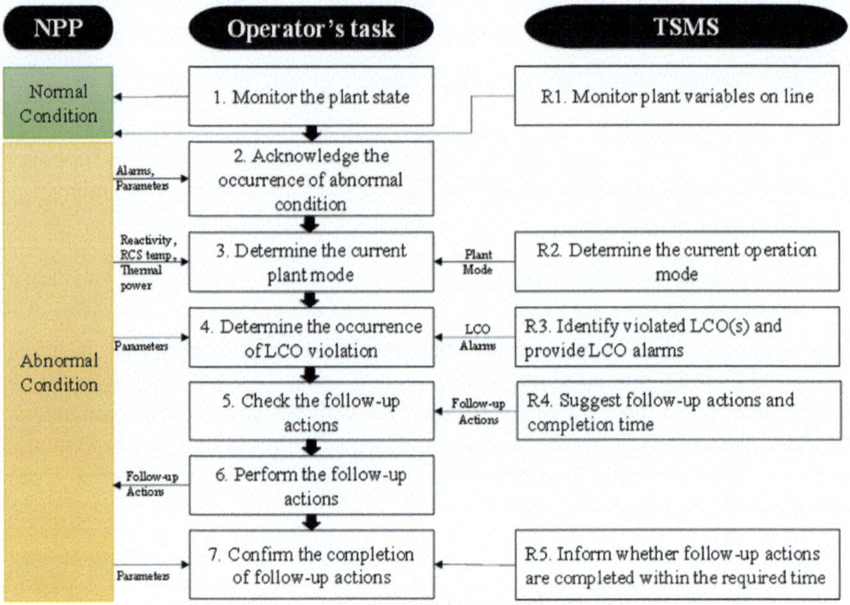

Fig. 7.19 Operator tasks in the Tech Spec and corresponding high-level functional requirements of the TSMS. Reproduced with permission from Lee and Kim (2022)

The task analysis results are used to determine the high-level requirements for the TSMS to support operators in Tech Spec applications, as listed here and as also depicted in Fig. 7.19.

- R1: The TSMS should continuously monitor the following plant variables:
  - R1.1: Parameters
  - R1.2: System and component statuses

  - R1.3: Calculated parameters
  - R1.4: Graph-related parameters
  - R1.5: Condition of instrumentation
  - R1.6: Operator inputs

- R2: The TSMS should determine the current operation mode of the plant.
- R3: The TSMS should identify the violated LCO(s) and provide LCO-related alarms upon the violation of any LCO.
- R4: The TSMS should propose follow-up actions and the required completion time.
- R5: The TSMS should monitor and inform whether a follow-up action is completed within the required completion time.

7.2.2 Conceptual Design of the TSMS

Figure 7.20 illustrates the overall TSMS architecture. The system comprises six sub-functions: operation mode monitoring, parameter calculation, graph-related parameter monitoring, instrumentation monitoring, LCO monitoring, and follow-up action monitoring. Table 7.10 provides a description of each function along with their input, output, and implementation method. Further descriptions of the TSMS functions are given in the following subsections.

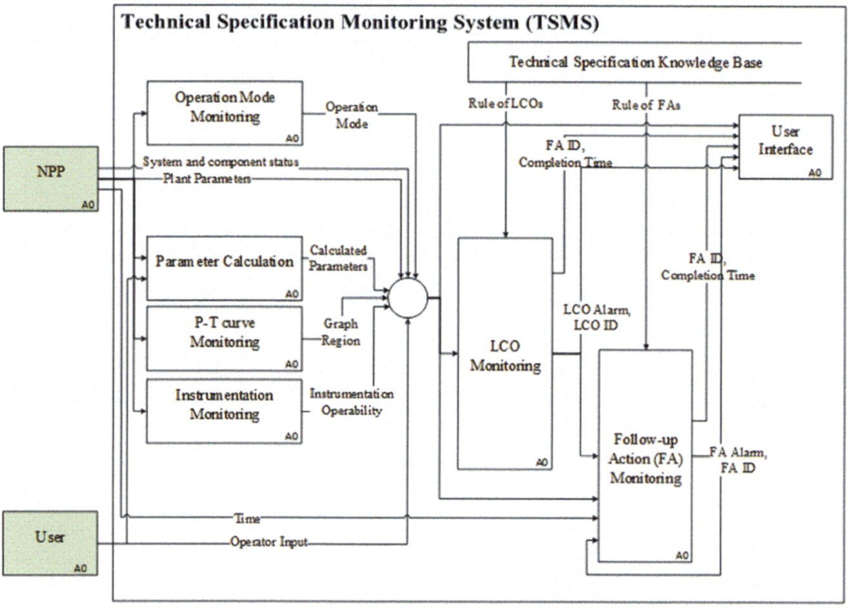

Fig. 7.20 Architecture of the TSMS. Reproduced with permission from Lee and Kim (2022)

Table 7.10 Summary of TSMS functions

Function	Inputs	Outputs	Functional requirements	Automation level	Implementation technique
Operation mode monitoring	Plant parameters (from NPP)	Operation mode (to LCO monitoring and follow-up action monitoring)	R1	Full automation	Rule-based system
Parameter calculation	Plant parameters (from NPP) Operator inputs (from user)	Calculated parameter (to LCO monitoring and follow-up action monitoring)	R1.3	Partial automation	Calculation algorithm
Graph-related parameter monitoring	Plant parameters (from NPP)	Graph region (to LCO monitoring and follow-up action monitoring)	R1.4	Full automation	Support vector machine
Instrumentation monitoring	Plant parameters (from NPP)	Instrumentation operability (to LCO monitoring and follow-up action monitoring)	R1.5	Full automation	Neural network—autoencoder
LCO monitoring	Operation mode (from operation mode monitoring) Plant parameters (from NPP) System and component status (from NPP) Calculated parameters (from parameter calculation) Graph-region (from graph-related parameter monitoring) Condition of instrumentation (from instrumentation monitoring) Operator inputs (from user)	LCO alarm and ID (to follow-up action monitoring and user interface) Follow-up action ID and completion time (to user interface)	R1 R2 R3 R4	Partial automation	Rule-based system

(continued)

Table 7.10 (continued)

Function	Inputs	Outputs	Functional requirements	Automation level	Implementation technique
Follow-up action monitoring	LCO alarm and ID (from LCO monitoring) Operation mode (from operation mode monitoring) Plant parameters (from NPP) System and component status (from NPP) Calculated parameters (from parameter calculation) Graph-region (from graph-related parameter monitoring) Condition of instrumentation (from instrumentation monitoring) Operator inputs (from user) Time	Follow-up action alarm and ID (to follow-up action and user interface) Follow-up action ID and completion time (to user interface)	R4 R5	Partial automation	Rule-based system

Reproduced with permission from Lee and Kim (2022)

7.2.2.1 Operation Mode Monitoring Function

The operation mode monitoring function continuously classifies the current operation mode of the plant into one of the six modes listed in Table 7.8. As shown above in Fig. 7.18, the operation mode corresponds to the applicability area of the Tech Spec. Three parameters are regarded as the inputs of this function: reactivity, percent thermal power, and RCS T_{avg}, as in Table 7.8.

A rule-based system is applied for this function because clear determination rules exist for the different operation modes. The decision rules in Table 7.8 are converted to if–then rules in the TSMS. For example, for a reactivity condition of greater than or equal to 0.99 and a rated percent thermal power of greater than or equal to 5%, the operation mode is classified as Mode 1, power operation.

7.2.2.2 Parameter Calculation Function

The parameter calculation function generates parameters that require complicated calculations using specific formulas. The Tech Spec instructs operators to manually compute certain parameters to be monitored, at times referring to figures showing trends and tables expressing plant parameters. As the calculations may entail complicated steps, this process may become long and time-consuming. Examples of parameters that require calculation are the shutdown margin (SDM), RCS operational leakage, and boron concentration, among others. For the parameter calculation function of the TSMS, calculation algorithms are developed that apply a specific formula to each parameter and, when necessary, also provide data tables and figures from the Tech Spec and other related documents.

7.2.2.3 Graph-Related Parameter Monitoring Function

The objective of the graph-related parameter monitoring function is to classify different regions of parameter graphs when the LCOs require the use of graphs. Here, different graph regions indicate the current plant condition. For monitoring certain LCOs, operators must use a graph and ensure that the current state is in the desired area of the graph. Examples of this type of monitoring include the P–T limit curve and the departure of nucleate boiling ratio. Figure 7.21 shows an example of a simplified P–T curve. Parameters falling into any of the unacceptable areas means that the RCS is being operated under conditions that can result in brittle failure, which has the potential to lead to a LOCA. Based on such graphs, operators must ensure that the current RCS pressure and temperature are in the acceptable area.

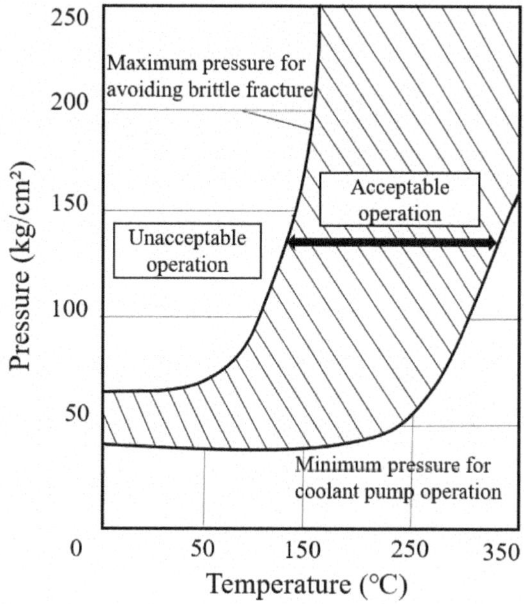

Fig. 7.21 Example pressure–temperature (P–T) limit curve. Reproduced with permission from Lee and Kim (2022)

An SVM was chosen as the method of implementation for the graph-related parameter monitoring function of the TSMS. Generally, SVMs provide good performance in sorting data into two categories with a limited number of attributes. Accordingly, the SVM is considered appropriate for this function as graph-related parameter monitoring generally distinguishes two regions—desired and undesired, for example—based on two or three characteristics.

7.2.2.4 Instrumentation Monitoring Function

The purpose of the instrumentation monitoring function is to test the process sensors of the instrumentation. The monitoring of some LCOs requires operators to test the instruments used in the safety-related systems such as the RPS, engineered safety feature actuation system, and control element assembly calculators. The tests are to verify that the main components of the instrumentation are in a healthy state to perform their function. An example can be seen in LCO 3.3.1 in the standard Tech Spec for a Westinghouse-type plant (NRC 2012), which dictates that the RPS instrumentation must be operable. In terms of components, the RPS instrumentation is made up of process sensors, bistables, logic matrices, and reactor trip switch gear. Testing of these components in actual NPPs is conducted by operators in the case of the process sensors, while on the other hand by software in the case of the bistables, logic matrices, and reactor trip switch gear (Lee et al. 2008).

As the implementation method for this function, an autoencoder is applied. As mentioned previously, neural network methods such as autoencoders provide good performance in processing large data and nonlinear data relationships. Furthermore, autoencoder models are well known for effectively reconstructing input values by adopting the same structure for the input and output layers. This makes an autoencoder a suitable approach for the instrumentation monitoring function as the sensor reconstruction model should estimate the normal behavior of the sensors.

7.2.2.5 LCO Monitoring Function

The LCO monitoring function determines if the plant status complies with the various LCOs dictated by the Tech Spec. In order to define the current plant status, this function tracks the six types of monitored variables (Table 7.9) and compares them with the relevant LCOs. As plant data, the six types of monitored variables are transmitted from the NPP, operators, and other TSMS functions (e.g., the operation mode monitoring, parameter calculation, graph-related parameter monitoring, and instrumentation monitoring functions). The moment any LCO is violated, the LCO monitoring function raises an alarm and provides operators with follow-up actions to take in order to return the plant to a safe status. The LCO alarms and follow-up actions are conveyed to operators through the user interface, and the determined follow-up actions are also passed to the subsequent function. For implementation, the LCO monitoring function applies a rule-based system because in this case, if–then rules are able to clearly define the conditions for the decisions. The LCOs for the minimum amount of equipment as well as the operating parameters are thoroughly detailed in the Tech Spec.

7.2.2.6 Follow-Up Action Monitoring Function

The purpose of the follow-up action monitoring function is to confirm that the follow-up actions previously called for are completed within the appropriate amount of time. When the plant status violates any LCO, the Tech Spec generally instructs operators to perform a follow-up action or actions to return the plant to a safe status. In this case, the operators must confirm the termination time of the follow-up action and complete it within the available time. If the follow-up action is not completed in the appropriate amount of time or otherwise not properly performed, the operator needs to perform an alternative action according to the current plant status. When the follow-up action monitoring function receives the LCO alarm and ID of the violated LCO, it first identifies the follow-up action(s) to be monitored. Then to confirm the completion of the given follow-up action(s), this function tracks the six types of monitored variables. If the follow-up action is not completed within the available time or otherwise fails, the function raises an alarm and outputs the failed follow-up ID and alternate action to the user interface. Then to monitor the completion of the alternative follow-up action or actions, follow-up action alarms and IDs are sent in turn. Like the LCO monitoring function, the follow-up action monitoring function applies a rule-based system since the decision condition (i.e., if) and follow-up action (i.e., then) are clearly delineated in the Tech Spec.

7.2.3 *Implementation of the TSMS*

This section details the implementation of the selected AI methods in Sect. 7.2.2 for the development of each function of the TSMS. For the implementation, a CNS developed by KAERI (KAERI 1990) for a Westinghouse 990 MWe three-loop PWR reference plant is employed as the testbed. For developing the software, Python 3.7 is used as the programming language, and for modeling the AI methods, the keras and scikit-learn software libraries are used.

7.2.3.1 Operation Mode Monitoring Function

As discussed in the previous section, the operation mode monitoring function applies a rule-based system and classifies the current operation mode into Modes 1–6. A rule-based system is a type of AI method composed of three main parts working together: a knowledge base, database, and inference engine. As the core of the system, the knowledge base contains a set of rules such as ‘if (condition)–then (action)’ to solve a particular problem. The database includes a set of facts used to compare the ‘if (condition)’ aspect of the rules in the knowledge base. The inference engine provides the reasoning by which the rule-based system reaches a solution from the knowledge base and database (Jadhav and Channe 2016).

Figure 7.22 illustrates a flowchart of the rules to determine the operation mode, which as stated above can be classified into six types, namely Modes 1–6: power operation, startup, hot standby, hot shutdown, cold shutdown, and refueling, respectively. These rules for determination are stored in the knowledge base of the rule-based system. The relevant plant data, in other words the reactivity (k-value), thermal power, and RCS temperature, are used to compare with the ‘if (condition)’ aspect of the rules in the knowledge base. From this comparison, the rule-based system determines the current operation mode via reasoning about the plant data and knowledge base.

The bolded path in Fig. 7.22 highlights an example. Here, the plant is in normal operation, the k-value is 1.0, the thermal power is 0%, and the reactor coolant temperature is 200 °C. Based on these, the operation mode monitoring function adopting a rule-based system determines the plant to be in Mode 3.

7.2.3.2 Parameter Calculation Function

The parameter calculation function uses specific calculation algorithms in place of manual calculations by operators. One typical calculated parameter is the SDM,

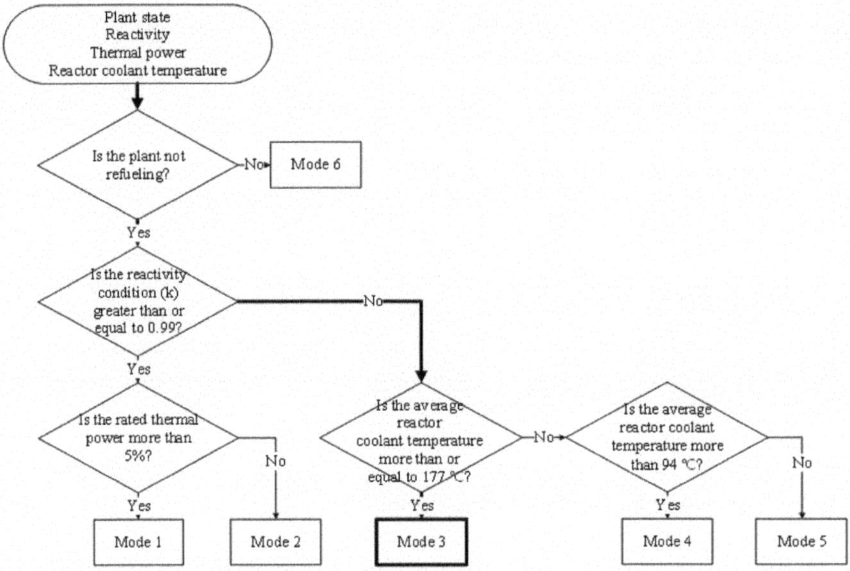

Fig. 7.22 Rules for operation mode monitoring function. Reproduced with permission from Lee and Kim (2022)

which is the instantaneous amount of reactivity to define the sub-criticality of the reactor. The complicated manual calculation process for the SDM can be eliminated by the parameter calculation function.

Table 7.11 lists the total 14 input data used for calculating the SDM, and Fig. 7.23 illustrates the SDM calculation procedure of the function. The steps in this series of calculations are as follows. First, the power defect is estimated in the current reactor power at the beginning of life (BOL) as in Eq. (7.3). Second, the power defect is estimated in the current reactor power at the end of life (EOL) as in Eq. (7.4). Third, the power defect is estimated in the current power at the current burnup as in Eq. (7.5). The total reactivity defect is then estimated as in Eq. (7.6), after which if there are any inoperable or abnormal control rods, the total inoperable and abnormal rod worth is calculated using Eq. (7.7). Lastly, the SDM can finally be calculated as in Eq. (7.10).

Figure 7.24 shows an example SDM calculation with the following initial conditions. The reactor is in normal operation with 90% thermal power for 10 min, and the current burnup is 4000 MWD/MTU. The reactor trips at 10 min due to a LOCA. In order for the parameter calculation function to work, the operators should input the current inoperable rod number, i.e., 1 in the example, the abnormal bank name C, and abnormal rod number 1. The other input variables are automatically determined by the TSMS. The Tech Spec recommends that the lowest limit of the SDM, shown as the red line in the plot of Fig. 7.24, should be 1770 pcm in Mode 1 and Mode 2. The calculated SDM from the TSMS is plotted as the blue line in the figure, showing a calculated SDM of 2568 pcm before the accident followed by an increase in the SDM to 4500 pcm after the trip.

Table 7.11 Inputs for the SDM calculation function

1	Hot full reactor power	8	Void content
2	Current reactor power	9	Worst stuck rod worth
3	Power defect (hot full power, BOL)	10	Inoperable rod number
4	Power defect (hot full power, EOL)	11	Bank worth
5	Current burnup	12	Abnormal bank name
6	BOL burnup	13	Abnormal rod number
7	EOL burnup	14	Total rod worth

Reproduced with permission from Lee and Kim (2022)

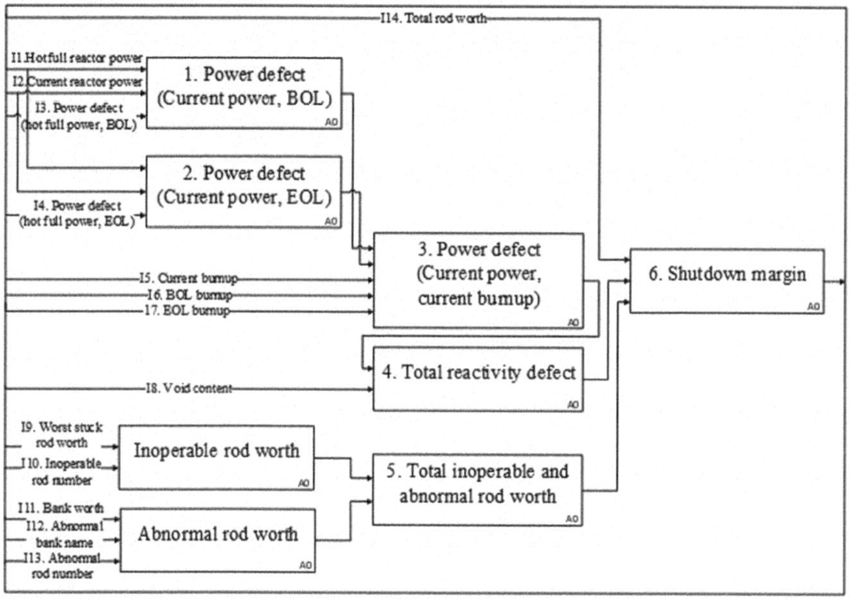

Fig. 7.23 Calculation algorithm of the SDM calculation function. Reproduced with permission from Lee and Kim (2022)

$$Power\ defect\ (BOL) = \frac{power\ defect\ (hot\,full\ power,\ \ BOL) \times current\ reactor\ power}{hot\,full\ reactor\ power} \tag{7.3}$$

$$Power\ defect\ (EOL) = \frac{power\ defect\ (hot\,full\ power,\ \ EOL) \times current\ reactor\ power}{hot\,full\ reactor\ power} \tag{7.4}$$

$$Power\ defect = \frac{(EOL - BOL) \times (current\ burnup - BOL\,burnup)}{(EOL\,burnup - BOL\,burnup)} + BOL \tag{7.5}$$

$$\text{Total reactivity defect} = \text{void content} + \text{power defect} \tag{7.6}$$

$$\begin{aligned}&\text{Total inoperable and abnormal rod worth}\\ &\quad = \text{inoperable rod worth} + \text{abnormal rod worth}\end{aligned} \tag{7.7}$$

$$\text{Inoperable rod worth} = \text{worst stuck rod worth} \times \text{inoperable rod number} \tag{7.8}$$

$$\text{Abnormal rod worth} = \frac{bank\ worth}{rod\ number} \times abnormal\ rod\ number \tag{7.9}$$

$$\begin{aligned}&\text{Shutdown margin (SDM)} = \text{total rod worth}\\ &\quad - \text{Total inoperable and abnormal rod worth} - \text{Total reactivity defect}\end{aligned} \tag{7.10}$$

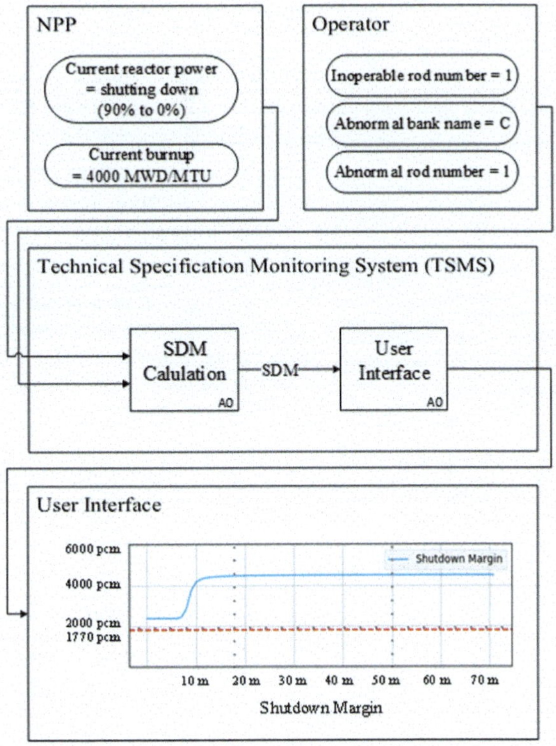

Fig. 7.24 Example SDM calculation via parameter calculation function. Reproduced with permission from Lee and Kim (2022)

7.2.3.3 Graph-Related Parameter Monitoring Function

The graph-related parameter function adopts an SVM to determine whether the graph-related parameters fall within the desired areas of the graph. An SVM can divide data into two different classes by identifying a hyperplane; Fig. 7.25 illustrates the different elements of the SVM. Support vectors here refer to data points located closely to the hyperplane (Berwick 2003), where the hyperplane is a decision factor dividing given data into two different classes. The largest distance between the hyperplane and the support vectors is known as the maximum margin. For more effective classification, the SVM attempts to find the hyperplane with the maximum margin. To establish a hyperplane to divide data classes, this method uses a kernel function to map the data onto a high-dimensional space, and subsequently determines the maximum margin hyperplane within that space. Several common kernel functions include the linear kernel, polynomial kernel, radial basis function (RBF), and sigmoid kernel (Chen et al. 2011). Once set, the hyperplane can be regarded as a classifier.

Figure 7.26 depicts the algorithm for the P–T curve monitoring function applying an SVM. Preprocessing of the two inputs, RCS pressure and RCS T_{avg}, is carried out to scale the values for application to the SVM model. For the input preprocessing, min–max normalization is utilized in this case. The minimum and maximum of each value are defined from the inputs. Then following the min–max normalization method, the input values are rescaled to the range of 0 to 1 using Eq. (7.11).

$$Xnor = \frac{(X - X_{min})}{(X_{max} - X_{min})} \tag{7.11}$$

The SVM model classifies the P–T curve as 1 or 0 representing the desired region and undesired region, respectively. To classify these P–T curve regions, the SVM model is pre-trained to generate different outputs from the two different classes based on the input values, i.e., RCS pressure and RCS T_{avg}. The initially prepared training dataset thus contains RCS pressure and RCS T_{avg} values, which are projected on the P–T limit curve. The regions are labeled by checking whether each data point representing the current RCS pressure and RCS T_{avg} is in the desired region or in the undesired region in the P–T limit curve. Then the datasets are divided into

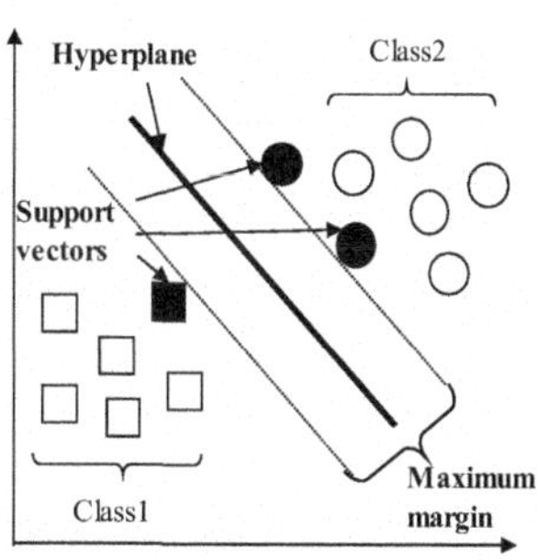

Fig. 7.25 Support vectors, hyperplane, and maximum margin. Reproduced with permission from Lee and Kim (2022)

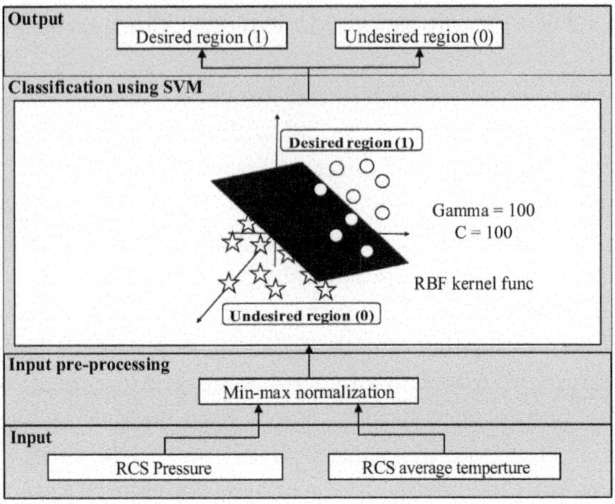

Fig. 7.26 Algorithm of the P–T curve monitoring function using SVM. Reproduced with permission from Lee and Kim (2022)

training and testing datasets. The training dataset contains the RCS pressure, RCS T_{avg}, and P–T curve region with 300 samples. During training, the SVM aims to find the optimal hyperplane dividing the data into two different classes, or in other words, desired and undesired regions on the P–T curve in the present case. To do so, the kernel function maps the data into a higher-dimensional space. Table 7.12 lists different model classification accuracies according to different kernel functions. For the P–T curve monitoring function of the TSMS, the RBF is selected as the kernel function as it shows the highest accuracy at 93.5%. With the RBF kernel function, the performance of the SVM depends on fine-tuning the C and gamma parameters that are hypermeters in SVM to control error. Table 7.13 shows the results of determining the optimal parameters. As a result, (C, gamma) = (100, 100) is selected as it achieved the highest performance, namely 100%.

Following successful training, the algorithm is then tested using the test dataset, which contains the RCS pressure, RCS T_{avg}, and P–T curve region with 100 samples. Test results confirm that the regions of the P–T curve can be classified at an accuracy of 99%. Figure 7.27 shows an example of the P–T curve monitoring function in case of a LOCA. In this event, the RCS pressure and temperature decrease sharply over time. Accordingly, through the interface, the P–T curve monitoring function indicates that the plant status moves into the undesired region at about 1000 s.

Table 7.12 Classification accuracy of the SVM model with different kernel functions

Kernel function	Classification accuracy
Linear kernel	0.634
Polynomial kernel	0.814
Radial basis function (RBF)	0.935
Sigmoid kernel	0.401

Reproduced with permission from Lee and Kim (2022)

Table 7.13 Classification accuracy of the SVM model with the RBF kernel function for different settings of C and gamma

Gamma/C	0.001	0.01	0.1	1	10	100
0.001	0.52	0.52	0.52	0.52	0.52	0.58
0.01	0.52	0.52	0.52	0.52	0.58	0.59
0.1	0.52	0.52	0.52	0.58	0.63	0.83
1	0.52	0.52	0.60	0.82	0.92	0.94
10	0.52	0.52	0.90	0.94	0.98	0.99
100	0.52	0.52	0.96	0.99	0.99	1.00

Reproduced with permission from Lee and Kim (2022)

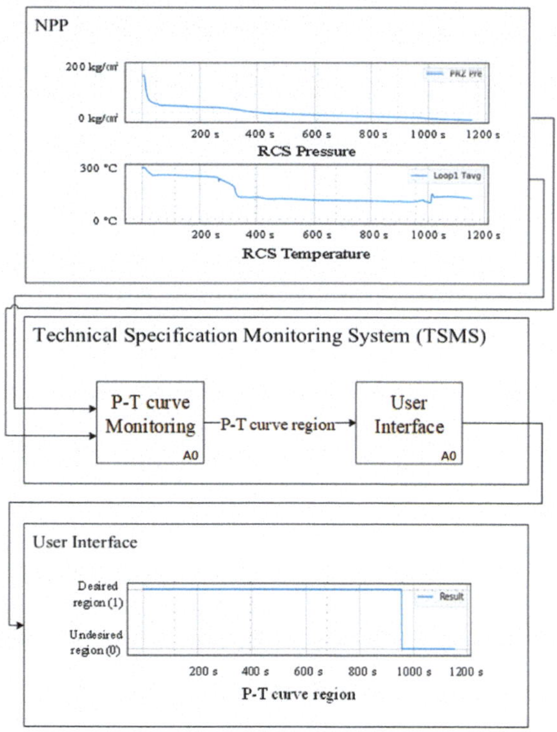

Fig. 7.27 Example result of the P–T curve monitoring function. Reproduced with permission from Lee and Kim (2022)

7.2.3.4 Instrumentation Monitoring Function

The instrumentation monitoring function adopts an autoencoder; Fig. 7.28 illustrates the general form of this type of neural network model. As can be seen in the figure, the autoencoder has the same number of inputs and outputs with encoder and decoder functions. For input x, the encoder transforms corrupted input data $\tilde{x}$ to a hidden representation, h, as described in Eq. (7.12), where f $(\cdot)$ is a nonlinear activation

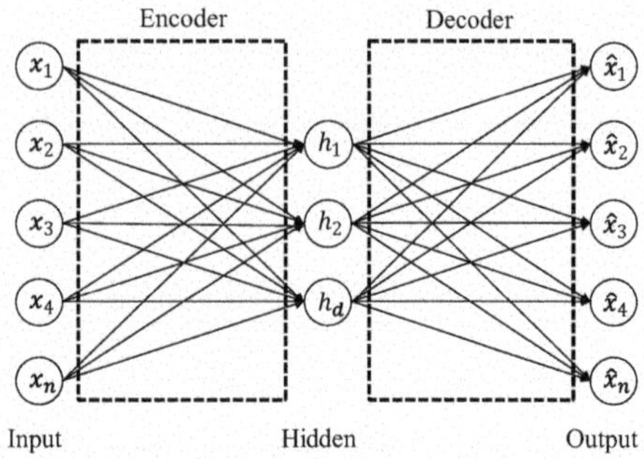

Fig. 7.28 Structure of an autoencoder. Reproduced with permission from Lee and Kim (2022)

function such as sigmoid. The sigmoid function $f(z) = \frac{1}{1+\exp(-z)}$ is commonly applied. For the other terms, $W_1 \in \mathbb{R}^{d \times m}$ is the weight matrix and $b \in \mathbb{R}^d$ is the bias vector to be optimized in the encoding function with d nodes in the hidden layer.

$$H = f(W_1\tilde{x} + b) \tag{7.12}$$

With parameters $W_2 \in \mathbb{R}^{m \times d}$ and $c \in \mathbb{R}^m$, the decoder subsequently attempts to map the hidden representation to the reconstructed vector $\hat{x}$ in the output layer using a nonlinear transformation, as in Eq. (7.13).

$$\hat{x} = g(W_2 h + c) \tag{7.13}$$

For an input training set of $\{x_i\}_{i=1}^n$, the reconstruction error can be computed via $\sum_{i=1}^n \left\| x_i - \hat{x}_i^2 \right\|$. The training objective of the autoencoder is to determine the optimal parameter $\theta = \{W_1, b, c\}$ of the encoder and decoder by minimizing the reconstruction error, as defined in Eq. (7.14) (Shaheryar et al. 2016).

$$\min_{\theta} \sum_{i=1}^{n} \left\| x_i - \hat{x}_i^2 \right\| \tag{7.14}$$

Figure 7.29 illustrates the RPS instrumentation monitoring function with the autoencoder. In this example, the sensor variables are RCS loop #1, 2, 3 T_{avg}, PZR pressure, feedwater line #1, 2, 3 flow, and SG loop #1, 2, 3 level, which are the RPS process sensors. Input preprocessing is first performed to normalize the sensor data, specifically with the min–max normalization method to properly scale the sensor data for the input layer of the autoencoder model. The minimum and maximum of each variable are determined from the sensor data, and then each is rescaled to the range of 0 to 1 following Eq. (7.11), similar to the previous section.

When normalized sensor data is used as inputs, the signal reconstruction process with an autoencoder produces outputs of reconstructed sensor data. For sensor data reconstruction, the autoencoder model is pre-trained with a training dataset

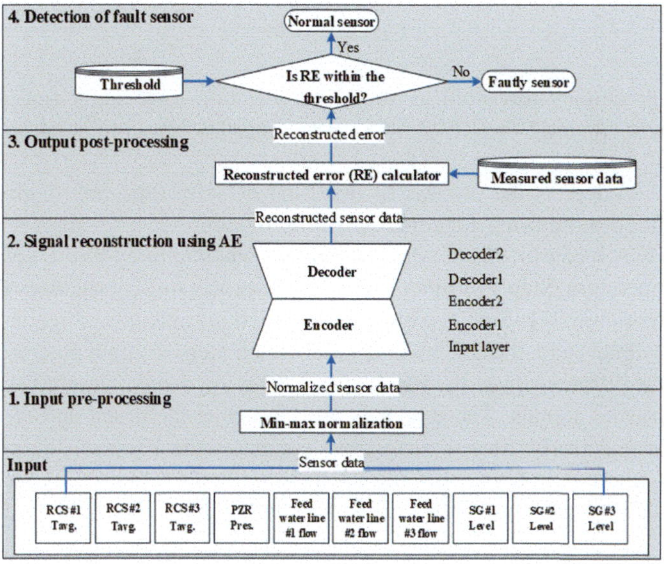

Fig. 7.29 Algorithm of the RPS instrumentation monitoring function with an autoencoder. Reproduced with permission from Lee and Kim (2022)

comprising 3354 data samples of selected sensor data. The autoencoder model has five layers, namely an input layer, two encoder layers, and two decoder layers, as shown in Fig. 7.28. The objective of training for this model is to minimize the difference between the input sensor data and the reconstructed sensor data.

As the next step shown in Fig. 7.29, output post-processing computes the RE, which is the square of the error between the input and reconstructed output obtained from a well-trained model. A large residual may indicate that a particular senor is faulty; for example, with a training dataset including only normal sensor data, a trained model will produce a low residual since the training resulted in a good ability to reconstruct the data. But with an input from a faulty sensor, the trained model will produce a high residual because, based on its training, it cannot reconstruct faulty sensor data well. It is therefore necessary to determine a threshold that distinguishes normal and faulty sensor data.

To define the threshold, a method suggested by Shewhart (Shewhart 1931) is applied based on the REs. The main elements of a Shewhart chart are the center line CL, the upper control limit UCL, and the lower control limit LCL, as shown in Eqs. (7.15) to (7.17). In these equations, μ and σ denote the mean and standard deviation of each residual.

$$\mathrm{UCL} = \mu + 3\sigma \tag{7.15}$$

$$\mathrm{CL} = \mu \tag{7.16}$$

$$\mathrm{LCL} = \mu - 3\sigma \tag{7.17}$$

With a predefined threshold as such, the detection of a faulty sensor can be achieved by comparing the RE. An RE exceeding the threshold suggests a faulty sensor.

Figure 7.30 depicts the process of detecting a faulty sensor through the RPS instrumentation monitoring function. First, the function receives signals as inputs during a LOCA scenario, one of which is faulty: a stuck failure (300 °C) of the RCS loop #1 temperature. Step 1 of function (input pre-processing) normalizes the signal data into the range 0–1. Step 2 (signal reconstruction) attempts to reconstruct the normalized signals with the autoencoder, after which Step 3 (output post-processing) calculates the difference, i.e., the RE, between the normalized measured signals and the reconstructed signals. These signals are plotted as the green and red lines in the figure, respectively. Step 4 (detection) compares the RE with the predefined thresholds, where it is found that the RE for the faulty signal is higher than the threshold. The algorithm thus determines that the RCS loop #1 temperature is faulty.

7.2.3.5 LCO Monitoring Function

The LCO monitoring function applies a rule-based system to confirm that the plant condition is in compliance with the LCOs given in the Tech Spec. The rule base for this function is constructed by extracting if–then rules for all of the LCOs. Updated plant information is continuously sent to the LCO monitoring function, which compares the plant information with the predefined rule base. Based on the comparison, the function raises an alarm if any LCO is violated. An example of building a rule base for an LCO is shown in Fig. 7.31, where LCO 3.4.2 of the standard Tech Spec for a Westinghouse-type plant (NRC 2012) indicates that the average temperature of each RCS loop should be 283 °C. This is converted into if–then rules as follows: if RCS $T_{avg} \geq 283$ °C, then LCO 3.4.2 is satisfied (Rule 1), and if RCS $T_{avg} < 283$ °C, then LCO 3.4.2 is violated (Rule 2).

Figure 7.32 illustrates an example of the LCO monitoring function process for LCO 3.4.1. The function first receives the PZR pressure, RCS cold-leg temperature, and RCS total flow rate from the NPP, and receives the current plant operation mode from the operation mode monitoring function of the TSMS. Then the LCO monitoring function compares the plant parameters with the predefined Rules 1 and 2 for this LCO. In this example, the current status matches Rule 2, which is a violation of the LCO, and accordingly the function generates an LCO alarm and provides operators with the ID of the violated LCO (LCO 3.4.1 in this case), the ID of the appropriate follow-up actions (FA341-1), and the required completion time (2 h). Table 7.14 lists the LCOs and corresponding rules implemented in the TSMS prototype.

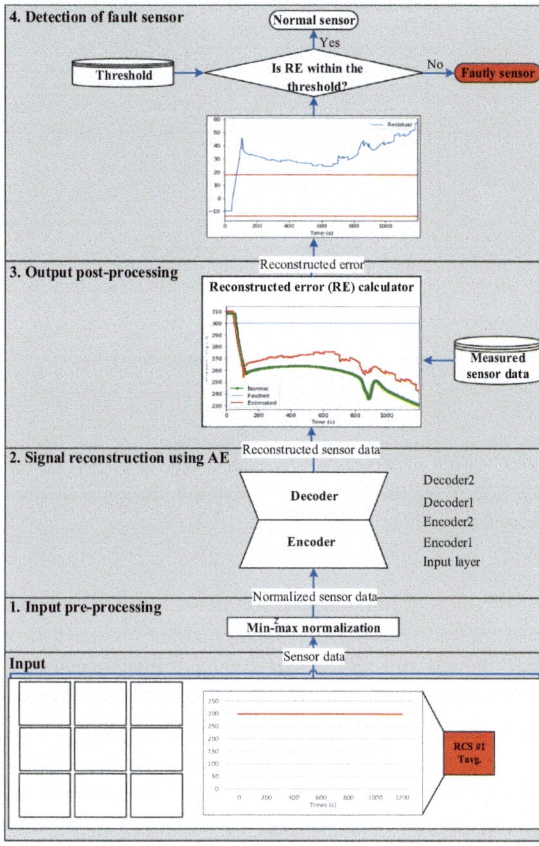

Fig. 7.30 Example of the RPS instrumentation monitoring function. Reproduced with permission from Lee and Kim (2022)

3.4 REACTOR COOLANT SYSTEM (RCS)

3.4.2 RCS Minimum Temperature for Criticality

LCO 3.4.2 Each RCS loop average temperature (T_{avg}) shall be ≥ 283 °C

LCO ID	Rules (if-then)
LCO 3.4.2	**if** RCS $T_{avg} \geq 283$ °C **then** LCO 3.4.2 satisfaction **if** RCS $T_{avg} < 283$ °C **then** LCO 3.4.2 violation

Fig. 7.31 Example extraction of an if–then rule for the LCO monitoring function. Reproduced with permission from Lee and Kim (2022)

7.2.3.6 Follow-Up Action Monitoring Function

As discussed in Sect. 7.2.2.6, the follow-up action monitoring function adopts a rule-based system to confirm that the follow-up actions are successfully conducted

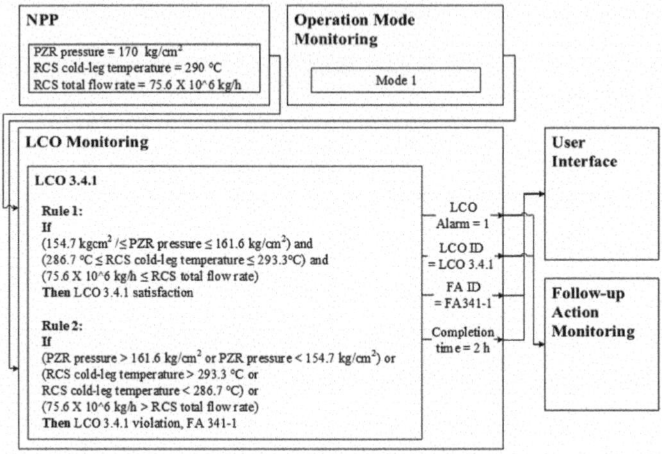

Fig. 7.32 Example of the LCO monitoring function process. Reproduced with permission from Lee and Kim (2022)

within the required completion time. Similar to the previous section, the rule base is constructed by extracting if–then rules for the follow-up actions given in the Tech Spec. When an LCO alarm is raised, this function performs a comparison between the current plant information and the rule base. If any rules are violated, the function outputs alarms indicating the failure of the follow-up action.

Figure 7.33 depicts an example of the follow-up action monitoring function implementation and process. If–then rules are first produced to build the rule base from the follow-up actions for each LCO listed in the Tech Spec. As seen in Fig. 7.33a, when one or more of the RCS departure from nucleate boiling parameters exceed the limits, operators must perform actions to restore the parameter(s) to under the limit(s) within 2 h. These operator tasks can be converted into if–then rules as shown in the figure.

Figure 7.33b illustrates an example of the process of the follow-up action monitoring function. The function is first activated by receiving an alarm for LCO 3.4.1 from the LCO monitoring function. Along with the alarm, the follow-up action monitoring function also receives parameter values of the PZR pressure, RCS cold-leg temperature, and RCS total flow rate from the NPP and the current operation mode from the operation mode monitoring function. Additionally, a timer is initiated to measure the elapsed time. The plant parameters and the required completion time are monitored based on Rules 1 and 2. As its output, the function provides operators either with information about the successful completion of the follow-up actions within the required completion time or with other follow-up actions suggested by the system in case the follow-up actions fail.

Table 7.14 Rule base for the LCO monitoring function

Rule ID	LCO ID	Rules	
		If (condition)	Then (action)
R1	LCO 3.1.1	Shutdown margin ≥ 1770 pcm	LCO 3.1.1 satisfaction
R2		Shutdown margin < 1770 pcm	LCO 3.1.1 violation, FA 311–1
R3	LCO 3.3.1	Condition of RPS instrumentation = normal sensor	LCO 3.3.1 satisfaction
R4		Condition of RPS instrumentation = faulty sensor	LCO 3.3.1 violation, FA 331–1
R5	LCO 3.4.1	(154.7 kg/cm^2 ≤ PZR pressure ≤ 161.6 kg/cm^2) and (286.7 °C ≤ RCS cold-leg temperature ≤ 293.3 °C) and (75.6 × kg/h ≤ RCS total flow rate)	LCO 3.4.1 satisfaction
R6		(PZR pressure > 161.6 kg/cm^2 or PZR pressure < 154.7 kg/cm^2 or (RCS cold-leg temperature > 293.3 °C or RCS cold-leg temperature > 286.7 °C) or (75.6 × 10^6 kg/h > RCS total flow rate)	LCO 3.4.1 violation, FA 341–1
R7	LCO 3.4.3	Region of P–T curve = desired region	LCO 3.4.3 satisfaction
R8		Region of P–T curve = undesired region	LCO 3.4.3 violation, FA 343–1
R9	LCO 3.4.4	Reactor coolant pumps 1, 2, and 3 are operable	LCO 3.4.4 satisfaction
R10		Reactor coolant pumps 1 or 2 or 3 is inoperable	LCO 3.4.4 violation, FA 344–1

Reproduced with permission from Lee and Kim (2022)

7.2.4 *TSMS Prototype*

Following the above function implementations, a full TSMS prototype is now developed. In brief, the system continuously compares the current plant condition with the Tech Spec of the NPP. In the event that any LCO is violated, the TSMS generates LCO alarms and recommends follow-up actions to the operators. It then monitors the operator performance of the follow-up actions in terms of the required completion time and identifies when the plant re-enters a safe condition.

An image of the HSI of the TSMS is shown in Fig. 7.34 for a test of the prototype simulating a LOCA scenario with a break size of 100 cm^2 in the loop 1 hot-leg. Figure 7.34a is the LCO alarm list for notifying operators of the LCO violations, where the first column shows the ID of the violated LCO and the second column

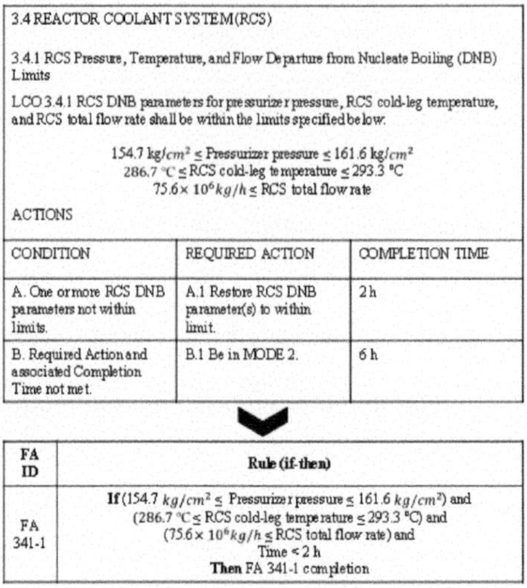

(a) Rule building from the Tech Spec.

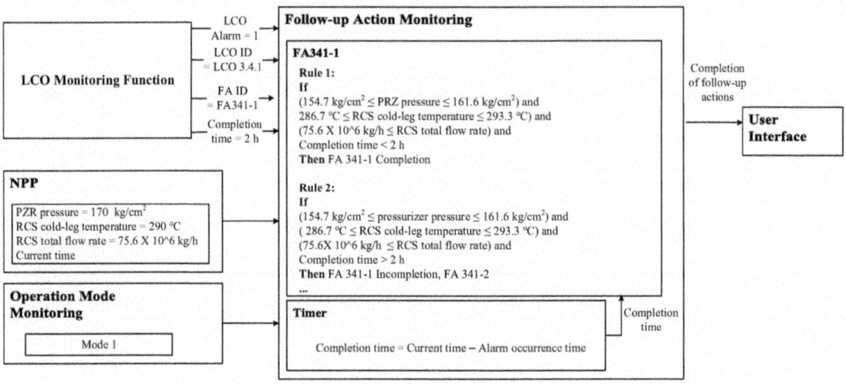

(b) Function process.

Fig. 7.33 Example of the follow-up action monitoring function. Reproduced with permission from Lee and Kim (2022)

gives a description of the relevant LCO with information about the violated NPP system and component. The third column presents the time of the alarm occurrence upon LCO violation, and the fourth column shows the end time of the alarm upon completion of the follow-up action to satisfy the LCO.

The test results indicate that LCO 3.4.1 and LCO 3.4.3 are violated at 37 s and at 2 min 25 s after LOCA occurrence, respectively. A violation of LCO 3.4.1 indicates that the RCS pressure, temperature, and flow departure from nucleate boiling have violated their specified limits. A violation of LCO 3.4.3 indicates that the RCS pressure and temperature have moved into the undesired region of the P–T limit

Fig. 7.34 Appearance of the human–system interface of the TSMS. Reproduced with permission from Lee and Kim (2022)

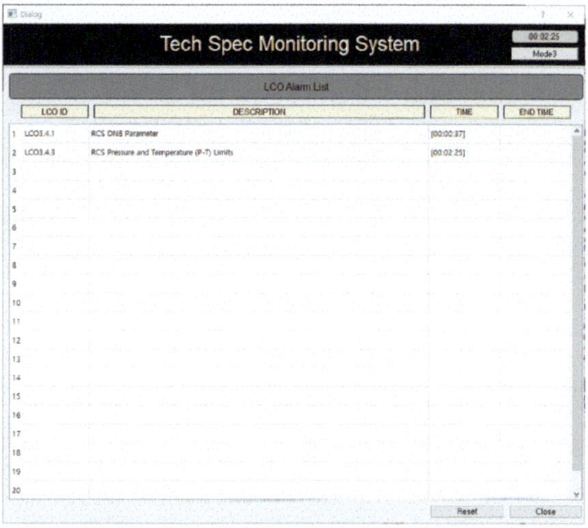

(a) LCO alarm list

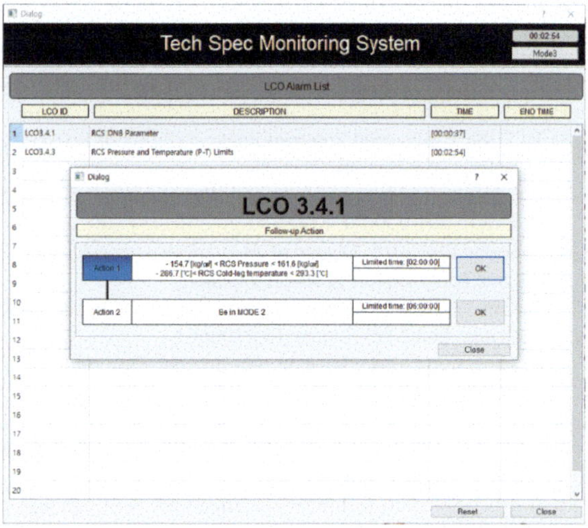

(b) Pop-up window representing follow-up ation

curve. By selecting the LCO in the alarm window, operators can obtain information on the appropriate follow-up action from a pop-up window.

An example of this pop-up window presenting information on the recommended follow-up action is shown in Fig. 7.34b for LCO 3.4.1. In this case, the operators must complete the first follow-up action to restore the RCS pressure, temperature, and flow rate to back within the normal operating ranges. As shown in the window, this action should be completed within 2 h. If the operators cannot complete the first follow-up action within this time limit, a second follow-up action is indicated as an alternative.

References

Ahn J, Bae J, Min BJ, Lee SJ (2022) Operation validation system to prevent human errors in nuclear power plants. Available at SSRN 4062799
Ahn J, Lee SJ (2020) Deep learning-based procedure compliance check system for nuclear power plant emergency operation. Nucl Eng Des 370:110868
Berwick R (2003) An Idiot's guide to support vector machines (SVMs). Retrieved on October, 21, pp 2011
Chen K-Y, Chen L-S, Chen M-C, Lee C-L (2011) Using SVM based method for equipment fault detection in a thermal power plant. Comput Ind 62(1):42–50
Endsley MR, Kaber DB (1999) Level of automation effects on performance, situation awareness and workload in a dynamic control task. Ergonomics 42(3):462–492
Endsley MR (1996) Automation and situation awareness. In: Automation and human performance: theory and applications. CRC Press
Gertman D, Halbert B, Parrish M, Sattison M, Brownson D, Tortorelli J (2002) Review of findings for human performance contribution to risk in operating events. IDAHO NATIONAL ENGINEERING AND ENVIRONMENTAL LAB IDAHO FALLS
Infield DG (1987) Summary report on the post-accident review meeting on the chernobyl accident: safety series No 75–INSAG–1. IOP Publishing
Jadhav SD, Channe H (2016) Comparative study of K-NN, naive Bayes and decision tree classification techniques. Int J Sci Res (IJSR) 5(1):1842–1845
Kaber DB, Endsley MR (2004) The effects of level of automation and adaptive automation on human performance, situation awareness and workload in a dynamic control task. Theor Issues Ergon Sci 5(2):113–153
KAERI (1990) Advanced compact nuclear simulator textbook. Nuclear Training Center in Korea Atomic Energy Research
Lee DY, Lee CK, Hwang IK (2008) Development of the digital reactor safety system (No. KAERI/RR–2914/2007). Korea Atomic Energy Research Institute
Lee SJ, Seong PH (2014) Design of an integrated operator support system for advanced NPP MCRs: issues and perspectives. Springer Japan
Lee S, Kim J (2022) Design of computerized operator support system for technical specification monitoring. Ann Nucl Energy 165:108661
Lee JD, See KA (2004) Trust in automation: Designing for appropriate reliance. Hum Factors 46(1):50–80
Lidsky L, Lanning D, Dobrzeniecki A, Meyers K, Reese T (1988) The use of prolog for computerized technical specifications. In: Artificial intelligence and other innovative computer applications in the nuclear industry. Springer
NRC (2012) Standard technical specifications westinghouse plants (NUREG-1431 Rev. 4). US Nuclear Regulatory Commission
NRC, U. S. (1979) TMI-2 Lessons Learned Task Force. Final report. NUREG-0585. Nuclear Regulatory Commission
NRC, U. S. (1980) Clarification of TMI action plan requirements. Technical report. NUREG-0737. Nuclear Regulatory Commission
OHara J, Higgins J, Fleger S, Barnes V (2010) Human-system interfaces for automatic systems. Brookhaven National Lab.(BNL), Upton, NY (United States)
Paiva GV, Schirru R (2015) Application of an expert system for real time diagnosis of the limiting conditions for operation in nuclear power plants
Parasuraman R, Riley V (1997) Humans and automation: use, misuse, disuse, abuse. Hum Factors 39(2):230–253
Ragheb M, Abdelhai M, Cook J (1988) Producton-rule analysis system for nuclear plant technical specifications tracking. In: Artificial intelligence and other innovative computer applications in the nuclear industry. Springer

Shaheryar A, Yin X-C, Hao H-W, Ali H, Iqbal K (2016) A denoising based autoassociative model for robust sensor monitoring in nuclear power plants. Sci Technol Nucl Installations
Shewhart WA (1931) Economic control of quality of manufactured product. Macmillan And Co Ltd., London
Stanton NA (1996) Human factors in nuclear safety. CRC Press

Chapter 8
Human–Autonomous System Interface and Decision-Making

8.1 Human–Autonomous System Interface

The goal of the human–autonomous system interface or HASI is to support the manual operations of NPP operators. Primarily, it provides the means by which operators can obtain information for monitoring and also perform control actions. For monitoring, the HASI presents operators with alarms and information displays. The information displays include the status of systems and components as well as process parameters. For the control actions, the HASI provides the controllers through which the operator actions are transferred to the system. These primary features are the same as those provided in the HSI of conventional NPPs.

Additionally, the second goal of the HASI is to support the interaction between the operators and the autonomous operation system. In order to design the HASI to support this interaction, two aspects need to be taken into account: the level of automation, and the issues of human performance in automation.

The roles of the operator and the system in an autonomous operation environment are based on the level of automation. Generally, the level of automation is defined on a continuum of fully manual operation to fully automated operation. While Chap. 1 introduced the Billings' levels of automation originally developed for the aviation industry, Table 8.1 in this chapter details five levels of automation better suited to NPP operation (O'Hara and Higgins 2010). O'Hara and Higgins (2010) specifically defined each level of automation and provided examples of each level for NPPs. The autonomous NPP of this book aims at Level 4, Operation by Exception.

Once the designer determines the level of automation for a particular operation or control, its various functions are allocated to the operators and to the automated system. Here, the HASI is related to supporting the functions of the operators. According to NUREG-0711 (NRC 2012), a task analysis should define the requirements for the operator functions, or in other words, the relevant tasks, information, and controls to perform the functions allocated to the operators. HASI design then follows by addressing the defined requirements to support the operators. That is, the

J. Kim et al., *Autonomous Nuclear Power Plants with Artificial Intelligence*, Lecture Notes in Energy 94, https://doi.org/10.1007/978-3-031-22386-0_8

Table 8.1 Levels of automation for NPP applications (O'Hara and Higgins 2010)

Level	Automation functions	Human functions
1. Manual operation	No automation	Operators manually perform all functions and tasks
2. Shared operation	Automatic performance of some functions/tasks	Manual performance of some functions/tasks
3. Operation by consent	Automatic performance when directed by operators to do so, under close monitoring and supervision	Operators monitor closely, approve actions, and may intervene with supervisory commands that automation follows
4. Operation by exception	Essentially autonomous operation unless specific situations or circumstances are encountered	Operators must approve of critical decisions and may intervene
5. Autonomous operation	Fully autonomous operation. System or function not normally able to be disabled, but may be manually started	Operators monitor performance and perform backup if necessary, feasible, and permitted

HASI should include all the information and controls necessary to perform the tasks assigned to the operators.

Many issues related to human performance based on interactions with automated systems have been reported. Although automation technology has introduced numerous benefits in many fields and applications, ill-designed automation may fail when the design does not consider the interactions between humans and automation. Lee well summarized the automation-related problems, including the following (Lee 2006).

- Out-of-the-loop unfamiliarity
- Clumsy automation
- Automation-induced errors
- Inappropriate trust
- Inadequate training and skill loss

Out-of-the-loop unfamiliarity refers to the diminished ability of operators to detect a failure of automation and to resume manual control (Endsley and Kiris 1995). This problem often occurs with a highly automated system in which the operators are removed from directly performing the control. In this case, the operators have difficulty in maintaining situational awareness about the plant status as well as the control activity by the autonomous system. Clumsy automation refers to the case when automation makes easy tasks easier and hard tasks harder. This can occur by automating easy tasks but leaving hard tasks to the operators. The third problem listed above, automation-induced errors, refers to error types newly introduced as a result of automation. A typical example is a mode error, in which the operator must take an action when the automation is in the wrong mode. The inappropriate trust issue includes both misuse and disuse. Misuse refers to when an operator over-relies on the automation and thus fails to detect a failure of the automation and intervene in the operation. Disuse refers to the situation in which the operator does not use the automation because of a low level of trust in its reliability. Lastly, the inadequate training and skill loss issue can appear in situations where automation reduces or eliminates the opportunities for operators to obtain skills by doing the job. More issues and explanations are presented in (Lee 2006).

One of the approaches to address the above issues is called human-centered automation. As Billings (2018) suggests, human-centered automation means that automation needs to be designed to work cooperatively with human operators toward a common goal. In other words, this concept suggests that automated systems should be designed to be team players (Wiener and Curry 1980; Woods 2018; Sarter and Woods 1997). Applying this concept to the design of HSIs in NPPs, the OECD Halden Reactor Project performed experimental studies to investigate the human–automation cooperation quality between two different interfaces (Massaiu et al. 2004). One was a conventional interface commonly applied in NPPs, while the other followed a human-centered design that improves the transparency of the systems through explicit representations of the activity of the automated systems. The results showed that the human-centered design achieved a clear improvement in human–automation cooperation.

O'Hara and Higgins (2010) also suggested that the ecological-interface design (EID) concept could support operators in the monitoring and understanding of automation. They identified two main points: (1) the work domain analysis used in the EID concept based on an abstraction hierarchy can identify the particular information that should be contained in interactions with automation, and (2) the EID principle can increase operator understanding of automation and situational awareness.

As the interest in automation is increasing in the nuclear industry, design guidelines have recently been published for improving human–automation interaction. For example, Naser (2005) provided guidance for function allocation and the design of personnel interaction with automation after surveying the contemporary state-of-the-art automation technology and its impact on personnel and integrated human–system performance. Moreover, the US Nuclear Regulatory Commission has added a new chapter in NUREG-0700, Human-System Interface Designed Review Guidelines, called Automation System (NRC 2002). This chapter provides design guidelines on automation display, automation levels, automation modes, and so on, based on the results from a previous study (O'Hara and Higgins 2010). Also, Anuar and Kim (2014) derived design requirements to address the issues of human performance in automation using a modeling method named the Itemized Sequence Diagram.

8.2 Decision-Making

Decision-making in the context of the autonomous operation of an NPP refers to the selection of the relevant operational strategies based on the current plant status and all forms of feedback. To facilitate this, the decision-making function communicates with the other functions and evaluates the plant status before selecting the operational strategies.

Figure 8.1 illustrates the interactions between the decision-making function and the other functions under the framework suggested in this book. From the diagnosis function, the decision-making function receives information about the occurrence of an abnormal or emergency state along with the diagnosis results of the plant state. From the prediction function, the decision-making function receives predicted trends of parameters as well as predictions about a reactor trip and potential threats. From the monitoring function, it receives information about any threats to the safety functions, violations of LCOs from the Tech Spec monitoring, and information about inappropriate controls by the autonomous system. From the signal validation function, the decision-making function receives information about any detected faulty signal and determines whether to replace the signal with a newly generated one or remove or ignore it in the system.

Based on all this information from the other functions, the decision-making function is able to select the appropriate operational strategy. To do so, the development of decision rules is necessary. Figure 8.2 shows an example flowchart of decision rules for an autonomous operation system. In general, NPP states can be divided into normal, abnormal, and emergency operation. Definitions of these states are given in IAEA Safety Reports Series No. 48 as follows.

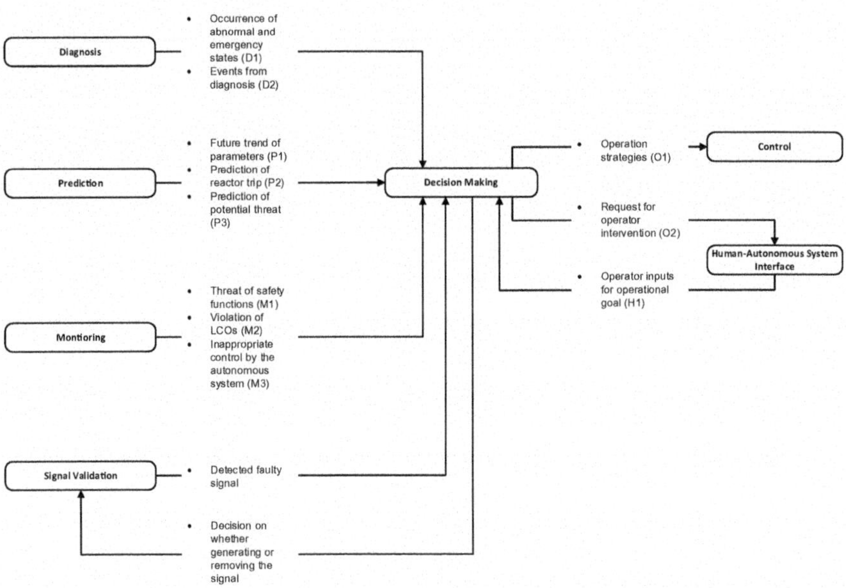

Fig. 8.1 Interactions between the decision-making function and other functions in the autonomous NPP

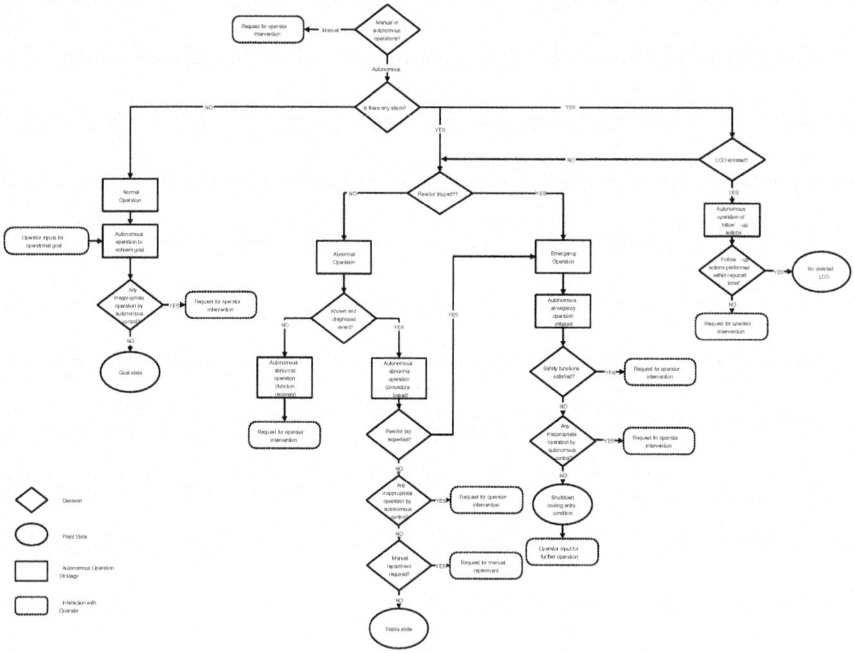

Fig. 8.2 Decision rules for strategy selection

> Normal operation is defined as plant operation within specified operational limits and conditions. Examples include starting up and shutting down the plant, normal power operation, shutdown, maintenance, testing and refueling.
>
> Abnormal operation or anticipated operational occurrence (AOO) is an off-normal operation state which would most likely not cause any significant damage to items important to safety nor lead to accident conditions. Examples of abnormal operation events include loss of normal electrical power and faults such a turbine trip, malfunctions of individual components of a normally running plant, failure to function of individual items of control equipment, and loss of power to the main coolant pumps.
>
> Accident conditions (i.e., emergency operations) are defined as deviations from normal operation more severe than AOOs, including design basis accidents, beyond design basis accidents and severe accidents. Examples of such deviations include loss of coolant accidents (LOCAs), complete loss of residual heat removal from the core, and anticipated transient with scram. (IAEA 2006)

Normal and off-normal (i.e., abnormal or emergency) operations are generally divided depending on alarm occurrence. In case of normal operation with no alarms, the operational goals are handled by the operators. Examples of operational goals include "increase the reactor power to 100% from 2%", "create a bubble in the PZR", or "shut down the reactor". The range of such operations depends on the level of automation. That is, at higher levels of automation, more extensive operations are covered by the autonomous operation system. Once any inappropriate operation by the autonomous system is detected via operation validation in the monitoring function, a request for operator intervention is generated and transferred to the HASI. Although Sect. 7.1 introduced the concept of operation validation only for emergency operation, the algorithm developed for this function can be extended to normal and abnormal operations.

In typical PWRs, abnormal or emergency operations are distinguished by the occurrence of a reactor trip. In abnormal operation in which the reactor is not tripped, the operational strategies can differ depending on whether the current status has been trained, meaning whether it is able to be identified by the diagnosis function or not. If the current status is trained and diagnosed, the control function is expected to manage the situation successfully based on its prior training for the specific event. On the other hand, if the status is evaluated as an unknown event, the strategy focuses on recovering the failed function and maintaining the plant in as stable a condition as possible until the operators intervene. In addition, many abnormal operation scenarios require manual control actions after stabilizing the plant. In this case, requests for the manual actions are generated. If the prediction function predicts that the reactor is about to trip under the current situation, the strategy is switched to reflect emergency operation.

In emergency operation initiated by a reactor trip, one general goal in PWRs is to cool down the reactor and depressurize the primary system to the entry condition for long-term cooling performed by the shutdown cooling system. During emergency operation, any failure of a safety function indicates that the autonomous operations are not working well. In this case, as above, requests for operator intervention are generated. Once the plant reaches the entry condition for long-term cooling, the operators decide upon further operations.

The Tech Spec monitoring function works in both abnormal and emergency operations. If any LCO is violated, the decision-making function provides the optimal strategy with follow-up actions to perform for the control.

References

Anuar N, Kim J (2014) A direct methodology to establish design requirements for human–system interface (HSI) of automatic systems in nuclear power plants. Ann Nucl Eng 63:326–338

Billings CE (2018) Aviation automation: the search for a human-centered approach. CRC Press

Endsley MR, Kiris EO (1995) The out-of-the-loop performance problem and level of control in automation. Hum Factors 37(2):381–394

IAEA (2006) Development and review of plant specific emergency operating procedures. Int At Energy Agency

Lee J (2006) Human factors and ergonomics in automation design. In: Salvendy G (ed) Handbook of human factors and ergonomics. Wiley

Massaiu S, Skjerve ABM, Skraaning Jr G, Strand S, Waeroe I (2004) Studying human-automation interactions: methodological lessons learned from the human-centred automation experiments 1997–2001 (No. HWR–760). Institutt for energiteknikk

Naser J (2005) Guidance for the design and use of automation in nuclear power plants. Electric Power Research Institute: California, USA

NRC (2002) Human-system interface design review guideline (NUREG-0700 Rev. 2). US Nuclear Regulatory Commission

NRC (2012) Human factors engineering program review model (NUREG-0711 Rev. 3). US Nuclear Regulatory Commission

O'Hara JM, Higgins J (2010) Human-system interfaces to automatic systems: review guidance and technical basis

Sarter NB, Woods DD (1997) Team play with a powerful and independent agent: operational experiences and automation surprises on the Airbus A-320. Hum Factors 39(4):553–569

Wiener EL, Curry RE (1980) Flight-deck automation: promises and problems. Ergonomics 23(10):995–1011

Woods DD (2018) Decomposing automation: apparent simplicity, real complexity. In: Automation and human performance. Theory and applications, pp 3–17. CRC Press

Chapter 9
Conclusion

This book introduced applicable AI techniques and examples related to the autonomous operation of NPPs. The utilization of AI is a recent trend in increasingly many industrial fields, stemming from the explosive growth in AI adaptation due to increased data processing and developments in hardware design, graphics processing units, and related methods. Although NPPs are already highly automated to reduce human error and ensure the reliability of the various system operations, the term *autonomous* in this context remains relatively unpopular because the nuclear industry strictly relies on proven technologies. But despite this, the adoption of AI techniques and the promotion of autonomous operations based on them seem unavoidable when considering their great advantages, especially for advanced reactors and small modular reactors.

Novel approaches and practical examples suggested by the authors were discussed in this book for the autonomous operation of NPPs aimed at minimizing human operator intervention. In addition to the high-level framework to develop an autonomous operation system, the topics covered all necessary areas including signal validation, diagnosis, prediction, control, monitoring, and the human–autonomous system interface. For all these various functions, the techniques addressed in this book covered a wide range of applicable methods such neural networks and reinforcement learning as well as traditional knowledge-based systems.

There are many potential areas to which the approaches and examples introduced in this book can be applied in addition to autonomous NPPs. For instance, the techniques for diagnosis and prediction can be used as an algorithm of operator support systems to help operators' diagnosis and decision making in existing NPPs. In addition, the operation validation system and tech spec monitoring system introduced in Chap. 7 may be installed as an independent support system in the current NPPs. The algorithms for the signal validation would be a part of the instrumentation and control system to detect failures of sensors and signals. Therefore, the algorithms and methods in this book could be used as key techniques for operator support systems in NPPs.

J. Kim et al., *Autonomous Nuclear Power Plants with Artificial Intelligence*, Lecture Notes in Energy 94, https://doi.org/10.1007/978-3-031-22386-0_9

The authors hope that this book can provide useful information for researchers and students who are interested in applying AI techniques in the nuclear field as well as other industries. Various potential areas of AI applications and available methods were discussed with examples. Traditional approaches to recent applications of AI were examined. In addition, the specific techniques and modeling examples provided would be informative for beginners in AI studies. While the focus of this book was the autonomous operation of NPPs with AI, the methods addressed here would also be applicable to other industries that are both complex and safety–critical.

Index

J. Kim et al., *Autonomous Nuclear Power Plants with Artificial Intelligence*, Lecture Notes in Energy 94, https://doi.org/10.1007/978-3-031-22386-0

The manufacturer's authorised representative in the EU is Springer Nature Customer Service Centre GmbH, Europaplatz 3, 69115 Heidelberg, Germany. If you have any concerns regarding our products, please contact ProductSafety@springernature.com

Printed and bound by CPI Group (UK) Ltd, Croydon, CR0 4YY
30/09/2026
02225299-0003